Modelling Archaeology and Palaeoenvironments in Wetlands

MODELLING ARCHAEOLOGY AND PALAEOENVIRONMENTS IN WETLANDS

the hidden landscape archaeology of Hatfield and Thorne Moors, eastern England

Henry P. Chapman and Benjamin R. Gearey

with contributions by

Michael Bamforth, Nóra Bermingham, Pete Marshall, Ian Powlesland, Maisie Taylor and Nicki Whitehouse

OXBOW BOOKS
Oxford and Oakville

Published by
Oxbow Books, Oxford, UK

© Oxbow Books and the individual authors, 2013

ISBN 978-1-78297-174-0

Cover image: Vincent van Gogh: 'Travaux aux Champs' ('Landscape with bog trunks' 1883)
Photograph © 2013 Museum of Fine Arts, Boston

This book is available direct from:

Oxbow Books, Oxford, UK
(Phone: 01865-241249; Fax: 01865-794449)

and

The David Brown Book Company
PO Box 511, Oakville, CT 06779, USA
(Phone: 860-945-9329; Fax: 860-945-9468)

or from our website

www.oxbowbooks.com

A CIP record for this book is available from the British Library
Library of Congress Cataloging-in-Publication Data

Chapman, Henry, 1973-
 Modelling archaeology and palaeoenvironments in wetlands : the hidden landscape archaeology of Hatfield and Thorne Moors, eastern England / Henry P. Chapman and Benjamin R. Gearey ; with contributions by Michael Bamforth, N?ra Bermingham, Pete Marshall, Ian Powlesland, Maisie Taylor, and Nicki Whitehouse.
    pages cm
 Includes bibliographical references and index.
 ISBN 978-1-78297-174-0
 1. Hatfield Moors (England)--Antiquities. 2. Thorne Moors (England)--Antiquities. 3. Water-saturated sites (Archaeology)--England. 4. Peatlands--England. 5. Landscape archaeology--England. 6. Excavations (Archaeology)--England. 7. Paleoecology--England. 8. Hatfield Moors (England)--Environmental conditions. 9. Thorne Moors (England)--Environmental conditions. I. Gearey, Benjamin R. II. Bamforth, Michael. III. Title. IV. Title: Modeling archaeology and palaeoenvironments in wetlands.
 DA670.H27C42 2013
 942.8'1--dc23
                                                            2013038878

Printed and bound in Great Britain by
Short Run Press, Exeter

For our Fathers

Contents

List of figures	xiii
List of tables	xv
Acknowledgments	xvii
Foreword, Robert Van der Noort	xix
Summary	xxi
Résumé	xxiii
Zusammenfassung	xxv

1. Landscape archaeology, space, chronology, palaeoenvironments and peatlands — 1

1.1 Introduction — 1
1.2 Space, time and landscape archaeology — 1
    Space and landscape archaeology — 1
    Time and landscape archaeology — 4
    Spatial and temporal resolution and the palaeoenvironmental record — 5
1.3 Peatlands and raised mires: a case study in space and time — 5
1.4 Outline of this book — 8

2. Raised mires and the Humber Peatlands — 9

2.1 Introduction — 9
2.2 Raised mires: location and formation — 9
2.3 Archaeological and palaeoenvironmental investigations of raised mires — 11
    Archaeological research and raised mires — 11
    Palaeoenvironmental research and raised mires — 14
    Integrated approaches to the archaeology of mire environments — 14
    Towards a 'hidden landscape archaeology' of raised mires — 15
2.4 Case studies: Hatfield and Thorne Moors — 17
2.5 Topographic, archaeological and palaeoenvironmental data for Hatfield and Thorne Moors — 21
    Historical mapping — 22
    Geological mapping — 22
    Three-dimensional surface data — 26
2.6 The archaeological record from Hatfield and Thorne Moors — 26
    Hatfield Moors — 27
    Thorne Moors — 28
2.7 Previous palaeoenvironmental research on Hatfield and Thorne Moors — 29
2.8 Models of peatland development — 37
2.9 Summary — 37

3. Building from the bottom up: aims and approaches — 38

3.1 The challenge: space, time and raised mires — 38

3.2	Key research objectives	39
	Objective 1: the morphology of the pre-peat landscapes	39
	Objective 2: wetland inception and peat spread	39
	Objective 3: modelling and interpreting palaeoenvironmental records: evidence for mire development, landscape change, human impact and the archaeological record	40
	Objective 4: assessing and quantifying the surviving peat resource	41
3.3	Approach and methods	41
	Geographical Information Systems: a platform for modelling past landscapes	41
	Methods for modelling pre-peat landscapes (objective 1)	42
	Methods for modelling wetland inception and peat spread (objective 2)	46
	Methods for interpreting palaeoenvironmental records: evidence for mire development, landscape change, human impact and the archaeological record (objective 3)	47
	Methods for assessing and quantifying the surviving peat resource (objective 4)	47
3.4	Summary	48

4. Laying the foundations: modelling pre-peat landscapes 49

4.1	Introduction	49
4.2	Hatfield Moors pre-peat landscape	49
	Assessing the value and coverage of legacy datasets for modelling the pre-peat landscape	49
	Ground-truthing and fieldwork locations	50
	Results from the gridded borehole surveys	52
	Comparing legacy datasets with data from the gridded borehole surveys	53
	Results from the borehole transects	57
	Summary of gridded borehole surveys and transects	65
	Peat stratigraphy on Hatfield Moors	65
	Generation of a new Digital Elevation Model of the pre-peat landscape of Hatfield Moors	65
4.3	Thorne Moors pre-peat landscape	67
	Assessing the value and coverage of legacy datasets for modelling the pre-peat landscape	67
	Testing the legacy data and determining fieldwork locations	67
	Results from the gridded borehole surveys	68
	Comparison of legacy datasets with the data from the gridded borehole surveys	69
	Results from the borehole transects	70
	Peat stratigraphy on Thorne Moors	75
	Generation of a new Digital Elevation Model of the pre-peat landscape of Thorne Moors	76
4.4	Summary	76

5. Modelling, dating and contextualising palaeoenvironmental records 77

5.1	Introduction	77
5.2	Radiocarbon dating peat and modelling chronologies	77
	Bayesian chronological modelling: assessing the synchronicity of palaeoenvironmental 'events'	80
	Chronological modelling: methods, results and discussion	80
5.3	Sampling strategy: radiocarbon dating of samples from Hatfield Moors	81
5.4	Results: Hatfield Moors radiocarbon dates	82
5.5	Contextualising palaeoenvironmental evidence for wetland inception on Hatfield Moors	85
	North of Lindholme Island (northern area of Hatfield Moors)	85
	West of Lindholme Island (western area of Hatfield Moors)	86
	Packards Southwest (southwestern area of Hatfield Moors)	87
	Packards South (southern area of Hatfield Moors	87
	Middle Moor (southeast of Lindholme Island)	87
	Southeastern area (north of Porters Drain)	88
	The Porters Drain area (southeastern area of Hatfield Moors)	88
5.6	Discussion: patterns of early wetland development on Hatfield Moors	88
5.7	Ombrotrophic mire development on Hatfield Moors	89
5.8	Wetland inception and ombrotrophic mire development on Thorne Moors	90

5.9 Modelling chronologies: assessing the evidence for climate change and human activity in previous palaeoenvironmental studies	89
The Humberhead Levels 'recurrence surfaces'	91
Modelling the temporal range of human impact identified in the palynological records from Hatfield and Thorne Moors	93
Discussion of spatial and temporal patterns of vegetation change	93
From local to regional patterns of change: interpreting BSW records and assessing implications for human activity	96
5.10 Summary and conclusions	97

6. Patterns of change: modelling mires in four dimensions — 99

6.1 Introduction	99
6.2 Modelling Hatfield Moors in four dimensions	100
Wetland inception on Hatfield Moors	100
Modelling peatland spread across Hatfield Moors	106
Modelling the surviving peatland resource on Hatfield Moors	108
Modelling the depth of surviving peat on Hatfield Moors	108
Modelling the date of the cut-over peat surface of Hatfield Moors	108
Summary: the peatland resource on Hatfield Moors	110
6.3 Modelling Thorne Moors in four dimensions	110
Wetland inception on Thorne Moors	110
Modelling the timing of wetland inception in relation to elevation on Thorne Moors	111
Modelling the timing of wetland inception in relation to proximity to watercourses on Thorne Moors	112
Modelling the timing of wetland inception in relation to proximity to surface run-off on Thorne Moors	112
Summary: patterns of wetland inception and peat spread on Thorne Moors	114
Modelling the surviving peatland resource on Thorne Moors	114
Modelling the depth of surviving peat on Thorne Moors	114
Modelling the date of the cut peat surface of Thorne Moors	117
Summary: the peatland resource on Thorne Moors	117
6.4 Summary: modelling Hatfield and Thorne Moors in four dimensions	117
Wetland inception and peat spread	117
Peatland resource modelling	118
Summary	118

7. Archaeological investigations of a late Neolithic site on Hatfield Moors — 119

7.1 Introduction	119
7.2 Methods	119
Borehole survey	119
Excavation, sampling and fieldwalking	119
7.3 Results	120
The trackway and platform	120
Site chronology	121
Finds	122
Coleoptera and macrofossils	124
7.4 Discussion	129
7.5 Conclusions	130

8. The 'hidden landscape archaeology' of Hatfield and Thorne Moors — 131

8.1 Introduction	131

8.2	An integrated landscape archaeology of Hatfield Moors	131
	The late Mesolithic period (pre-4000 BC)	131
	The earlier Neolithic period (c. 4000–3300 BC)	133
	The later Neolithic period (c. 3300–2500 BC)	134
	The earlier Bronze Age period (c. 2500–1500 BC)	138
	The later Bronze Age period (c. 1500–700 BC)	139
	The Iron Age and later	140
8.3	Thorne Moors: complex patterns of mire development	141
	The Mesolithic period (pre-4000 BC)	141
	The Neolithic and Bronze Age periods (c. 4000–700 BC)	141
	After the Bronze Age	147
8.4	Peat inception and mire development: the regional context	147
8.5	Palaeoenvironmental evidence for human activity, the Humberhead Levels 'recurrence surfaces' and the identification of Holocene climate change	149
8.6	The regional archaeological context of Hatfield and Thorne Moors	150
	The Mesolithic period	150
	The earlier Neolithic period	150
	The later Neolithic period	150
	The Bronze Age period	152
	The Iron Age period	152
8.7	Summary	152

9. Conclusions: themes in the archaeo-environmental study of peatlands — 153

9.1	Introduction	153
9.2	Hatfield and Thorne Moors: modelling mires	153
	Exploring and understanding pre-peat landscapes	155
	Modelling wetland inception and peat spread	155
	Reconstructing raised mire development: spatial variations in the transition to ombrotrophy	156
	Radiocarbon dating and Bayesian modelling: exploring spatial and temporal dimensions of palaeoenvironmental records	158
	Landscape archaeology and the palaeoenvironmental record	160
9.3	Peatlands, space and time: reflections	160
	How much data is enough? Resolution and deposit modelling	160
	Quantifying the archaeo-environmental record of peatlands for cultural resource management	161
9.4	The heritage management of peatlands in the 21st century	162
9.5	Landscape archaeology and peatlands: developing a 'tool-kit' for future archaeo-environmental investigations	163
9.6	Peatlands, palaeoecology and archaeology	164

Appendix 1. Radiocarbon dating protocol and Bayesian chronological modelling methodology	167
Radiocarbon dating programme	167
Calibration	167
Chronological modelling of the legacy palaeoenvironmental datasets	167
Hatfield Moors	168
Hatfield Moors Site 1 (HAT1)	168
Hatfield Moors Site 2 (HAT2)	168
Hatfield Moors Site 3 (HAT3)	168
Hatfield Moors Site 4 (HAT4)	168
Lindholme Bank Road (LIND_B)	169
Porters Drain	169
Hatfield Moors 'recurrence surfaces'	170

Thorne Moors	170
Crowle Moor Site 1 (CLM1)	170
Crowle Moor Site 2 (CLM2)	172
Goole Moor Site 1 (GLM1 area)	173
Goole Moor Site 2 (GLM1 area)	173
Goole Moor Site 3 (GLM1 area)	173
Rawcliffe Moor Site 1 (RWM1)	173
Thorne Moors Site 1 (TM1 area)	173
Thorne Moors Site 2 (TM1 area)	173
Thorne Moors Trackway site (TM1 area)	173
Thorne Waste Site 1	173
Thorne Waste Site 2	176
Appendix 2. Coleoptera from the Hatfield trackway and platform site	185
Bibliography	188
Index	201

List of Figures

Figure 1.1 Location map showing the positions of Hatfield and Thorne Moors in relation to the Humber lowlands
Figure 1.2 Restoration measures on Hatfield Moors
Figure 1.3 Re-wetting of Hatfield Moors
Figure 1.4 Tollund Man
Figure 1.5 The flat and featureless landscape of Hatfield Moors
Figure 2.1 The distribution of lowland and upland peat in England
Figure 2.2 Simplified model of hydroseral succession and raised mire development
Figure 2.3 Detail of an Iron Age trackway from the peatland at Corlea, Co. Longford, Ireland
Figure 2.4 Series of 'time slice' models showing the development of Derryville Bog, Co. Tipperary, Ireland
Figure 2.5 Plot of radiocarbon and dendrochronological dates of wetland archaeological structures from Derryville, Co. Tipperary, Ireland, in relation to shifts in the bog's surface wetness
Figure 2.6 Detailed location map of Hatfield and Thorne Moors in relation to modern settlements
Figure 2.7 The modern landscapes of Hatfield and Thorne Moors are largely the product of post-Medieval industrial activity
Figure 2.8 The end of peat cutting on Hatfield Moors
Figure 2.9 The flooding of Hatfield Moors
Figure 2.10 Principal areas of Hatfield Moors
Figure 2.11 Principal areas of Thorne Moors
Figure 2.12 Section of the 1st edition 1:2500 scale Ordnance Survey mapping of Hatfield Moors (1855)
Figure 2.13 Section of the 1st edition 1:2500 scale Ordnance Survey mapping of Thorne Moors (1853–1855)
Figure 2.14 Drift geology of the Humber Peatlands
Figure 2.15 The LIDAR data for Hatfield Moors
Figure 2.16 Plan of the Thorne Moors trackway excavation trenches
Figure 2.17 The location of the Thorne Moors Bronze Age trackway following re-wetting, looking east
Figure 2.18 Spatial distribution of palaeoenvironmental study sites on Hatfield Moors
Figure 2.19 Spatial distribution of palaeoenvironmental study sites on Thorne Moors
Figure 2.20 Models of mire evolution
Figure 3.1 Landscape archaeology approaches can be challenging within peatland environments
Figure 3.2 How three-dimensional space is represented within a Digital Elevation Model
Figure 3.3 Pre-peat landscape of Hatfield Moors
Figure 3.4 Pre-peat landscape of Thorne Moors
Figure 3.5 Extent of the GPR survey data used for Hatfield Moors showing the resulting modelled pre-peat land-surface
Figure 3.6 Excavating boreholes on Hatfield Moors
Figure 4.1 Pre-peat sands forming a low dune revealed in a drain section on Hatfield Moors
Figure 4.2 Locations of gridded borehole surveys on Hatfield Moors
Figure 4.3 Locations of borehole transects on Hatfield Moors
Figure 4.4 DEM derived from borehole Grid 5
Figure 4.5 DEM derived from borehole Grid 4
Figure 4.6 DEM derived from borehole Grid 6
Figure 4.7 DEM derived from borehole Grid 8
Figure 4.8 DEM derived from borehole Grid 7
Figure 4.9 DEM derived from borehole Grid 1
Figure 4.10 DEM derived from borehole Grid 2
Figure 4.11 DEM derived from borehole Grid 3
Figure 4.12 Borehole transects to the north of Lindholme Island (transects 1–4)
Figure 4.13 Borehole transect to the southwest of Lindholme Island (transect 5)
Figure 4.14 Borehole transects in the Porters Drain area of Hatfield Moors (transects 6–10)
Figure 4.15 DEM of the pre-peat landscape of Hatfield Moors
Figure 4.16 Locations of gridded borehole surveys on Thorne Moors
Figure 4.17 Locations of borehole transects on Thorne Moors
Figure 4.18 DEM derived from borehole Grid 1
Figure 4.19 DEM derived from borehole Grid 2
Figure 4.20 DEM derived from borehole Grid 3
Figure 4.21 DEM derived from borehole Grid 4
Figure 4.22 Borehole transects in the areas of Cottage Dyke and Birtwistle on Thorne Moors
Figure 5.1 Locations of previous radiocarbon dated sequences on Hatfield Moors
Figure 5.2 Locations of previous radiocarbon dated sequences on Thorne Moors
Figure 5.3 Radiocarbon sampling site locations on Hatfield Moors
Figure 5.4 Elevations of radiocarbon samples taken from Hatfield Moors
Figure 5.5 Probability distribution of radiocarbon dates from Hatfield Moors
Figure 5.6 Offsets between radiocarbon measurements on the humic acid and humin fractions of bulk sediment samples
Figure 5.7 Replicate radiocarbon determinations on humic acid and humin bulk fractions and waterlogged plant remains from sediment samples
Figure 5.8 Probability distributions of radiocarbon dates associated with the Humberhead Levels (HHL) 'recurrence surfaces' on Hatfield and Thorne Moors
Figure 5.9 Probability distributions of radiocarbon dates associated with *Tilia* declines
Figure 5.10 Probability distributions of radiocarbon dates associated with increased values of *Plantago lanceolata*
Figure 5.11 Probability distributions of radiocarbon dates associated with reductions in *Plantago lanceolata* values
Figure 6.1 The cut-over peatland landscape of Hatfield Moors
Figure 6.2 Age-altitude model for Hatfield Moors (humic fractions)
Figure 6.3 Watercourse proximity model with locations of radiocarbon dates
Figure 6.4 Age-distance to watercourse model of wetland inception on Hatfield Moors (macrofossils and humic fractions)

Figure 6.5 Age-distance to watercourse model for Hatfield Moors (macrofossils and humin fractions)
Figure 6.6 How flow-accumulation is calculated in a GIS
Figure 6.7 Flow-accumulation model of Hatfield Moors based on the pre-peat land-surface DEM and showing the locations of basal radiocarbon samples
Figure 6.8 Flow-accumulation in relation to calibrated radiocarbon dates in the area north of Lindholme Island
Figure 6.9 Flow-accumulation model for Hatfield Moors (macrofossils and humic fractions)
Figure 6.10 Flow-accumulation model for Hatfield Moors (macrofossils and humin fractions)
Figure 6.11 Flow-accumulation model with values converted to predicted dates for wetland inception across Hatfield Moors
Figure 6.12 Resulting GIS model of wetland inception by cultural period on Hatfield Moors
Figure 6.13 Modelled depths of surviving peat across Hatfield Moors
Figure 6.14 Resulting model of the dates of the surface peat across Hatfield Moors in relation to calibrated radiocarbon dates from surface peat samples
Figure 6.15 The modelled dates of surface peat across Hatfield Moors defined by cultural period
Figure 6.16 Watercourse proximity model for Thorne Moors with the locations of radiocarbon dates
Figure 6.17 Flow-accumulation model of Thorne Moors based on the pre-peat land-surface DEM and showing the locations of basal radiocarbon samples
Figure 6.18 Resulting GIS model of wetland inception by cultural period on Thorne Moors
Figure 6.19 Modelled depths of surviving peat across Thorne Moors
Figure 7.1 Location map of the discovery of a cluster of worked wooden poles on the northern side of Lindholme Island on Hatfield Moors
Figure 7.2 The initial discovery of wooden poles on Hatfield Moors
Figure 7.3 The gridded borehole survey in relation to the current land-surface, the pre-peat land-surface sands (in yellow-brown) and the locations of the initial discovery, looking northwest
Figure 7.4 The excavation trenches
Figure 7.5 Fieldwalking survey area in relation to the site and the location of the concentration of worked lithics
Figure 7.6 Overall site plan drawing together all phases of excavation
Figure 7.7 Profile through the site showing the exposed area of the structure following peat cutting, with the locations of samples (S1–6) taken for Coleopteran analysis
Figure 7.8a and b One of the birch bark layers laid onto the surface of the peat between the horizontal poles of the superstructure
Figure 7.9a and b Two of the vertical *Pinus sylvestris* pegs identified within the structure showing worked ends
Figure 7.10 Calibration of the radiocarbon dates from the five samples taken from the trackway and platform structure
Figure 7.11 Selection of lithics discovered within the proximity of the site during fieldwalking
Figure 7.12 Fossil beetles from the Hatfield trackway and platform site
Figure 7.13 Model of the site in its contemporary local environmental context
Figure 8.1 Wetland inception on Hatfield Moors towards the end of the Mesolithic period in the second half of the sixth millennium BC
Figure 8.2 Wetland inception and peat spread on Hatfield Moors by the early Neolithic period in the mid to late fourth millennium BC

Figure 8.3 Wetland inception and peat spread on Hatfield Moors by the later Neolithic period in the first half of the third millennium BC
Figure 8.4 Location of the Hatfield trackway and platform in relation to wetland inception and peat spread within the local area
Figure 8.5 The Hatfield trackway and platform site
Figure 8.6 *Pinus* trunk preserved towards the base of the peat on Hatfield Moors
Figure 8.7 Wetland inception and peat spread on Hatfield Moors by the earlier Bronze Age in the second half of the third millennium BC
Figure 8.8 Summary of selected previous palaeoenvironmental evidence for mire development on Thorne Moors
Figure 8.9 Wetland inception on Thorne Moors by the earlier Neolithic period around the middle of the fourth millennium BC
Figure 8.10 Wetland inception and peat spread on Thorne Moors by the later Neolithic period in the first half of the third millennium BC
Figure 8.11 Wetland inception and peat spread on Thorne Moors by the early Bronze Age period in the second half of the third millennium BC
Figure 8.12 Wetland inception and peat spread on Thorne Moors by the middle Bronze Age period in the middle of the second millennium BC
Figure 8.13 Distribution of sites dating to the Mesolithic period from the local region
Figure 9.1 The landscape on the western side of Hatfield Moors in 2012
Figure 9.2 *Posterior density estimates* for peat inception and subsequent raised (ombrotrophic) mire development on Hatfield and Thorne Moors
Figure 9.3 Offsets between radiocarbon measurements of the humic acid and humin fractions of bulk sediment samples
Figure 9.4 Sunset on Hatfield Moors
Figure A1.1 Probability distributions of dates from Hatfield Moors Site 1 (HAT1) *P-Sequence* model
Figure A1.2 Probability distributions of dates from Hatfield Moors Site 2 (HAT2) *P-Sequence* model
Figure A1.3 Probability distribution of dates from Hatfield Moors Site 3 (HAT3)
Figure A1.4 Probability distributions of dates from Hatfield Moors Site 4 (HAT4) *P-Sequence* model
Figure A1.5 Probability distribution of dates from Lindholme Bank Road (LIND_B)
Figure A1.6 Probability distribution of dates from Porters Drain
Figure A1.7 Probability distributions of dates from Crowle Moor Site 1 (CLM1) *P-Sequence* model
Figure A1.8 Probability distributions of dates from Crowle Moor Site 2 (CLM2) *P-Sequence* model
Figure A1.9 Probability distributions of dates from Goole Moor Site 1 (GLM1) *P-Sequence* model
Figure A1.10 Probability distributions of dates from Goole Moor Site 2 (GLM2) *P-Sequence* model
Figure A1.11 Probability distributions of dates from Goole Moor Site 3 (GLM3) *U-Sequence* model
Figure A1.12 Probability distributions of dates from Rawcliffe Moor Site 1 (RAW1) *P-Sequence* model
Figure A1.13 Probability distributions of dates from Thorne Moor Sites 1 and 2 (TM1 and TM2) *P-Sequence* model
Figure A1.14 Probability distributions of dates from Thorne Moor trackway site
Figure A1.15 Probability distributions of dates from Thorne Waste Site 2

List of Tables

Table 2.1 Available topographic and archaeological datasets for Hatfield and Thorne Moors
Table 2.2 Archaeological finds from Hatfield Moors
Table 2.3 Archaeological finds from Thorne Moors
Table 2.4 Summary of the Humberhead Levels 'Recurrence Zones'
Table 2.5 Summary of the Humberhead Levels 'Regional Pollen Assemblage Zones'
Table 2.6 Summary of vegetation changes at Porters Drain and comparison with other palaeoenvironmental studies on Hatfield Moors
Table 2.7 Summary of palaeohydrological changes at the Porters Drain (Hatfield Moors) and Middle Moor (Thorne Moors) study sites
Table 4.1 Comparison of statistical results from the various surface datasets for Hatfield Moors
Table 4.2 Borehole transects on Hatfield Moors
Table 4.3 Basal topography statistics from the values of boreholes within the grids for Hatfield Moors
Table 4.4 Values from the GIS models for each of the gridded borehole survey areas in comparison with the results from other datasets for the same areas on Hatfield Moors
Table 4.5 Statistical results from the borehole transects on Hatfield Moors
Table 4.6 Basal topography statistics from the values of boreholes within grids for Thorne Moors
Table 4.7 Comparative statistics from the GIS models of each of the gridded areas
Table 5.1 Summary of the results of Bayesian modelling of palaeoenvironmental sequences
Table 5.2 Radiocarbon dates from Hatfield Moors
Table 5.3 Chi-squared test results of radiocarbon dates from Hatfield Moors
Table 5.4 *Posterior density estimates* for 'recurrence surfaces'
Table 5.5 Probability distributions of dates associated with falls in *Tilia* values
Table 5.6 Probability distributions of dates associated with rises in *Plantago lanceolata* values
Table 5.7 Probability distributions of dates associated with reductions in *Plantago lanceolata* values
Table 6.1 Basal dates from the area north of Lindholme Island in relation to elevation
Table 6.2 Basal radiocarbon dates in relation to distance to watercourses
Table 6.3 Flow-accumulation values in relation to basal radiocarbon dates
Table 6.4 Comparing modelled dates with calibrated radiocarbon dates from legacy datasets
Table 6.5 Radiocarbon dates from surface deposits across Hatfield Moors
Table 6.6 Basal dates from Thorne Moors in relation to elevation
Table 6.7 Basal radiocarbon dates in relation to distance to watercourses
Table 6.8 Flow-accumulation values in relation to basal radiocarbon dates
Table 6.9 Comparison of depth statistics from the boreholes and Natural England data
Table 7.1 Radiocarbon dates from the site
Table 9.1 *Posterior density estimates* for peat inception and raised (ombrotrophic) mire development on Hatfield and Thorne Moors
Table A1.1 Hatfield Moors Site 1 radiocarbon results
Table A1.2 Hatfield Moors Site 2 radiocarbon results
Table A1.3 Hatfield Moors Site 3 radiocarbon results
Table A1.4 Hatfield Moors Site 4 radiocarbon results
Table A1.5 Lindholme Bank Road (LIND_B) radiocarbon results
Table A1.6 Porters Drain radiocarbon results
Table A1.7 Hatfield Moors 'recurrence surfaces' radiocarbon results
Table A1.8 Crowle Moor Site 1 radiocarbon results
Table A1.9 Crowle Moor Site 2 radiocarbon results
Table A1.10 Goole Moor Site 1 radiocarbon results
Table A1.11 Goole Moor Site 2 radiocarbon results
Table A1.12 Goole Moor Site 3 radiocarbon results
Table A1.13 Rawcliffe Moor Site 1 radiocarbon results
Table A1.14 Thorne Moor Site 1 radiocarbon results
Table A1.15 Thorne Moor Site 2 radiocarbon results
Table A1.16 Thorne Moor trackway site radiocarbon results
Table A1.17 Thorne Waste Site 1 radiocarbon result
Table A1.18 Thorne Waste Site 2 radiocarbon results
Table A2.1 Coleoptera from the Hatfield trackway and platform site

Acknowledgements

The research on Hatfield and Thorne Moors, and the excavations of the Hatfield Trackway and Platform were made possible by the encouragement, support and endeavour by numerous individuals and organisations. Those who contributed to the final report are mentioned by name in this publication. Most importantly of all, we wish to acknowledge the unwavering support of English Heritage who funded the project, but also provided specialist advice. Specifically, we acknowledge the guidance and assistance of Inspector of Ancient Monuments Keith Miller, Project Officer Helen Keeley, and Senior Investigator Magnus Alexander. Likewise, we would also like to acknowledge the assistance provided by Natural England who funded phases of excavation and provided access to Hatfield and Thorne Moors throughout the project, and we would particularly like to thank Tim Kohler and Kevin Bull who supplied data and also arranged fieldwork including the fieldwalking of the landscape. Scotts Ltd. provided additional advice and was extremely supportive through the supply of data from their own surveys, and we are grateful for the access to LIDAR data supplied by the Environment Agency.

Throughout the project, the members of the Thorne and Hatfield Conservation Forum shared in depth knowledge of these landscapes in addition to support both on and off site. In particular we acknowledge the assistance of Helen Kirk. We would also like to acknowledge our debt to Mick Oliver who made the initial discovery of the Hatfield Trackway and Platform, and to Paul Buckland who provided initial comment and advice. Assistance both on site and off was provided by Peter Robinson of Doncaster Museum and Art Gallery, and additional advice was given throughout the project by South Yorkshire Archaeology Service, and particularly by Dinah Saich and Jim McNeil.

We are grateful for the help we had in the field during both excavation and coring, and would particularly like to acknowledge the assistance provided by Nóra Bermingham, Will Fletcher, Kristina Krawiec and Emma Tetlow, as well as Ian Stead for metal detector surveys and local knowledge. We would also like to thank Derek Hamilton for assistance during the formulation of the radiocarbon strategy.

Finally, we would like to thank the publishers, Oxbow Books, for their patience, guidance and support throughout the publication process. In particular we would like to thank Julie Gardiner and Julie Blackmore for making this happen.

Foreword

My days wandering across Thorne and Hatfield Moors go back to the 1990s when, as part of the English Heritage funded Humber Wetlands Project, I led the team that undertook an archaeological and palaeoenvironmental survey of both Moors. We found very little in terms of archaeology, were more successful where it concerned palaeoenvironmental research, but being out on the Moors left a sense of awe that I will never completely forget. In the 1990s, Thorne and Hatfield Moors were more-or-less completely overcut for peat milling, and looked desert-like. Following the decision by the UK Government in 2002 to buy the Moors for the benefit of the nation, Natural England has commenced the long-term conservation of the peatlands, and the landscape is slowly being transformed into a mire and a raised bog.

Thorne and Hatfield Moors are exceptional landscapes, the last remnants of what once was an extended wetland wilderness in the area covered by Lake Humber in the last Post-Glacial. We found it was nigh on impossible to read the landscape, something that we had been trained to do. Mires are not simple palimpsests, where different generations leave their imprint on the landscape for archaeologists to decipher. Instead, peat growth obliterates these imprints, and no geophysical technique can reveal them. Even then, it was clear that only the application of new techniques, methods and ideas could do justice to this landscape.

The current book is just doing that: it employs high-resolution chronological modelling and methodological advances in Geographical Information Systems (GIS) to attain a new level of understanding in the origin and development of Hatfield and Thorne Moors. It has succeeded in this completely. This book presents a 'time-lapse' series of the development of the mires, which provides a framework for furthering our understanding of the origin and growth of the peat, shows us where opportunities for future discovery are located, and can even assist nature conservation managers in their quest to restore the raised bogs. In achieving such a high-resolution picture of the dynamic natural development, especially for Hatfield Moors, Ben Gearey and Henry Chapman have set new standards in wetland archaeology.

But this book is much more than making progress in the mapping of peat-landscape development over long periods of time. By drawing on and synthesizing all previous research undertaken on Thorne and Hatfield Moors, much of this not readily available for the public before this publication, this book serves as the new starting point for anybody who is more than a little curious about Thorne and Hatfield Moors.

And then there is the trackway. The discovery of a late Neolithic trackway on Hatfield Moors in 2005 on the edge of Lindholme Island, leading into the mire, presents an important turning point in the archaeological study of the Humberhead Peatlands. This is the first prehistoric site found *in situ* on the Moors, and important for that fact alone. More importantly, however, this trackway presents us with an incredible insight into the interaction of our prehistoric ancestors with Thorne and Hatfield Moors. What the evidence seems to suggest is that this was not simply a trackway across the expanding mire, but one that led into the bog. Whatever the reason for its construction, it implies that the fascination with Thorne and Hatfield Moors in the Neolithic, more than 4500 years ago, was not dissimilar from our fascination with Thorne and Hatfield Moors today.

Professor Robert Van de Noort FSA

Summary

This monograph describes an integrated approach to modelling patterns and processes of landscape change in raised mire (peatland) environments with the aim of contextualising past human activity and the archaeological record. Chapter 1 presents a summary of the archaeological and palaeoenvironmental potential of raised mire environments, where the waterlogged, anoxic conditions enable the preservation of organic cultural remains alongside sub-fossil material such as pollen, insect and plant remains which provide the source material for the reconstruction of past environmental change. The study areas of two of the largest lowland raised mire systems in the UK, Hatfield and Thorne Moors, east England are introduced in Chapter 2. These peatlands have been the subject of extensive previous palaeoenvironmental research, but most archaeological discoveries relate to Antiquarian reports with very few securely provenanced finds in recent years. The sites have also been heavily cut for peat, not only destroying an unknown extent of the resource, but also complicating and inhibiting understanding of the evolution of these landscapes. A series of hypotheses regarding the formation processes and patterns of the development of the peatlands have previously been advanced, but these have not been tested. The available archaeological, palaeoenvironmental, cartographic and topographical data are summarised. Chapter 3 outlines a series of outstanding questions regarding the archaeological and palaeoenvironmental records from the study areas. Chapter 4 assesses the available legacy data in terms of its spatial extent and detail and describes fieldwork in the form of the excavation of borehole grids and transects, aimed at determining the robustness of the different datasets which are then used to generate digital elevation models (DEM) of the pre-peat landscapes. Chapter 5 presents the results of chronological (Bayesian) modelling of the radiocarbon dates from previous palaeoenvironmental study of the peatlands, with the aim of determining the robustness of the inferred chronology of 'events' interpreted from the records, including human impact on the vegetation and evidence for palaeohydrological changes in the peatland potentially caused by climatic change. The results of the radiocarbon dating of samples from the base and top of peats on Hatfield Moors are also detailed. The DEMs of the pre-peat landscape generated in the previous chapter are also used to contextualise previous palaeoenvironmental investigations on the peatlands in terms of the relationship between the topography and the timing and character of wetland development and peat inception. The Bayesian chronological modelling is used to compare the timing of palynological evidence for human impact across the study area and also to assess the synchronicity of evidence for climatic forcing of the peatlands in the form of the 'Humberhead levels 'recurrence surfaces'. Chapter 6 outlines the testing of the hypotheses of mire inception processes described in Chapter 2, using a GIS-based approach integrating the chronological, spatial and topographic data presented earlier in the monograph. The results of these analyses suggest that local microtopographic variation in the pre-peat landscape of Hatfield Moors identified using a 'flow-accumulation model', may have been highly significant in terms of early processes of paludification and subsequent peat growth. The relationship between the flow-accumulation model and the age of basal peat deposits is used to generate a further GIS model which provides an estimation of the chronology of peat inception across the study area. The relationship between the age and elevation of the cut-over peatland surface is also investigated, which indicates that much of the surviving peat dates to the Bronze Age, with very few intact deposits which preserve a later Holocene record surviving. The same analyses are also presented for Thorne Moors, but limitations in the data and the complexity of the pre-peat landscape rather limit the outputs for this peatland. Chapter 7 describes the excavation of a late Neolithic platform and trackway site on Hatfield Moors, utilising the models of peat inception and spread to contextualise this site and human activity within local and landscape scale processes of environmental change during this period. Chapter 8 presents an integrated 'hidden landscape archaeology' of Hatfield and Thorne Moors, using the results of the GIS-based and chronological modelling to describe spatial and chronological patterns and processes of landscape change across the mid to late Holocene and to suggest possible implications for human activity and the archaeological record. It is suggested that the Hatfield trackway and platform site can only be understood within the context of these patterns of local environmental change over time, and it is hypothesized that the construction of site may be related to the perception and response of human communities to landscape change and in particular the flooding and loss of land during the later Neolithic. The relationship of processes of peat inception on Hatfield and Thorne Moors to regional 'drivers' including rises in relative sea level and the impact on local fluvial systems is also considered and the problems and potential of using palaeohydrological records from raised mires to identify climate change and its relationship with human activity is discussed. Chapter 9 provides a summary and synthesis of the study, considering the potential of the approach for investigations of other peatland landscapes. The value of Bayesian approaches for assessing the robustness of chronologies and also for comparing the relative timing of different palaeoenvironmental and archaeological 'events' is highlighted. The problems of radiocarbon dating of peat deposits are considered in terms of the difference between the fractions of sediment which may be used for dating associated palaeoenvironmental sequences. The importance of modelling the surviving depth and extent of peat across the two peatlands is highlighted, in terms of its potential to preserve *in situ* archaeological sites and deposits within the context of heritage management and current initiatives for peatland conservation and restoration. Methodological issues concerning data quantity and quality for spatial modelling of 'hidden landscapes' are discussed. The significance of mapping and modelling peat deposits for future archaeological understanding and input into recent peatland restoration initiatives within broader frameworks such as 'ecosystem services' is outlined. The monograph concludes with reflections on research at the interface between palaeoecology and archaeology and how investigating the 'hidden landscapes' of peatlands may enhance our understanding of past landscapes, archaeology and human activity more broadly.

Résumé

Cette monographie décrit une approche intégrée de la modélisation des variations et des procédés d'évolution du paysage dans un environnement de hauts marais (tourbière) dans le but de replacer dans leur contexte activités humaines passées et vestiges archéologiques. Le chapitre 1 présente un résumé du potentiel archéologique et paléoenvironnemental des environnements de hauts marais où des conditions de sol gorgé d'eau, anoxique permettent la préservation de restes culturels organiques à côté de matériaux sub-fossiles tels que pollens, insectes et vestiges de plantes qui fournissent les matériaux de base pour la reconstruction des changements environnementaux passés. On introduit au chapitre 2 les zones d'études, deux des plus grands systèmes de tourbières bombées de basses terres du Royaume-Uni, Hatfield et Thorne Moors, dans l'est de l'Angleterre. Ces tourbières ont fait l'objet de vastes recherches paléoenvironnementales auparavant, mais la plupart de ces découvertes archéologiques relèvent de comptes rendus d'amateurs d'histoire avec très peu de trouvailles de provenance confirmée ces dernières années. On a aussi fortement exploité les sites pour leur tourbe, non seulement détruisant on ne sait quelle étendue de cette ressource, mais compliquant et inhibant aussi la compréhension de l'évolution de ces paysages. Une série d'hypothèses sur les procédés de formation et les étapes du développement des tourbières ont été émises dans le passé mais n'ont pas été testées. On résume les données archéologiques, paléoenvironnementales, cartographiques et topographiques disponibles. Le chapitre 3 passe en revue une série de questions non résolues sur les vestiges archéologiques et paléoenvironnementaux des zones d'étude. Le chapitre 4 évalue les anciennes données disponibles en matière de leurs étendue et coordonnées spatiales et décrit la prospection de terrain sous la forme d'excavation de quadrillages et transects de forage dans le but de déterminer la robustesse des divers ensembles de données qui sont ensuite utilisées pour générer des modèles numériques d'élévation (MNE) des paysages d'avant la tourbe. Le chapitre 5 présente les résultats d'une modélisation chronologique (bayésienne) des dates au carbone 14 d'une précédente étude paléoenvironnementale des tourbières, afin de déterminer la robustesse de la chronologie déduite des 'événements' interprétés à partir de vestiges, y compris l'impact de l'homme sur la végétation et les témoignages de changements paléohydrologiques dans la tourbière éventuellement causés par des changements climatiques.

Les résultats de la datation au C14 des échantillons de la base et du sommet de la tourbe de Hatfield Moors sont également détaillés. Les MNE du paysage d'avant la tourbe générés dans le chapitre précédent sont aussi utilisés pour replacer dans leur contexte les précédentes investigations paléoenvironnementales des tourbières en matière de relations entre la topographie et la temporalité et le caractère du développement des hauts marais et la formation de la tourbe. La modélisation chronologique bayésienne est utilisée pour comparer la temporalité des indices palynologiques pour l'impact humain sur l'ensemble de la zone d'étude et aussi pour évaluer le synchronisme des témoignages de pression climatique sur les tourbières sous la forme de 'surfaces de récurrence des marais de Humberhead'. Le chapitre 6 donne les grandes lignes de l'évaluation des hypothèses des procédés de formation des hauts marais décrits au chapitre 2, utilisant une approche basée sur SIG qui intègre les données chronologiques, spatiales et topographiques présentées plus haut dans cette monographie. Les résultats de ces analyses donnent à penser que la variation microtopographique locale du paysage d'avant la tourbe d'Hatfield Moors identifiée grâce à un modèle par 'accumulation de flux', pourrait avoir été extrêmement significative en matière de procédés anciens de paludification et par la suite de croissance de tourbe. La relation entre le modèle par accumulation de flux. et l'âge des dépôts à la base de la tourbe est utilisée pour générer un autre modèle SIG qui fournit une estimation de la chronologie du développement de la tourbe dans la zone étudiée. On a également analysé la relation entre l'âge et l'élévation de la surface coupée de la tourbière, ceci indique qu'une grande partie de la tourbe qui a survécu date de l'âge du bronze, très peu de dépôts intacts préservant des témoignages de l'holocène tardif ayant survécu. Nous présentons également ces mêmes analyses pour Thorne Moors mais les limitations dans les données et la complexité du paysage pré-tourbe limitent plutôt les résultats pour cette tourbière. Le chapitre 7 décrit l'excavation d'une plateforme et d'un chemin du néolithique tardif sur Hatfield Moors, utilisant les modèles de formation et de propagation de la tourbe pour remettre dans son contexte ce site et l'activité humaine dans le cadre de procédés de changements environnementaux à l'échelle locale et à celle du paysage pendant cette période. Le chapitre 8 présente une archéologie intégrée du 'paysage caché' de Hatfield et Thorne Moors en utilisant les résultats d'une modélisation chronologique et basée sur SIG pour décrire les variations et procédés de changement du paysage au cours de l'Holocène moyen et tardif et pour suggérer d'éventuelles implications pour l'activité humaine et les indices archéologiques. On avance que le site à plateforme et chemin d'Hatfield ne peut se comprendre que dans le contexte de ces schémas de changements environnementaux locaux au fil du temps et on émet l'hypothèse que la construction du site pourrait avoir un lien avec la perception et la réaction de communautés humaines aux changements du paysage et en particulier aux inondations et à la perte de terres au cours du néolithique final. On considère également la relation entre les procédés de formation de tourbe sur les landes de Hatfield et Thorne et des `conducteurs' régionaux y compris l'élévation du niveau de la mer relatif et l'impact sur les systèmes fluviaux locaux et on discute des problèmes et du potentiel de l'utilisation de vestiges paléohydrologiques des hauts marais pour identifier le changement climatique et son lien avec l'activité humaine. Le chapitre 9 fournit un résumé et une synthèse de l'étude, considérant le potentiel de cette approche pour des investigations d'autres paysages de tourbière. On souligne la valeur des approches bayésiennes pour évaluer la robustesse des chronologies et aussi pour comparer la relative chronologie des divers `événements' paléoenvironnementaux et archéologiques. On considère les problèmes de datation au carbone 14 des dépôts de tourbe en matière de différence entre les fractions de sédiments qui peuvent être utilisées pour dater des séquences paléoenvironnementales et archéologiques associées. On souligne l'importance de modéliser l'épaisseur et l'étendue de la tourbe qui subsiste sur la totalité des deux tourbières en matière de son potentiel de préservation de dépôts et sites archéologiques *in situ* dans le contexte de gestion du patrimoine et d'initiatives actuelles pour la conservation et la restauration des tourbières. On discute des questions de méthodologie concernant la quantité et la qualité des données pour une modélisation des `paysages cachés'. On met en évidence l'importance de cartographier et modéliser les dépôts de tourbe pour la compréhension archéologique dans l'avenir et la contribution aux récentes initiatives de restauration des tourbières dans des cadres plus étendus tels que les `services écosystémiques'. La monographie se termine par une réflexion sur l'interface entre paléoécologie et archéologie et comment une investigation des `paysages cachés' des tourbières peut renforcer et élargir notre compréhension des paysages, de l'archéologie et des activités humaines passés.

Traduction Annie Pritchard
08/10/13

Zusammenfassung

Diese Monographie stellt einen integrierten Ansatz vor, durch den Strukturen und Prozesse des Landschaftswandels in Hochmooren modelliert werden können, mit dem Ziel, vergangene menschliche Aktivitäten und die archäologische Überlieferung zu kontextualisieren. Kapitel 1 fasst das historische und landschaftsarchäologische Potenzial von Hochmooren zusammen, deren wassergesättigte und deshalb luftsauerstofffreie Erhaltungsbedingungen die Konservierung organischer materieller Kultur sowie subfossiler Materialien wie Pollen und Resten von Insekten und Pflanzen ermöglichen; diese kulturellen und natürlichen Hinterlassenschaften bilden das Quellenmaterial für die Rekonstruktion des historischen Wandels der Umwelt. Die Untersuchungsgebiete der vorliegenden Arbeit, zwei der größten Hochmoorregionen im Tiefland Großbritanniens, werden in Kapitel 2 vorgestellt: Hatfield Moors und Thorne Moors im östlichen England. Diese Torfgebiete waren bereits extensiv landschaftsgeschichtlich untersucht worden, jedoch resultieren die meisten archäologischen Nachrichten aus antiquarischen Berichten, während Funde mit gut dokumentierter Provenienz in den letzten Jahren nur selten geborgen wurden.

In den Untersuchungsräumen fand zudem intensiver Torfabbau statt, was nicht nur zur Zerstörung von Quellen in unbekanntem Ausmaß führte, sondern auch das Verständnis der Entwicklung dieser Landschaften verkomplizierte und behinderte. Eine Reihe von Hypothesen in Bezug auf die Entstehungsprozesse und die Entwicklungsstrukturen der Moorlandschaften wurden in jüngerer Zeit entfaltet, jedoch wurden diese nie getestet. Die bislang vorhandenen archäologischen, landschaftsgeschichtlichen, kartographischen und topographischen Daten werden zusammengefasst. Kapitel 3 benennt eine Reihe von vordringlichen Fragen mit Bezug auf die archäologische und landschaftsgeschichtliche Datengrundlage in den Untersuchungsgebieten. Kapitel 4 bewertet die verfügbaren Altdaten hinsichtlich ihres räumlichen Umfangs und ihrer Detailgenauigkeit und beschreibt die durchgeführten Feldarbeiten wie die Ausgrabung von Profilen und eines Rasters von Bohrlöchern, deren Ziel es war zu bestimmen, wie robust die verschiedenen Datensets sind; diese werden dann angewandt um digitale Höhenmodelle (digital elevation models, DEM) der Landschaften vor der Entstehung der Torfmoore zu generieren. Kapitel 5 legt die Resultate der chronologischen (Bayes'schen) Modellierung der Radiokarbondaten aus früheren landschaftsgeschichtlichen Untersuchungen der Torfgebiete vor, ebenfalls mit dem Ziel zu bestimmen, wie robust der aus der vorhandenen Datengrundlage erschlossene zeitliche Ablauf der „Ereignisse" sein kann, einschließlich des menschlichen Einflusses auf die Vegetation und der Hinweise auf paläohydrologische Veränderungen in den Torflandschaften, die möglicherweise durch klimatischen Wandel hervorgerufen worden waren. Ebenso werden Details zu den Radiokarbondatierungen von Proben aus den oberen und unteren Lagen der Torfe von Hatfield Moors diskutiert.

Die digitalen Höhenmodelle der Landschaften der Vor-Torf-Zeit werden schließlich genutzt um frühere landschaftsgeschichtliche Untersuchungen der Torfgebiete zu kontextualisieren in Bezug auf die Beziehungen zwischen Topographie und Art und Ablauf der Entwicklung der Feuchtböden und dem Beginn der Torfbildung. Die Bayes'sche chronologische Modellierung wird angewandt um den Ablauf der menschlichen Einflussnahme, wie er sich in den palynologischen Daten zeigt, in den Untersuchungsgebieten zu erfassen und zu vergleichen und um die Synchronizität der Hinweise auf klimatischen Druck auf die Torfe zu bewerten, wie sie in Form der „Humberhead Levels recurrent surfaces" (alte, durch unterschiedliche Humifizierung im Torf entstandene Schichtoberflächen) dokumentiert wurden.

Kapitel 6 behandelt wie die Hypothesen der Moorbildungsprozesse, die in Kapitel 2 beschrieben werden, getestet wurden unter Anwendung einer GIS-basierten Methodik, die die chronologischen, räumlichen und topographischen Daten zusammenführt, die weiter oben in dieser Monographie vorgestellt worden sind. Die Ergebnisse dieser Untersuchungen legen nahe, dass lokale mikrotopographische Unterschiede in der vor-torf-zeitlichen Landschaft von Hatfield Moors, die durch ein „flow-accumulation model" erkannt werden konnten, von besonderer Bedeutung waren für die frühen Prozesse der Versumpfung und die anschließende Torfbildung. Die Zusammenhänge zwischen „flow-accumulation" Modell und dem Alter der untersten Torfschichten werden genutzt um ein weiteres GIS-Modell zu generieren, das eine ungefähre Chronologie der Torfentstehungsprozesse in den Untersuchungsräumen ermöglicht. Ebenso wird der Zusammenhang von Alter und Höhe der abgebauten Torfgebiete untersucht, was darauf hinweist, dass ein großer Teil des noch vorhandenen Torfs in die Bronzezeit datiert werden kann; es sind noch wenige intakte Schichten erhalten, die somit eine spätholozäne Quelle konservieren. Die gleichen Analysen werden auch für Thorne Moors präsentiert, wobei jedoch Einschränkungen bei den Daten und die Komplexität der Landschaft vor der Torfbildung die Ergebnisse für dieses Torfgebiet eher begrenzen.

Kapitel 7 beschreibt die Ausgrabung eines spätneolithischen Fundplatzes mit Plattform und Bohlenweg in Hatfield Moors, bei der die Modelle für Torfbildung und -ausbreitung genutzt wurden um den Fundplatz und die dortigen menschlichen Aktivitäten mit Prozessen des jungsteinzeitlichen Landschaftswandels auf lokaler und regionaler Ebene zu kontextualisieren. Kapitel 8 legt eine integrierte Archäologie „verborgener Landschaften" von Hatfield Moors und Thorne Moors vor und wendet dabei die Ergebnisse der GIS-basierten und chronologischen Modellierungen an um die räumlichen und zeitlichen Strukturen und Prozesse des Landschaftswandels vom mittleren bis ins späte Holozän zu beschreiben und mögliche Implikationen für menschliche Aktivitäten in der archäologischen Überlieferung aufzuzeigen. Es wird deutlich, dass der Fundplatz von Hatfield mit Bohlenweg und Plattform nur im Kontext dieser Strukturen lokalen Landschaftswandels im Verlauf der Zeit verstanden werden kann, und die Hypothese wird diskutiert, dass die Anlage des Platzes in Beziehung steht mit der Wahrnehmung von und dem Umgang mit dem Landschaftswandel durch örtliche Gemeinschaften, insbesondere als Reaktion auf die Überflutung und den Verlust von Flächen während des jüngeren Neolithikums.

Ebenso wird das Verhältnis von Torfbildungsprozessen in Hatfield Moors und Thorne Moors zu regionalen "Triebkräften" erörtert, wie z.B. Erhöhungen des relativen Meeresspiegels und der Einfluss auf lokale Flusssysteme, und werden die Probleme und Möglichkeiten der Nutzung paläohydrologischer Daten aus Hochmooren zur Identifizierung von Klimawandel und dessen Relation zu menschlicher Aktivität diskutiert. Kapitel 9 bildet eine Zusammenfassung und Synthese der Untersuchung, in der das Potenzial dieses Ansatzes für die Erforschung weiterer Torfbodenlandschaften erörtert wird. Der Wert Bayes'scher Ansätze für die Bewertung der Robustizität von Chronologien und für den Vergleich der relativen Zeitabläufe verschiedener landschaftsgeschichtlicher und archäologischer „Ereignisse" wird ebenfalls beleuchtet. Die Probleme der Datierung von Torfböden mit C14 werden in Bezug auf die Unterschiede in den Sedimentfraktionen besprochen, die für die Datierung verschiedener landschaftsgeschichtlicher Sequenzen genutzt werden können. Es wird die Notwendigkeit deutlich, die erhaltenen Tiefen und Ausdehnungen der Torfe in den beiden Torfgebieten zu modellieren, auch in Bezug auf die Möglichkeiten archäologische Fundplätze und Befunde *in situ* zu erhalten im Zusammenhang mit denkmalpflegerischen und landschaftspflegerischen Aktivitäten für die Bewahrung und Wiederherstellung von Torfgebieten. Darüber hinaus werden methodologische Fragen der Quantität und Qualität der Daten für eine räumliche Modellierung „verborgener Landschaften" diskutiert. Die Bedeutung der Kartierung und Modellierung von Torfböden für ein künftiges archäologisches Verständnis von Moorgeschichte und einen Beitrag zu gegenwärtigen Initiativen zur Wiederherstellung von Torfgebieten innerhalb breiter Themengebiete wie den „Ökosystem-Leistungen" wird dargestellt. Die Monographie schließt mit Überlegungen zu Forschungen an der Schnittstelle zwischen Paläoökologie und Archäologie und zur Frage, wie die Erforschung von „verborgenen Landschaften" von Torfgebieten unser Verständnis von historischen Landschaften, Archäologie und menschlicher Aktivität auf breiter Basis verbessern kann.

1. Landscape archaeology, space, chronology, palaeoenvironments and peatlands

1.1 Introduction

This book details an approach to the study of past spatial and temporal patterns of environmental change within the broader framework of landscape archaeology. It focuses on a particular form of wetland landscape and uses quantitative and qualitative approaches to different records, using GIS (Geographical Information System) modelling to investigate spatial and temporal patterns of Holocene landscape change for two raised mires in South Yorkshire, Hatfield and Thorne Moors (Figure 1.1). Whilst it is concerned in part with specific aspects of landscape evolution including rates and patterns of peat growth and spread, which are of course a feature of wetland environments, it aims to illustrate the synergy which may be generated through linking different datasets to better understand past landscapes, human activity and the archaeological record. Furthermore, it explores how, through the generation of spatial models integrated with chronological information, data can be generated that has practical applications for current management concerns, such as *in situ* preservation, heritage management and policy. The 'modelling' of the book's title therefore refers to the GIS based methodology for manipulating, analysing and presenting data from topographic, stratigraphic, cartographic, palaeoenvironmental and archaeological sources, and also to the application of Bayesian approaches for assessing the robustness of radiocarbon chronologies.

The book builds on a rich legacy of previous palaeoenvironmental research on Hatfield and Thorne Moors, using these data as a means to explore certain questions situated at the interface between themes in 'traditional' palaeoecological study, wetland archaeology and landscape archaeology (Gearey and Chapman 2004). Wetland inception and peat growth, for example, is a subject of predominantly palaeoecological interest (e.g. see Charman 2002), but little attention has been paid to the implications of peatland spread in relation to past human activity and the character and location of archaeological remains within a pre-peat landscape. It has been estimated that the total area of peatlands in Europe is around 515,000km^2 (JNCC 2011) amounting to a significant area of corresponding pre-peat landscape which are rarely exposed and hence generally remain 'out of sight and mind'. Many of these 'hidden' landscapes may date back to the early Holocene but, a few notable execeptions aside, have seen little or no archaeological investigation.

This book incorporates methodological advances in the areas of GIS (e.g. Chapman and Gearey 2003, Chapman 2006) and chronological modelling (e.g. Bayliss *et al.* 2007, Bronk Ramsey 2008, Blockley *et al.* 2008, Gearey *et al.* 2009) and focuses primarily on Hatfield Moors. This is because of the variable access to both landscapes that resulted from the timing of the implementation of peatland restoration measures undertaken by Natural England (Figure 1.2), significantly restricting access (Figure 1.3), although this was more of a problem for Thorne Moors, where restoration commenced earlier and with greater impact on the initial stages of fieldwork on which this book is based. The primary focus on Hatfield Moors is reflected both in the content of this monograph and also in the inversion of the 'traditional' nomenclature 'Thorne and Hatfield Moors' in the title and throughout this book. Collectively, these peatlands are commonly referred to as 'the Moors' or 'the Humber peatlands'.

1.2 Space, time and landscape archaeology

Space and landscape archaeology

The measurement and analysis of space is central to archaeological study, such as through the recording of site plans and sections, and the interpretation of spatial relationships between artefacts and features. Certain spatial relationships, such as the investigation of stratigraphy, enable broad inferences to be drawn regarding relative chronological sequencing. It has long been recognised that archaeological landscapes represent the product of multiple layers of human activity from different periods. For example, in his book *Archaeology in the Field,* O.G.S. Crawford introduced the concept of the 'palimpsest', using

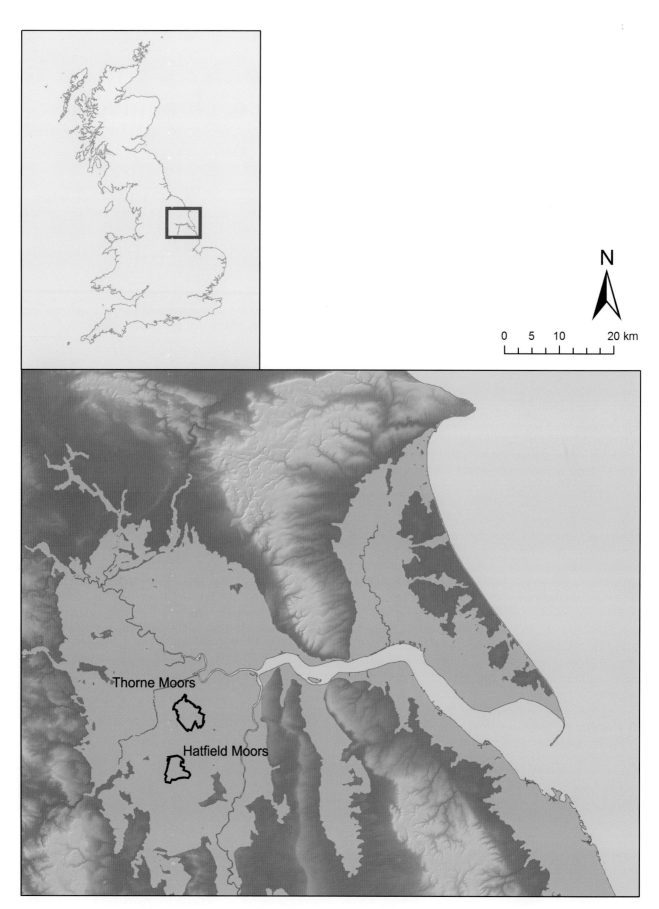

Figure 1.1 Location map showing the positions of Hatfield and Thorne Moors in relation to the Humber lowlands (area below the 10m contour shown in brown)

Figure 1.2 Restoration measures on Hatfield Moors

Figure 1.3 Re-wetting of Hatfield Moors

the analogy of layers of text on a re-used parchment to describe the manner in which the remains of successive waves of past human endeavour physically 'over write' each other in the landscape (Crawford 1953). Many of these ideas were expanded in the work of the landscape historian W.G. Hoskins and others over successive decades (e.g. Hoskins 1955, Aston and Rowley 1974, Taylor 1984, Aston 1985, Rackham 1986, Bowden 1999, Muir 2000; cf. Johnson 2007).

The general notion that the landscape is the product of previous layers of human activity has been applied in various ways, such as to the interpretation of prehistoric landscapes (Bradley 1993, 1998) and to the management of modern landscapes, particularly in the frameworks of historic landscape characterisation (Fairclough 1999, Fairclough and Rippon 2002, Rippon 2012). The idea of the palimpsest has also attracted significant reflection and discussion (e.g. Bailey 1981, 2007, Binford 1981, Foley 1981). Lucas (2012, 118) observed the different theoretical implications of the concept as: 'For Crawford, palimpsests contained the possibility of dissection and thus reconstructed sequences, whereas for Bailey, Binford and Foley it was the very impossibility of dissection that justified an archaeology of the longer term ...'

Approaches to the concept of 'space' in archaeology are problematic and contested; space has been variously described as both a 'backcloth ... of human activity' (Appleton 1975, 2) on the one hand, and as socially constructed on the other, for example in the role of locales (at a variety of spatial scales) in the processes of social production and reproduction (Giddens 1984). The definition of 'space' as the context for particular 'places' of human activity underlies a framework for considering spatial experience, using the individual human body as a vehicle for engaging with the external world (Tuan 1977). However, there has been much discussion and debate focusing on the nature of space in relation to themes including culture, environment, human perception and interpretation, and these emphasise the lack of a 'neutrality' of space (e.g. Ingold 1992, 1993) and the primacy of human sensory experience (e.g. Thomas 1993). Thus, related themes include somatic, perceptual, existential, architectural and cognitive spaces (Tilley 1994, 15).

These various concepts are encapsulated by different methodological approaches: for example, site location analyses might focus on factors such as topography and proximity to particular resources in terms of interpreting the factors that influence settlement choice in the past (e.g. Mehrer and Wescott 2006). In such cases, space is conceived of objectively and represented cartographically. In contrast, Tilley (1994) explored landscape interpretation largely on the basis of visual and physical relationships that would have been experienced by people in the past, aiming to 'describe the character of human experience, specifically the ways in which we apprehend the material world through direct intervention in our surroundings' (Brück 2005, 46). Such approaches highlight the variable and changing resolutions of an individual's perception of environments which is most apparent in terms of the visual (e.g. Fraser 1983). Following this, the resolution of space has been classified on the on the appearance of trees in the short-distance (or foreground), middle-distance and long-distance (or background) (Higuchi 1983, 14–15); a concept that was subsequently extended for the analysis of visibility patterns in the past using digital approaches (Wheatley and Gillings 2000).

The growth in the use of such digital technologies in archaeology has facilitated the integration, visualisation and analysis of spatial datasets (cf. Wheatley and Gillings 2002, Chapman 2006, Conolly and Lake 2006), although its use in this manner has been critiqued (e.g. Massey 1999). Space within a GIS environment is most commonly represented as an abstract series of regular square blocks, or 'cells' of a given size. Where a surface is derived from high resolution survey data, a single cell may represent a surface area of less than $1 \times 1m$, for example. However, irrespective of cell resolution, GIS provides a uniform simulation of space, which does not mirror the multiple resolutions at which people engage with the physical world. The use of GIS within archaeology has therefore been an area of significant debate and discussion (see Hacıgüzeller 2012 for a recent review). Thomas (2004), for example, regarded GIS as 'an illegitimate off-spring of modernism'. However, as Sturt (2006) has pointed out '... within GIS we are simply detailing our conceptualisations of space – not necessarily perceived or lived space'. Recognition of the methodological and theoretical issues of space and spatial representation should thus allow such technology to be used in a reflexive and critical manner. Hence, this book is not 'about' GIS, although it does rely in part on these technologies to analyse, manipulate and present complex data.

Time and landscape archaeology

Chronology is of course central to both archaeological (Whittle et al. 2011) and palaeoenvironmental (e.g. Telford et al. 2004) interpretation and understanding, although some assumptions which underlie approaches to time in archaeology are problematic (e.g. see Bailey 2007). Even apparently relatively straightforward concepts such as 'tensed' and 'tenseless' descriptions of time (the 'A' and 'B' series; see McTaggart 1908) continue to generate significant philosophical discussion (e.g. Zimmerman 2005). Whilst reflections on the nature of time itself (e.g. Ricoeur 1980, Brockmeier 2000) might not necessarily be regarded as central to archaeological practice, it has been noted that understanding chronology has long been a challenge (Whittle et al. 2011, 1; Bayliss et al. 2007). In general, there has been more theoretical reflection and consideration of time and chronology from an explicitly archaeological perspective (e.g. Lucas 2005, Bradley 1991) compared to palaeoenvironmental approaches (but see Caseldine 2012).

Time can only be 'seen' through the lens of chronology and, such theoretical discussions aside, various issues arise from the varying resolutions of the different methods that archaeologists use to construct chronologies. Even with 'absolute dates', such as those provided by radiocarbon dating, precision and accuracy can be highly variable. For example, dendrochronology might provide a felling date for an archaeological timber, accurate to a particular season of a particular year, which contrasts with that provided by even a very precise calibrated radiocarbon date. Composite archaeological chronologies often combine such varying forms and resolutions of chronological information and recent developments in Bayesian chronological modelling (Bayliss *et al.* 2007, Buck *et al.* 1996) now permit the formal modelling of such data.

Bayesian chronological modelling has recently fundamentally altered the approaches to, and the understanding of, archaeological chronologies (e.g. Whittle *et al.* 2011) and methods are now being employed for palaeoenvironmental chronologies, certain of which may achieve precision of ±50 years at two standard deviations (e.g. Swindles *et al.* 2007). In addition, such approaches permit the formal correlation of archaeological and palaeoenvironmental chronologies (e.g. Gearey *et al.* 2009). The work outlined in this book used Bayesian methodologies to test the robustness of radiocarbon dated palaeoenvironmental sequences, and to refine, compare and integrate different dated sequences (see Chapter 5).

Spatial and temporal resolution and the palaeoenvironmental record

The terms 'grain' (the fundamental unit by which a phenomenon is measured or described) and 'extent' (the spatial area or temporal duration of a phenomenon) are often used in discussions of scale (e.g. Allen and Hoekstra 1991). Cao and Lamb (1997) classified scales into four categories: cartographic (the spatial relation between a landscape and its map or model), observational (extrinsic extent), operational (the characteristic scales of different processes) and resolution (equivalent to grain), and further observed that operational scales are not necessarily equivalent to observational scales.

The latter point in particular has relevance to palaeoenvironmental interpretation, as the scales over which past processes operated do not correspond to how such a process is 'observed' (operational scale) in a given dataset. For example, palynological research aims to determine the spatial and temporal patterns of past vegetation change, but in terms of the former must rely: '... heavily on intuitive reconstruction founded in current ecological understanding and a qualitative estimation of relationships between landscape elements (geology, soils, aspect etc) and plant communities' (Caseldine *et al.* 2007, 135). In other words, deriving an explicit 'operational scale' from most palaeoenvironmental proxy records is highly problematic especially given the focus on single core studies (see Fyfe and Woodbridge 2012 for a recent discussion).

This presents a range of interpretative issues regarding understanding of past processes of environmental change, which may be particularly confounding at the interface between archaeological and palaeoenvironmental data (e.g. Chapman and Gearey 2000). Recently, there has been significant recent progress in deriving quantitative reconstructions of past landscapes from palynological data which effectively provide improved resolution at the observational scale and use, in part, GIS based approaches (e.g. Fyfe 2006, Gaillard *et al.* 2008). Inferring or reconstructing the operational scale of a record is not just a methodological problem in terms of palynological data, but arguably also for understanding other processes of environmental change, such as extrapolating local patterns from regional sea level curves (Gearey *et al.* in press). For example, whilst the rate of early Holocene relative sea level rise in the southern North Sea might be well constrained over a centennial scale, the pattern and process of this at a finer spatial and chronological scale will be less well resolved due to the grain or resolution of the various analytical methods employed.

This contrasts with archaeology and its intrinsic emphasis and concern with spatial patterns and scales and focus on contrasting observational scales. In the case of early Holocene sea level rise, this is encapsulated by questions regarding the rate and extent at which coastal areas might have been lost to the sea and whether these changes in sea level could have been perceived on the scale of a human lifetime (e.g. Chapman and Lillie 2004, Leary 2009; see also Sturt 2006). From a methodological perspective at least, this necessitates an understanding of rates of change on a very fine analytical scale.

These questions of varying scale and resolution in different records are especially pertinent when the focus is on linking environmental change with possible cultural response, such as in the context of debates regarding the interface between Holocene climate change and the archaeological record (e.g. Van de Noort 2004, Amesbury *et al.* 2008, Turney and Brown 2007, Berglund 2003, Bonsall *et al.* 2002, Ingram *et al.* 1981, Van Geel *et al.* 1996). This disjuncture presents a clear challenge to studies which seek to reconstruct environmental changes in three and four dimensions and to understand and situate past cultural activity within this context. Linking palaeoenvironmental and archaeological datasets centres on an understanding of the scale, pattern, process, and rates of change which in part entails developing methodologies to 'close the gap' between observational and operational scales. What was the spatial scale of a past process of change, or more precisely, what is the spatial grain at which such a process can be meaningfully identified or analysed? Did an event occur over a single lifetime, or over a series of generations? Can chronological resolution ever be attained that even permits such questions to be usefully asked?

1.3 Peatlands and raised mires: a case study in space and time

The importance of wetland environments for archaeology has long been recognised (e.g. Coles and Coles 1996,

Bernick 1998, Van de Noort and O'Sullivan 2006, Menotti 2012), generally centred on the fact that the anoxic conditions of peatlands facilitate the survival of organic remains (e.g. Chapman and Cheetham 2002), ranging from human remains or 'bog bodies' (e.g. Van der Sanden 1996; Figure 1.4) through to trackways (e.g. Raftery 1996) and entire settlements (e.g. Coles and Minnitt 1995). In addition, the peat matrix itself provides a rich source material for palaeoenvironmental reconstruction and dating of the archaeological context (e.g. Coles 1995). The recognition of the international importance of wetlands for the preservation of archaeological and palaeoenvironmental material has been highlighted by a number of statutes, recommendations and guidelines (e.g. Department of the Environment 1990, ICOMOS 1996, Willems 1998, Matthews 1993, Ramsar Convention Secretariat 2006). However, the value of wetland environments extends beyond such potential for remarkable preservation.

This book details an exploration of a range of spatial and chronological datasets relating to process and patterns of change of a specific type of landscape, and explores the ways in which we can integrate the two in both a quantitative and a qualitative manner, to provide a more comprehensive perspective on the study of past landscapes in general. The *foci* of study are peatland environments, and in particular a form of peatland called raised mires. Raised mires, like certain other wetland environments, are archaeologically distinctive in that they may be described as four dimensional archives of environmental change (*sensu* Brown 2008a). In other words, they hold the potential to provide spatial and temporal data at multiple resolutions which represents significant potential for exploring how archaeology might engage with the study of space, time, people and environments in the past. These data ultimately represent the only way in which we can explore the relationship between rates and processes of change and hence human perception and possible response and reaction to change.

The diversity of wetland ecosystems has been noted as having implications for archaeological study (e.g. Mitsch and Gosselink 1993, Dinnin and Van de Noort 1999, Mitsch and Gosselink 2000). These environments present specific practical challenges for archaeological study. They are 'hidden' landscapes in that the accretive nature of their growth conceals earlier landscapes, from their pre-peat land-surfaces through to the layers of peat that overlie them. As a result, many landscape archaeology approaches cannot be used directly 'off the shelf'.

For example, the 'regression' of terrestrial landscapes considers the modern landscape as the 'sum' of the previous phases. Morphological characteristics of certain archaeological features, such as field boundaries, might be typical of a particular period (Muir 2000) and a process of 'map regression' can be carried out to 'reveal' earlier phases of a landscape. This concept of a 'palimpsest' is closer to a cumulative palimpsest (see above, Bailey 1981, 2007) in which layers of sediment and any associated archaeological sites or artefacts generally overlie each other in sequence, preserving in sequence a record which may cover several millennia.

Furthermore, traditional methods of archaeological survey are generally less applicable in these environments compared with dryland landscapes. For example, aerial photography is largely dependent upon the natural drainage potential of soils (*cf.* Riley 1944, Wilson 1982), but cropmark formation is rarely a feature of wetlands (but see Cox *et al.* 2001). The applicability of geophysics and other forms of remote sensing is also severely restricted in wetland environments (*cf.* Schleifer *et al.* 2002), although there has been some progress on this front with developments in techniques such as Light Detection and Ranging (LIDAR) and Ground Penetrating Radar (GPR) (e.g. Carey *et al.* 2006).

The limitations of the standard methodological approaches used in landscape archaeology means that alternative approaches are necessary. Studies commonly incorporate a combination of ditch survey (to assess stratigraphy), excavation and borehole survey in order to access and sample earlier landscapes, and a heavy reliance on palaeoenvironmental data (see below, Chapter 2). Significant resources are required to investigate the evolution of raised mire landscapes (e.g. see Korhola *et al.* 1996, Newnham *et al.* 1995), with considerable investment in survey, radiocarbon dating and palaeoenvironmental analyses necessary for the comprehensive investigation of a mire system in four dimensions (e.g. see Mäkilä 1997 for the resources used in the palaeoenvironmental investigation of a relatively small mire system, and Gowen *et al.* 2005 for the integrated archaeo-environmental investigation of a slightly larger one).

However, few pristine peatlands survive and in the case of the many sites which have been drained and cut for peat, the landscape has effectively already been 'stripped backwards' in time, destroying a significant part of the archaeo-environmental record (e.g. the archaeological and palaeoenvironmental resource that *in situ* peat deposits represent) in the process. Somewhat ironically, given the difficulties of remote prospection outlined above, the excavation of drainage ditches and the cutting of peat are generally the agents of discovery of many wetland archaeological sites and artefacts (e.g. Buckland 1993). It has been estimated that, in 1997 alone, 0.95 million cubic metres of peat were extracted from peatlands in England (Van de Noort *et al.* 2002, Gearey *et al.* 2010) using mechanised methods which provide few opportunities for the identification of archaeological sites compared with hand-cutting (Coles and Coles 1996). For the Somerset Levels, of a total of 175 known sites, 59 had been partially or totally destroyed by peat cutting during the past 150 years (Brunning 2001). Following the extensive mechanized peat extraction which has impacted on many lowlands mires, the surviving landscape is flat and largely featureless, further inhibiting archaeological investigation (Figure 1.5). In general, little work has looked to assess the extent of the loss of peat from a specifically heritage management perspective.

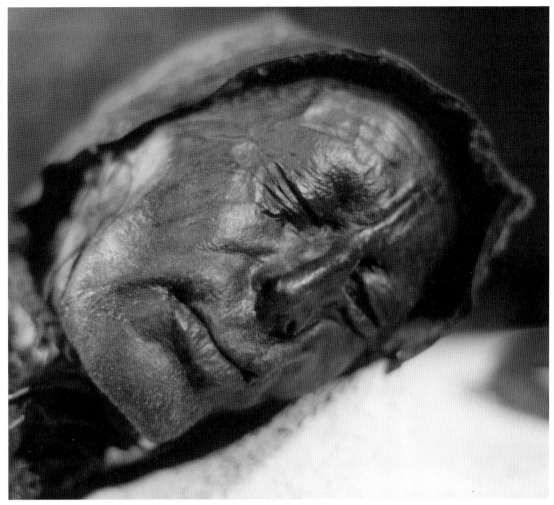

Figure 1.4 Tollund Man – an Iron Age bog body from Bjældskovdal peat bog near the village of Tollund in Denmark (photograph by Sven Rosborn)

Figure 1.5 The flat and featureless landscape of Hatfield Moors

The inability of traditional methods to identify archaeological remains within raised mire landscapes is in stark contrast with the demonstrated potential for the exceptional preservation of organic archaeological remains, the density of which has been estimated at perhaps 2.2 wet-preserved sites per km^2 in England (Van de Noort et al. 2002). Hence, peatlands provide an exceptional archaeo-environmental resource, but one which is hard to realise using the traditional methods of landscape archaeology. Without the investment of significant resources in terms of both fieldwork, palaeoenvironmental analyses and radiocarbon dating programmes (see below, Chapter 2), for example, it is difficult to assess their archaeo-environmental potential or begin to interpret them as archaeological landscapes. The 'tool kit' for this must draw on different methodological strands, including palaeoenvironmental research, field archaeology and landscape archaeology.

1.4 Outline of this book

This volume explores some of the issues discussed in the previous section: spatial and temporal patterns of environmental change, landscape archaeology, human activity and the archaeological record through the case study of peatlands and, specifically, of raised mire landscapes. As outlined above, it focuses on the two largest surviving areas of raised mires in the UK, Hatfield and Thorne Moors, which are situated predominantly in South Yorkshire. The work stems from earlier studies of these landscapes that began with the English Heritage funded Humber Wetlands Project, which investigated these landscapes in 1995–1996 (Van de Noort and Ellis 1997), and continued through commercially funded archaeological work (Gearey 2005). It also aims to provide a synthesis of this with the substantial and comprehensive body of palaeoenvironmental and archaeological work undertaken by other researchers such as Brian Smith, Paul Buckland, Nicki Whitehouse, Tessa Roper, Gretel Boswijk, Mark Dinnin and their collaborators on Hatfield and Thorne Moors over the years.

Chapter 2 begins with a detailed outline of the character of raised mires and previous archaeological and palaeoecological research within these environments. Within this context it introduces the two case studies: Hatfield Moors and Thorne Moors, outlining previous research on each, the available datasets from this work, and the various resulting hypotheses for the drivers of environmental change which offer starting points for analysing wetland inception and peat spread across the landscapes, in addition to outlining the current management challenges for them. Chapter 3 details the objectives and the methodologies applied to address them, integrating legacy data from previous studies and the strategy for obtaining new, bespoke datasets. It outlines the approaches used to generate models of the pre-peat landscapes, and to test the hypotheses outlined in Chapter 2 to generate models of wetland inception, peat spread and landscape development. It also describes the methodology applied for quantifying and understanding the peatland resources of Hatfield and Thorne Moors, both spatially and chronologically.

The following three chapters outline the results of the approaches detailed in Chapter 3. The first of this series, Chapter 4, provides the results of the spatial modelling of the pre-peat landscapes of Hatfield and Thorne Moors, combining legacy datasets with the results of fieldwork. In Chapter 5 the palaeoenvironmental record of landscape change is explored in terms of chronological robustness and modelling in order to interpret patterns of change, from wetland inception through to ombrotrophic mire development, including patterns of climate change and human activity. The final chapter in this series, Chapter 6, brings together the spatial modelling of the pre-peat landscape with the chronological modelling of palaeoenvironments in order to generate models of wetland inception and peat spread across Hatfield and Thorne Moors. Here, the results of testing the various hypotheses of the drivers for environmental change outlined in Chapter 2 are outlined, and the subsequent models are presented for both Moors. This chapter also provides the results of the quantification of the peatland resource for both landscapes, exploring both the volumetric modelling of the peat in physical terms, and the likely chronological span represented by different areas across the Moors.

Within the context of the modelled patterns of wetland inception and peat spread, Chapter 7 provides the results of archaeological excavations of the site of a late Neolithic trackway and platform from Hatfield Moors. The combination of data from these excavations with the models of environmental change illustrates an informed and integrated approach to the interpretation of this site, demonstrating the importance of integrating spatial and chronological datasets. Chapter 8 brings together the overall results, providing a summary of the landscape archaeology of Hatfield and Thorne Moors, a contextualisation of archaeological remains, and an assessment of the surviving peatland resource in terms of the potential for the preservation *in situ* of deposits and archaeological sites. Chapter 9 concludes this book with a discussion of the results within the context of the broader themes of the archaeology of raised mires, methodological approaches to their study, and the implications for the integrated investigation of the spatial and temporal within landscape archaeology.

2. Raised mires and the Humber Peatlands

with contributions by Nóra Bermingham

2.1 Introduction

As discussed in the previous chapter, peatlands, and specifically raised mires, provide the potential for generating records of four-dimensional landscape change. The land-surfaces beneath the peatlands are concealed and therefore preserved, whilst the peat itself contains a record of environmental changes through time, and the potential for preserving archaeological remains. This chapter begins with a general discussion of raised mires and outlines the archaeo-environmental research that has been carried out. In the second half of the chapter, the specific case studies of the raised mires of Hatfield and Thorne Moors are presented, including a consideration of the range of past work and the available datasets. Finally, this chapter discusses a series of testable hypotheses that have been formulated to explain the processes of peat inception for both sites.

2.2 Raised mires: location and formation

In the United Kingdom lowland raised (ombrotrophic, or precipitation fed, see below) mires are predominantly located in cool, humid regions, such as the northwest lowlands of England, the central and northeast lowlands of Scotland, Wales and Northern Ireland, but remnants also occur in some southern and eastern localities, for example Somerset, South Yorkshire and Fenland. Ombrotrophic mires are distributed across northern Europe from Estonia, the Ukraine, and Russia to northern Germany, Denmark, Finland and Scandinavia as well as Ireland and Britain (Moore 1984, Pfadenhauer *et al.* 1993). Extensive ombrotrophic peatlands exist in America and Canada (Zoltai 1988) and in the southern hemisphere in Australasia, South America and New Zealand (Charman 2002, Wilmshurst *et al.* 2003). The raised bogs of the Irish midlands represent the greatest northwest extent of European raised bog formation, ranked tenth globally in terms of total national peatland area, third in proportional terms to Finland (33.5%) and Canada (18.2%) (Taylor 1983).

Few of these peatlands in northwest Europe are pristine or undisturbed and the lowland peatlands of Britain in particular were once much more widespread. It has been estimated that only 4% of mires which were 'intact' in 1840, a critical period in the reclamation of this habitat, survive today, whilst only 6% of England's raised mires survive overall (Middleton and Wells 1990). Whilst this figure is dependent upon the accuracy of estimates drawn from the mid-19th century 1st edition Ordnance Survey mapping, the total loss of raised mire has nevertheless been extensive over the last 150 years (Hodgkinson *et al.* 2001). More recent reports have demonstrated the continuing depletion of the peatland resource through processes including peat extraction and wastage (Van de Noort *et al.* 2002). Recent initiatives are looking to restore damaged peatlands within a framework of 'ecosystem services' (see Cris *at al.* 2011), which should bring new opportunities for the protection and promotion of the archaeological resource (Gearey *et al.* 2010).

Mires are located in both upland and lowland contexts in the United Kingdom (Figure 2.1) and may be very approximately subdivided into three main types: blanket mires, transitional mires and raised mires (Moore 1984). Blanket mires form on upland slopes where precipitation is high; in excess of 1250mm/yr. Lindsay *et al.* (1988) suggest a minimum annual rainfall of 1000mm and a minimum 160 wet days per year are necessary to allow the formation of blanket bog. Transitional mires are where a combination of ombrotrophic (precipitation fed) and minerotrophic (fen) conditions are present suggesting the mire system is moving from a fen to a raised mire and they can have a similar distribution to raised mires (Heathwaite *et al.* 1993).

As stated above, raised (ombrotrophic) mires are defined as peatlands which rely solely on atmospheric precipitation for their primary water and nutrient supply (Moore 1986, 1990). This concept is central to the derivation of palaeohydrological records from raised mire systems and the use of such records to generate proxy data regarding past climatic changes. The anoxic nature of the mire in which the usual processes of decay are slowed down,

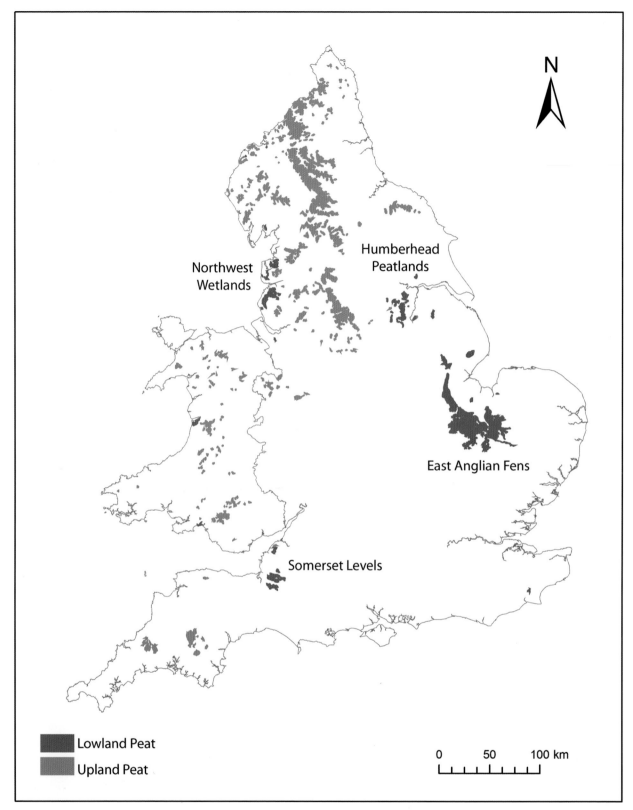

Figure 2.1 The distribution of lowland and upland peat in England and Wales

or even effectively halted, is likewise the reason why organic material – including palaeoenvironmental source material (sub-fossil pollen, plant remains and Coleoptera), archaeological artefacts and sites – may be preserved within ombrotrophic peats.

Precipitation and temperature are the main controls governing ombrotrophic mire formation, with a correlation between the occurrence of certain types of mire and climatic conditions (Moore and Bellamy 1973, Charman 2002). Raised mires develop in situations where there is a current excess of rainfall over evapo-transpiration in the order of 700–1250mm per year. Generally, three major processes of mire formation are recognised (Mitsch and Gosselink 1993): *terrestrialisation*, or the infilling of a basin via classic processes of hydroseral succession (Walker 1970), *paludification*, whereby blanket mires spread beyond their basins and subsume surrounding dryland areas; and *flow-through succession* or *topogenous development*, where the accumulation or development of peat modifies the pattern of water flow.

The vegetation of raised mire systems consists of plants which both control, and are adapted to, the physical and chemical environment of the system (Moore and Bellamy 1973). The raised surface may support a patterned mosaic of pools, hummocks and hollows, and lawns; microtopography created in part by the growth of the plants themselves, primarily bryophytes especially *Sphagnum*. *Sphagnum* mosses tend to be more resistant to decay than vascular plant tissue and thus are the primary constituent of peat (Lindsay 2010). Other plants including sedges, such as *Eriophorum angustifolium* (common cottongrass), *E. vaginatum* (hare's-tail cottongrass) and *Trichophorum cespitosum* (deergrass) are commonly found, with *Vaccinium oxycoccos* (Cranberry) and *Andromeda polifolia* (bog rosemary) less common associates, whilst *Rubus chamaemorus* (cloudberry) may forms dense patches locally. The nutrient-poor conditions are the main environment in the UK for several species of carnivorous plant, such *Drosera* spp. (sundews) and *Pinguicula* spp. (butterworts). Dwarf shrubs, especially *Calluna vulgaris* (heather), *Erica tetralix* (cross-leaved heath), *Vaccinium myrtillus* (bilberry) and *Empetrum nigrum* (crowberry) can be found on the drier areas.

The accumulation of peat in mire systems reflects vegetation succession, or the replacement of plant species in an orderly sequence (Mitsch and Gosselink 2000, Figure 2.2) which may occur in response to autogenic (internal) or allogenic (external) factors. In systems such as raised mires, the principal autogenic factor is the accumulation of the partially decayed remains of plants as peat growth raises the level of the growing surface (Walker 1970). However, this relatively simple concept hides some complexity in terms of process.

Tansley's (1939) original model of succession in hydroseres (the succession of plants over time in a freshwater environment) in a sub-oceanic climate suggested that open water led to reedswamp, followed by fen and fen carr and culminating in a 'climax' of *Quercus* (oak) woodland. In a wetter and cooler oceanic climate, the 'climax' community was regarded as *Sphagnum* ombrotrophic bog, which may follow directly from the open water (Figure 2.1) or even a paludified substrate in an extreme oceanic climate. Walker (1970) modified Tansley's (1939) model using stratigraphical and palynological data from various Holocene sequences, demonstrating many variations on the original pathways, with stages missed and sometimes reversed. Moreover, it was hypothesised that ombrotrophic bog was the 'climax' community of autogenic hydroseres in the British Isles and the transition from fen carr to *Quercus* woodland as proposed by Tansley (1939) was unsubstantiated.

The precise nature of hydrological processes within mires is complex and in many ways poorly understood (e.g. Holden and Burt 2003) and hence a detailed discussion is outside the scope of this chapter. However, certain general aspects of their hydrology are relatively well established. The 'classic' model of raised mires sees such systems as convex in cross-section, having a domed central massif elevated above the groundwater table and level of the surrounding landscape (Paasio 1933, 1934) and hence hydrologically separated from the adjacent dryland, although the margins of the system or 'lagg zone' will be influenced by groundwater (Ingram 1983). Large mire complexes may have more than one 'cupolum' or dome, with combinations of environmental gradients and corresponding site types (Seppa 2002).

A growing mire system is typically divided into the catotelm, the waterlogged and anoxic body of peat, and the acrotelm, the upper layer of living vegetation in which both recharge (water input) and discharge (water output) processes take place (e.g. Holden and Burt 2003). Hence the watertable is largely controlled within the acrotelm; as this is where new organic material is deposited, changes in the wetness of the acrotelm will be reflected in the character and composition of the accumulating peat deposit. Accumulation rates of peat vary, but can be comparatively rapid with estimates of up to 1.13mm per year (Borren *et al*. 2004).

2.3 *Archaeological and palaeoenvironmental investigations of raised mires*

Archaeological research and raised mires

The formation processes of raised mires, with continuous or near continuous accumulation of peat over what can be multi-millennial timescales, provide exceptional opportunities for both archaeological and palaeoenvironmental investigation. Some of the most evocative archaeological discoveries of the last century come from peatlands, such as the Neolithic 'Sweet Track' and other prehistoric trackways of the Somerset Levels (Coles and Coles 1986). This latter area is the most important in terms of the recorded peatland archaeological record in England. Some of the earliest reports of wet-preserved archaeological remains concern the Somerset Levels (e.g. Stradling 1849); one quarter of the surviving wet-preserved sites in England are found here and there are more scheduled examples of such sites than in the whole of the rest of the country combined (Jones *et al*. 2007).

The Sweet Track is a substantial Neolithic monument around 2km long, joining the 'island' of Westhay to the higher ground of Shapwick. Dendrochronological (tree ring) dating has provided a very precise date for the construction of the site in 3806 or 3807 BC (Hillam *et al*.

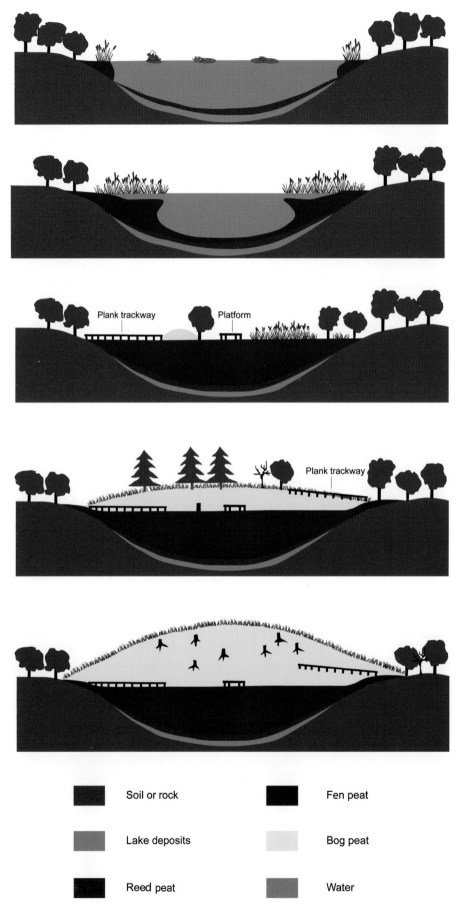

Figure 2.2 Simplified model of hydroseral succession and raised mire development

Figure 2.3 Detail of an Iron Age trackway (Corlea 1) from the peatland at Corlea, Co. Longford, Ireland (photography by Ingo Mehling reproduced under Creative Commons licensing: http://creativecommons.org/licenses/by-sa/3.0/deed.en)

1990). However, it is only one of a complex of prehistoric sites in the Somerset Levels area, with others including the Honeygore, Abbotts Way, Bells, Bakers, Westhay and Nidons trackways indicating human activity on and around the peatland throughout prehistory. Generally, these wooden trackways have been regarded as demonstrating the status of the peatland as an obstacle to be crossed and/or resource to be accessed, although recent research has begun to re-assess the possible significance of these monuments with respect to later Neolithic concepts of space, place and spirituality (see Bond 2004, 2007, 2010). The Someset Levels has also produced evidence for human settlement on the wetland areas in the form of the Glastonbury and Meare 'lake villages'. The range of material culture recovered from these Iron Age sites includes stone (quern stones, grinders and hammer stones), bone, horn and antler (combs, awls and needles), lead and tin (fishing weights and spindle whorls), bronze (rings, broaches, tweezers and needles), glass and amber beads, pottery and wooden containers providing insights into the full span of everyday life in the Iron Age (Coles and Minnitt 1995).

This rich archaeological record of the Somerset Levels can be contrasted with that of other lowland peatlands in the United Kingdom. This correlates in part with the extent to which different wetlands have been disturbed; for example, Fenns and Whixall Mosses on the Shropshire-Clwyd border remain comparatively undisturbed and have produced only one Bronze Age axe, a coin and a bog body of unknown date (Coles 1994). On the other hand, areas such as Hatfield and Thorne Moors, the case studies presented in this book, have seen significant peat extraction but have produced relatively few sites or artefacts in recent times at least (see below).

Archaeological finds from raised mires, and indeed other peatland environments, are therefore closely associated with physical disturbance, usually in the form of drainage and peat cutting. Hence, the cutting of peat in the raised mires of the midlands of Ireland has led to the discovery of a remarkably rich and diverse archaeological record. In Ireland raised bogs occupy much of the central lowlands, many of which initially developed over postglacial lakes, subsequently expanding and coalescing to form extensive complexes (Feehan and O'Donovan 1996). Trackways have been a feature of the peatland archaeological record in Ireland since at least the middle of the 20th century (Tohall and Van Zeist 1955, Rynne 1965a, 1965b), but there are accounts of sites discovered during the 19th century (Lucas 1985, Raftery 1999). By the late 1980s the potential of Irish raised bogs as rich repositories of material culture was becoming widely recognised following the survey and excavation of 58 trackways in Corlea Bog, Co. Longford (Raftery 1990) (Figure 2.3).

This was the first systematic archaeological survey of a bog in Ireland and this work prompted Coles and Coles (1989, 159) to suggest that '… there is some argument for saying that the Irish bogs still hold more information about the past than any other wetland in Europe …' A recent review has estimated that nearly 4000 sites have been identified in the peatlands of Ireland that have been drained and cut for peat since 1990 (Gearey et al. 2012). Of these sites the majority represent trackway structures (38%). Many of the bogs of northwest Europe have also produced well preserved human remains, so called 'bog bodies' (Briggs and Turner 1986, Ó Floinn 1995, van de Sanden 1996, Glob 1998). Whilst the majority of these bog bodies date to the later prehistoric period (van der Plicht et al. 2004), Neolithic, Bronze Age, and historic instances are also known. However, there is localised variation in this pattern with the majority of Irish bodies dating to the Medieval period (Bermingham and Delaney 2006).

Archaeological sites and finds are also known from other parts of Europe. Coles and Coles (1989) observed that there were around 1000 trackway structures from mire sites in Ireland, Britain, the Netherlands, Denmark and Germany. Surveys in Lower Saxony between from the 1950s to *c.* 1990 had produced somewhere between two and three hundred sites dating from the middle Neolithic to the Middle Ages (Casparie 1986, 1987, Coles and Coles 1989). Hayen (1987) developed a typology for trackways in the raised mire complexes of Lower Saxony based on the nature of construction rather than chronology.

Palaeoenvironmental research and raised mires

The value of raised mires as archives for palaeoenvironmental study has long been recognised, with some of the earliest pollen-based investigations of peat deposits focussing on the relationship between environmental changes and archaeological sites (e.g. Godwin 1960). Whilst palynological research continues to be an important area of study, recent years have seen an increased focus on the value of ombrotrophic peat deposits as providing records of changes in Holocene palaeoclimate (e.g. Aaby 1976, Barber 1981, 1982, 1994, Barber *et al.* 1994, 1998, Blundell and Barber 2005, Chambers *et al.* 1997, Charman *et al.* 1999, Hendon *et al.* 2001, Mauquoy *et al.* 2008, Swindles *et al.* 2007). As outlined above, this is based on the premise of a direct coupling of the surface wetness of ombrotrophic mire systems to the atmosphere via the precipitation:evaporation ratio. The identification and quantification of such changes have been approached mainly through the application of the analysis of peat humification (i.e. the extent to which the peat has decayed), plant macrofossils and testate amoebae (unicellular shelled animals, Protozoa: Rhizopoda) to derive 'Bog Surface Wetness' (BSW) records with chronological control generally supplied by radiocarbon dating (see Charman 2010 for a recent summary).

Research on sites across northwest Europe indicates a strong degree of synchronicity between certain periods of increased BSW over the mid- to late Holocene in particular (e.g. Chambers *et al.* 1997, Hughes *et al.* 2000, Charman *et al.* 2001, Charman 2010, Hendon *et al.* 2001, Langdon and Barber 2001). Charman (2010) has recently highlighted three 'key' wet-shifts from multiple records, dated to around 2250–2050 BC, 1650 BC and 810 BC and three other wet shifts of smaller magnitude at 1100 BC, 100 BC and AD 690. It has been hypothesised that some of these phases correlate with climatic shifts identified through in other proxy climate records, including the Greenland Ice cores (O'Brien *et al.* 1995), tree line and tree ring studies (e.g. Gear and Huntley 1991, Baillie and Munroe 1988), lake level and sedimentation data (e.g. Yu and Harrison 1995), and historical sources (Lamb 1977).

However, the relationship between peatland growth, proxy records and past climatic change is far from straightforward. The relative significance of temperature and precipitation in driving changes in surface wetness is uncertain, although research by Charman *et al.* (2001) has concluded that the peat record may be more strongly influenced by precipitation than by temperature changes. Hence the BSW record might be best interpreted as reflecting the past precipitation:evaporation ratio and for Northern Europe at least the: '... strength and position of summer westerly airflow, rather than temperature and precipitation *per se*' (Charman *et al.* 2009, 1817).

The causes of such 'sub-Milankovitch' scale climatic fluctuation are relatively poorly understood, although it is hypothesised that in northwest Europe at least, may be linked to solar forcing (changes in the amount of radiation emitted by the sun, e.g. Blaauw *et al.* 2004, Mauquoy *et al.* 2008, 2004, Van Geel *et al.* 1996) or shifts in ocean circulation (Barber *et al.* 1994, Andersen *et al.* 1998). The precise links between these various factors and the timing and nature of hydrological changes in mires remains an issue of some complexity (e.g. Swindles *et al.* 2007, Plunkett 2006). Väliranta *et al.* (2007) have, for example, highlighted the difficulties in identifying the 'forcing factors' behind hydrological changes in raised mires.

Integrated approaches to the archaeology of mire environments

The work of The Somerset Levels Project demonstrated the importance of the coherent integration of palaeo-environmental analyses into wetland archaeological investigations (see Coles and Coles 1986) and provided a model for much subsequent work in this arena, including management of the historic environment resource (Coles and Hall 1997). Other work has focussed on the dendrochronological record (e.g. Baillie 1992, Baillie and Brown 1996). Baillie and Brown (2002) suggested that the correlation between the start date of clusters of wetland sites and growth anomalies in Irish 'bog' oaks might reflect a cultural response linked to environmental deterioration. However, Swindles and Plunkett (2010) failed to establish any consistent correspondence between changes in tree populations and BSW records from Northern Ireland, concluding that further study of the patterns and processes of tree growth and hydrology were required.

Whilst the distribution and, in some instances, dates of clusters of sites in raised mires have been well established (e.g. Raftery 1996), there are often few coherent data relating to the evolution of the peatland systems in space and time. Discerning shifts in hydrology over time is important in terms of understanding episodes of past human activity on and around peatlands, but less work has considered the archaeological value of site-based palaeohydrological records from raised mires (Gearey and Caseldine 2006).

The potential of integrated archaeological and palaeo-environmental study in Ireland was initially demonstrated by the work at Corlea and other sites in the mires of Mountdillon, Co. Longford (Raftery 1996). The research carried out on these projects by the late Casparie, arguably represents some of the most innovative approaches to wetland archaeology, demonstrating the synergy which can result from combining palaeoenvironmental and archaeological interests and approaches (see below). Casparie carried out a series of highly detailed studies of mire development, hydrology and archaeology in the 1970s and 1980s in the Netherlands (Casparie 1972, 1982, 1984, 1986 and 1987, Casparie and Streefkerk 1992).

Later as part of the excavations in Irish peatlands at Corlea (Casparie and Moloney 1996) and Derryville (The Lisheen Archaeological Project, 2005), Casparie used extensive stratigraphic recording of sections and cores to reconstruct the development of mire systems through time. These studies demonstrated the complexity of pattern and process of mire development, the intricate relationship between internal (autogenic) and external forcing (allogenic) and the potential role of human activity within peatland development. The latter factor in particular is possibly one of the most intriguing and arguably underappreciated aspects of his work (see below). However, despite comprehensive survey and some excavation and analysis of the extensive archaeo-environmental resources in the midlands of Ireland outlined above, for example, relatively little recent study has considered the relationship between patterns of mire development and the associated archaeological record on a landscape scale.

The archaeological sequences excavated by the Lisheen Archaeological Project (Gowen *et al.* 2005) were also remarkable, including dense complexes of sites, such as that of the Cooleeny Complex in the southwest of the basin, which comprised a minimum of 48 wooden structures, a combination of well defined *toghers* (or trackways) and platforms lying within an area 100 × 100m. Archaeological strata lay in places up to 2m deep on the lower reaches of the gently sloping upland, whilst other parts of the mire revealed complexes of platforms, burnt mounds with wood-lined troughs and stone, plank and hurdle trackways (Cross *et al.* 2001). Bronze Age round house settlements and a cremation cemetery were identified on the lower slopes adjacent to the wetland, providing arguably the strongest evidence to date of a direct link between dryland occupation and human activity at the margin of a peatland in prehistory. An apparent association was also identified between shifts in the palaeohydrological record from relatively wet to dry mire surface and increased human activity, as reflected by prehistoric wetland site construction during the Bronze and Iron Ages in particular (Caseldine and Gearey 2005, Gearey and Caseldine 2006).

Detailed recording of the peat stratigraphy in drain faces and archaeological trenches carried out by Casparie (2005), coupled with high resolution survey of the basin and the radiocarbon dating of critical horizons, resulted in the creation of a series of 'time slice' models of the growth of the raised mire system (Figure 2.4). These show the development of a fen system during the Neolithic followed by the spread of raised mire from two cupola (or 'domes') across the entire basin by the Iron Age. The mapping of the evolution of the peatland in four dimensions in turn permitted an understanding of the relationship between the distribution of archaeological sites and the local wetland landscape context.

Casparie (2005) also presented compelling evidence that the construction of prehistoric trackways across Derryville Bog during the Bronze Age and Iron Age resulted in hydrological changes within the mire system (Figure 2.5; see also Cross *et al.* 2001), leading directly in one instance to a rapid shift in the central area of the mire from fen to ombrotrophic peat growth during the Bronze Age. Another timber trackway which was constructed across a discharge channel during the Iron Age appears to have had the effect of damming the flow of water leading to a 'bog burst', or catastrophic failure of the mire system which washed away significant areas of peat and resulted in a marked drop in the watertable and associated reduction in peat accumulation.

Casparie's (2005, 25) identification of such 'bog bursts' and contention that such events should be regarded as a 'normal' feature of raised mire growth have potential implications for the interpretation of palaeoclimate records from ombrotrophic mires. The estimated dates for 'bog burst' events at Derryville closely match independently identified and dated abrupt falls in the reconstructed watertable in sequences from across the mire (Caseldine and Gearey 2006). Despite the burgeoning of research into 'drivers' of peatland palaeohydrology and Holocene climate change in recent years (as discussed above), little subsequent attention has been given to the potential role and impact that autogenic events such as 'bog bursts' or human activity might have on palaeoclimate records from peatlands.

Towards a 'hidden landscape archaeology' of raised mires

The archaeological and palaeoenvironmental (archaeo-environmental) potential of raised mires is thus well established. Previous work, such as that outlined above, by the Somerset Levels Project and the Lisheen Archaeological Project, has demonstrated the remarkable synergy that can be derived from integrated approaches to peatlands. However, very little subsequent work research has been carried out aimed at investigating how mire systems might be conceptualised as archaeological landscapes worthy of study in their own right. Although recent work is beginning to address this imbalance more generally (Van de Noort and O' Sullivan 2006), peatland and wetland archaeology still tend to be viewed as somehow separate to the study of the archaeological record of drylands, with relatively little of these data making its way into broader syntheses of prehistory (e.g. Bradley 2007).

Gearey and Chapman (2004, 205) have discussed the value of a more nuanced approach to the archaeo-environmental conceptualisation of raised mires in particular, arguing that '... the full archaeoenvironmental potential of raised mires has yet to be fully realised'. Associated with this is the question of how palaeoenvironmental data, such as that from pollen diagrams or evidence for palaeoclimatic changes from peatlands (as discussed above), might be

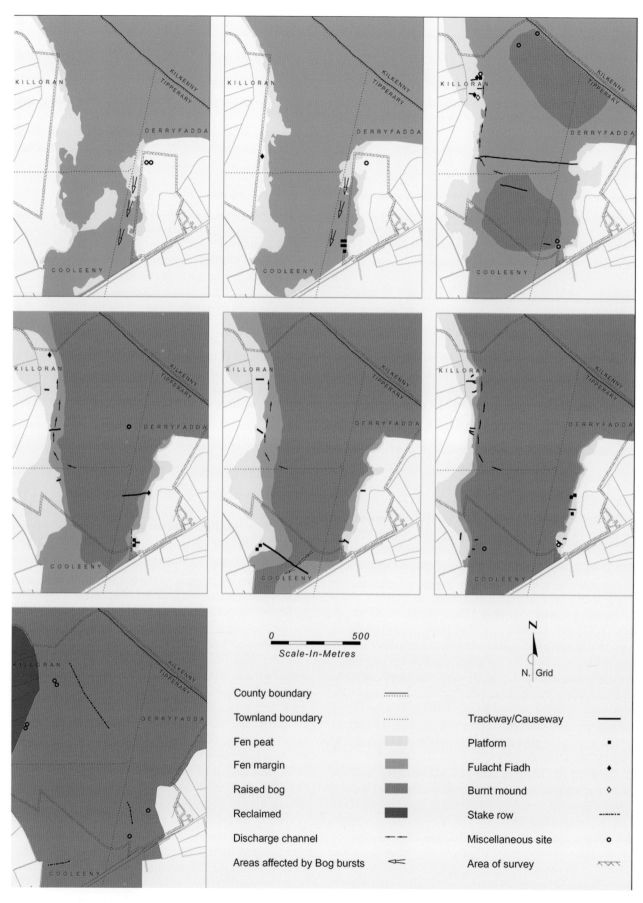

Figure 2.4 Series of 'time slice' models showing the development of Derryville Bog, Co. Tipperary, Ireland, based on detailed survey by Wil Casparie and Bernie Owens. These images show the development of the peatland from the Neolithic (top left) to the Medieval period (bottom left)

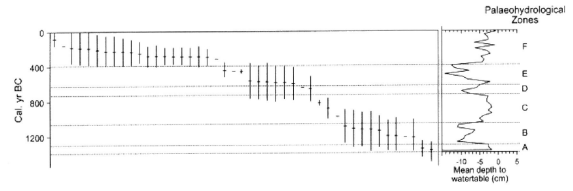

Figure 2.5 Plot of radiocarbon and dendrochronological dates of wetland archaeological structures from Derrryville, Co. Tipperary, Ireland, in relation to shifts in the bog's surface wetness, letters A–F indicate marked falls in watertable related to 'bog bursts' (after Gearey and Caseldine 2006)

more meaningfully integrated into narratives of past cultural change. In particular, how can palaeoenvironmental data, regarding rates, patterns and processes of past change, be better used to understand the complex relationship between people and their environment?

As Brown (2008a, 280) has observed, there is also the issue of the '… integration of very different spatial scales and meaningful interpolation in three dimensions.' Whilst this comment was made in the context of investigating alluvial wetland landscapes, it applies equally to peatlands. In other words, synthesising spatial and temporal datasets is in part a theoretical issue but is also a significant methodological challenge, especially as the resources required to carry out detailed archaeo-environmental investigations of peatlands can be considerable What are the data requirements for investigating raised mires as archaeological landscapes, particularly given that traditional methods of prospection, such as aerial photography and geophysics, are of limited value? The next section introduces the study areas that were the starting point for addressing these questions and previous research relating to them.

2.4 Case studies: Hatfield and Thorne Moors

The raised mires of Hatfield and Thorne Moors (Figure 2.6) were primarily chosen as case studies as they are the largest surviving areas of lowland raised mire in the United Kingdom, with a considerable legacy of previous research dedicated to them. They are situated in the broader region known as the Humber Wetlands which has a corresponding legacy of archaeological and palaeoenvironmental study (Van de Noort 2004) and are the remnants of a previously more extensive complex of wetlands on the lower floodplain of the River Trent which were once the largest examples of such wetlands in England (Lindsay *et al.* 1992).

The Humber peatlands fall partially under the rain shadow of the Pennines and hence receive a reduced westerly oceanic influence. Average annual precipitation is less than 600mm/year; with a potential evaporation of 510mm, mean January temperatures of 3.0°C and mean July temperatures of 16.3°C. These peatlands are thus unusual in that they are located in a region with a relatively warm, dry climate not generally associated with the development of lowland raised mire systems, which tend to be located in the western parts of Britain, that benefit from moist southwesterly airflows and orographic (resulting from the passage of moist air masses over hills or mountains) rainfall. As such, they are often compared to continental examples of mire systems (Dinnin 1993).

In their intact state both mires probably had 'classic' raised dome profiles; Woodruffe-Peacock (writing in 1920–21, quoted in Limbert 1987) stated how the entire complex of Thorne Moors was some 10km across and described the dome as rising 7.5–10.5m above the surrounding plain and fluctuating in height (*Moor-atmung*, rises and falls in the bog surface caused by climate, gas production and hydrology; Blaauw and Mauquoy 2012) by as much as 2m depending on seasonal rainfall. Rogers and Bellamy (1972) suggested that in this pristine state both mires may have had liquid cores. Although the mires would have been ombrotrophic systems, it is possible that run-off from the calcareous deposits of Lindholme Island (see below) may have had some influence on the hydrological regime of Hatfield Moors. Hatfield and Thorne Moors were once separated by a branch of the River Don which flowed eastwards towards the Trent being joined by the River Idle on the way (Figure 2.6). The latter once formed the eastern boundary of Hatfield Moors. The branch of the Don was stopped at Thorne and diverted eastwards in the 17th century, whilst the Idle was diverted eastwards to the Trent (Dinnin 1997a). The former courses of the River Torne and the River Idle are located to the south and east of Hatfield Moors respectively.

Although it has been suggested that drainage of the peatlands may have begun in the Roman period (Buckland and Sadler 1985), the current form of both landscapes is very much a product of later and Post-Medieval land-use (e.g. Limbert 1985, 1986, 1987), including drainage, reclamation for agriculture, peat cutting (for stable litter and later for horticulture) and other schemes of land 'improvement', such as 'warping' (the deposition

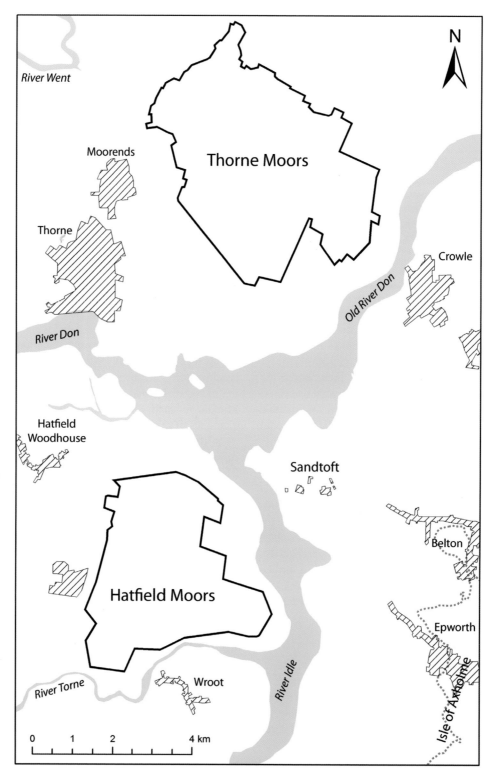

Figure 2.6 Detailed location map of Hatfield and Thorne Moors in relation to modern settlements (the areas in blue indicate the approximate extents of the river floodplains. The areas of hatching indicate current areas of settlement)

of alluvial sediments to improve the fertility of soils). Hence the current extents of the peatlands are certainly much reduced, although their precise maximum extents are unclear. Reclamation of the Humberhead levels began in earnest during the Middle Ages, but became more extensive following the passing in 1600 of *An Act for the recovery and inning of drowned and surrounded grounds and the draining dry of watery marshes, fens, bogs, moors and other grounds of like nature* (Dinnin 1997a). However, drainage of the area did not commence on a large scale until the 1620s with the work of Cornelius Vermuyden with drainage and peat cutting, particularly

Figure 2.7 The modern landscapes of Hatfield and Thorne Moors are largely the product of Post-Medieval industrial activity

Figure 2.8 The end of peat cutting on Hatfield Moors

between Tween Bridge Moor and Sand and Nun Moors on the southwestern side of Thorne. In contrast, the east, north and west sides of Thorne Moors were flood-warped from the late 18th to the early 20th century (Lillie 1997, 1998, 1999). On Hatfield Moors a large area of peatland east of Lindholme was reclaimed by cart-warping during the 19th century (Lillie 1997).

Peat continued to be cut on the un-reclaimed parts of both Hatfield and Thorne Moors between the late 19th and early 20th centuries. In the latter part of the 19th century intensive industrial peat cutting began on Thorne Moors and to a lesser degree on Hatfield Moors. Peat production, for use as stable litter, peaked in about 1910 (Eversham 1991) and was followed by a reduction in demand between about 1920 and 1950. Large scale extraction for the horticultural market began during the 1950s and was mechanised the following decade. Since then, increasingly effective drainage and extraction techniques have been developed, including the introduction of industrial surface milling during the 1980s, initially on Hatfield Moors and later on Thorne Moors (Figure 2.7).

Part of Hatfield Moors was notified as a site of Special Scientific Interest (SSSI) under the 1949 *National Parks and Access to the Countryside Act*, and re-notified in 1986 with boundary notifications under the 1981 *Wildlife and Countryside Act*. The boundary was extended in 1988 to include the whole peatland. Thorne, Goole and Crowle Moors were notified as an SSSI in 1970, when they were combined together as 'Thorne Moors'. In 1985, the then Nature Conservancy Council purchased a block of the peatland to form the National Nature Reserve. On the 27 February 2002, it was announced that the responsibility for the management of peatlands would transfer from the peat extractors, the Scotts Company, to Natural England (then English Nature). When peat cutting ended on Thorne Moors that same year, much of the southern, western and eastern parts of the Moors existed as abandoned peat cuttings in various stages of re-colonisation by vegetation, with a complex pattern of mire, wet and dry heath, scrub, fen and fen meadow communities.

Figure 2.9 The flooding of Hatfield Moors

The environmental management plan agreed between English Nature (now Natural England, the landowners) and The Scotts Company (the tenants) had included a clause that the basal 0.5m of peat should not be extracted in order to form a basis for future peatland rehabilitation (Wheeler and Shaw 1995). Peat cutting on Hatfield Moors finally ceased in 2004 (Figure 2.8). The cessation of peat cutting on Hatfield and Thorne Moors ended some 200 years of the removal of peat for domestic and commercial uses. Natural England has now embarked upon a programme of restoration which has focussed on the re-wetting of the remaining core areas of the peatlands in the hope that they will eventually regenerate into raised mires (Figure 2.9).

The current extents of each of the Moors (Hatfield Moors, 1360ha and Thorne Moors, 1760ha) have therefore been considerably reduced over the last 400 years through the combination of drainage, peat cutting and land improvement, particularly through 'warping'. The basal sediments under much of Hatfield Moors consist of blown sand deposits which reach nearly 5m deep in places and are underlain in parts by the lacustrine clay-silts of pro-glacial Lake Humber (Whitehouse, Buckland, Boswijk *et al.* 2001). Lindholme Island in the centre of Hatfield Moors (Figure 2.10) has always been an area of 'dryland' and is one of several discontinuous deposits of sand and gravel stretching southeast from near Selby to the southern part of the Isle of Axholme, deposited as an ice marginal morainic

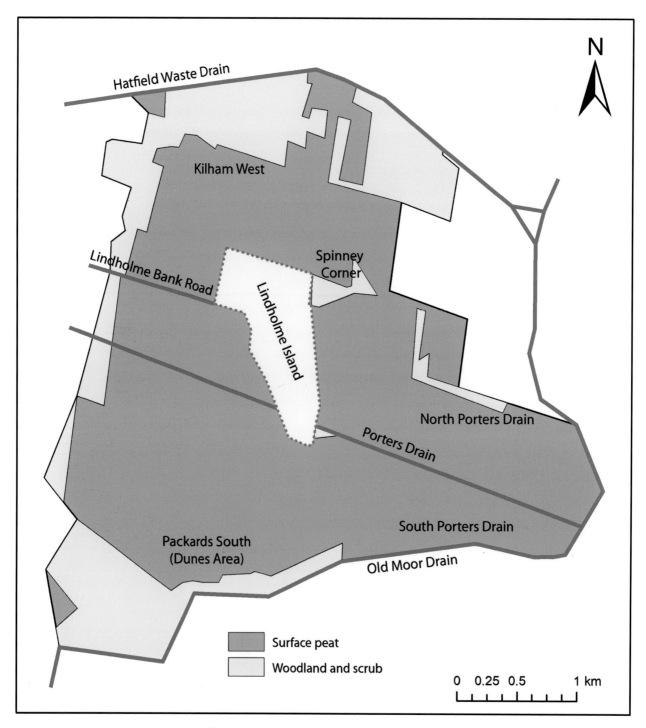

Figure 2.10 Principal areas of Hatfield Moors

sediment during the high level phase of 'Lake Humber' in the Late Devensian (Whitehouse, Buckland, Wagner *et al.* 2001). The presence of limestone in these deposits (see above) permitted a calciophile flora and fauna within an otherwise acidic environment. Thorne Moors (Figure 2.11) comprises the areas of Thorne Waste, Snaith and Cowick Moors, Rawcliffe Moor, Goole Moor and Crowle Moor. The site has been cut-over for peat and the network of ditches, canals and drains across the area are largely related to this process.

The study areas for both peatlands were initially defined by the surviving extents of cut-over peat not under warp or agricultural land. In practice the presence of dense vegetation in some parts of the Moors, as well as the progress in measures associated with Natural England's restoration of the peatlands, including the flooding of certain compartments and consequent limited access, reduced the area available for field work and survey. This was a particular problem on Thorne Moors which ultimately restricted the programme of work carried out on this site.

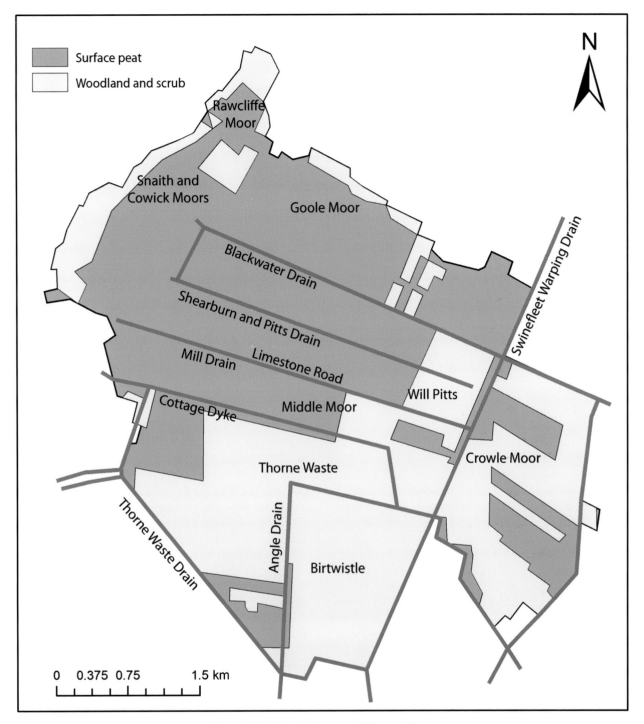

Figure 2.11 Principal areas of Thorne Moors

2.5 Topographic, archaeological and palaeoenvironmental data for Hatfield and Thorne Moors

Previous work by the Ordnance Survey and the British Geological Survey generated a range of cartographic datasets, including historical and modern mapping, three-dimensional topographical data and geological mapping. In addition, surveys undertaken by the Environment Agency and Natural England have produced information relating to the surface topography of Hatfield Moors, habitat distributions on Thorne Moors and low resolution peat depth data from across both landscapes. The peat extractors, the Scotts Company, made available ground penetrating radar (GPR) data relating to peat depths on Hatfield Moors. In addition to published research, archaeological and palaeoecological data were available through a variety of local and national databases. A summary of these datasets is provided in Table 2.1, and a discussion of the cartographic datasets follows.

Table 2.1 Available topographic and archaeological datasets for Hatfield and Thorne Moors

2D mapping and land cover	1:10,000 mapping
	1:25,000 mapping
	1:50,000 mapping
	Landline
	Mastermap
	Natural England Habitat distribution: Thorne Moors (2002)
	Natural England Regeneration land usage mapping: Thorne Moors (2002)
Historic mapping	Ordnance Survey 25in 1st edn County Series (1855)
	Ordnance Survey 25in 2nd edn County Series
	Ordnance Survey 10in
Geological mapping	British Geological Survey 1:50,000 scale mapping
	Sheet 079 Goole (Thorne Moors)
	Sheet 088 Doncaster (Hatfield Moors)
3D landscape surface data	Landform Panorama (50m resolution DTM)
	Landform Profile (10m resolution DTM)
	Environment Agency LIDAR for Hatfield Moors (2m resolution)
	DSM (Digital Surface Model – including tree tops)
	DTM (Digital Terrain Model – bare earth)
Other stratigraphic data	Natural England contour maps of peat depth, base of peat and peatland land-surface (1996)
	The Scotts Company GPR data for southern part of Hatfield Moors
HER data	South Yorkshire Archaeology Service SMR (Sheffield)
	North Lincolnshire SMR (Scunthorpe)
	Humber Archaeology Partnership SMR (Hull)
	The National Monuments Record (NMR)
	The Archaeology Data Service (ADS) online catalogue
	The Humber Wetlands Project database

Historical mapping

The early work by Stonehouse (1839) and Dunston (1909) relates to the study areas, although the resolution of information for each of the Moors is limited. However, the 1st and 2nd edition Ordnance Survey 1:2500 scale maps provide a 'snapshot' of land-use in the recent past. The 1855 1:2500 scale mapping of Hatfield Moors shows the predominantly northwest by southeast alignment of the drainage across the Moors and presumably reflects the intensification of peat cutting during the 19th century (Figure 2.12). The drains around Lindholme Island are less regular and presumably topographically determined.

Following the principal alignment defined by the drainage are a number of footpaths and the 'Holme Bank Road' from the western edge of Lindholme Island, which are likely to be contemporary with the laying out of the drainage pattern. However, a number of other pathways are depicted, some of which are respected by the drainage pattern and thus presumably pre-date it. These include 'David Bank', leading northwest across the Moors from the buildings at Lindholme, and three footpaths to the south of Lindholme Island leading towards Moor Drain at the northern edge of the old River Torne. The principal path leads south-southeast from the southern tip of Lindholme Island towards a bend in Moor Drain. To the west, and partially following the alignments of the drainage pattern, is a second footpath leading towards a 'wooden bridge' across Moor Drain. There is a third footpath crossing the drainage fields obliquely to the east.

The first edition Ordnance Survey mapping of Thorne Moors was conducted between 1853 and 1855 at a scale of 1:2500 (Figure 2.13) and depicts a predominantly west-northwest by east-southeast aligned drainage pattern with very few sub-divided fields. The density of drainage is low across large areas of the northern part of the Moors, although the pattern is very regular. The relatively featureless landscape of Thorne Moors during the mid-19th century is reflected in the lack of routeways across the peatland. The exception is in the south of the study area where a 'Causeway Bank' is depicted as a linear earthwork running broadly east–west connecting 'Moor Pits' and 'Brown's Well' on the western side of the Moors to the southern area of Crowle Moor to the east. This feature does not follow the general drainage alignment and therefore it probably pre-dates the intensification of peat extraction in the 19th century.

The historical mapping therefore provides a 'snapshot' view of the 19th century land-use of Hatfield and Thorne Moors, including some indication of relative chronology in terms of drainage patterns. The maps also depict post Medieval routeways across the Moors which may fossilise much earlier routeways. Whilst the drains were excavated to facilitate the cutting of peat, it is not clear from the mapping how much of the resource had been cut by this period. The value of these maps is thus in the identification of later land-use. For information regarding the 'hidden landscapes' within and beneath the peat, other data must be sought.

Geological mapping

Geological information (Figure 2.14) is available from the British Geological Society (BGS) at 1:50,000 scale. The

2. Raised mires and the Humber Peatlands

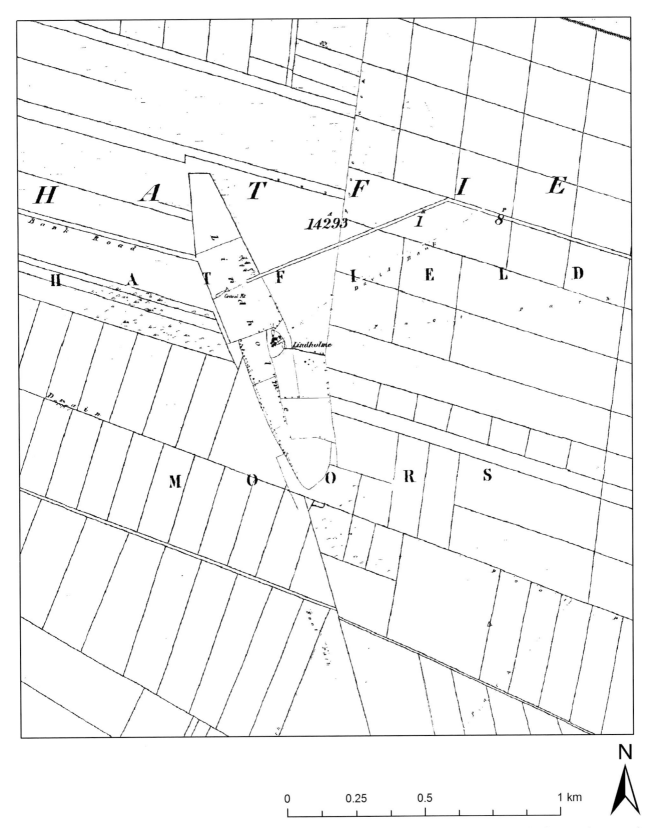

Figure 2.12 Section of the 1st edition 1:2500 scale Ordnance Survey mapping of Hatfield Moors (1855), showing the area of Lindholme Island

study area extends across two adjacent geology maps: Hatfield Moors (Sheet 088 Doncaster) and Thorne Moors (Sheet 097 Goole). The underlying geology is primarily Sherwood Sandstone, with the eastern edges of both Moors underlain by Mercia Mudstone. This solid geology is overlain by the silts and clays of the 'Lake Humber' deposits (Gaunt 1994). These sediments are part of the larger spread of laminated clays and sands which cover

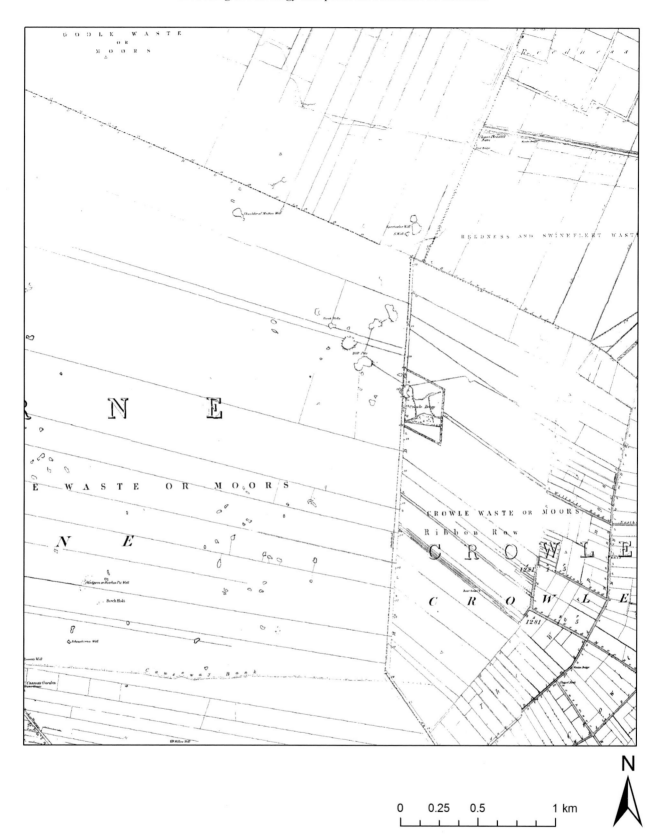

Figure 2.13 Section of the 1st edition 1:2500 scale Ordnance Survey mapping of Thorne Moors (1853–1855), showing the eastern area

much of the Vale of York below an elevation of 7.6m (or 25 ft, hence the name '25 foot drift') and have previously been identified as the basal deposit below Thorne Moors as a whole (see Whitehouse, Boswijk and Buckland 2001). Their geomorphological and lithological characteristics suggest lacustrine deposition in the pro-glacial 'Lake Humber', which was created when Late Devensian ice blocked the Humber gap, impounding the glacial outwash

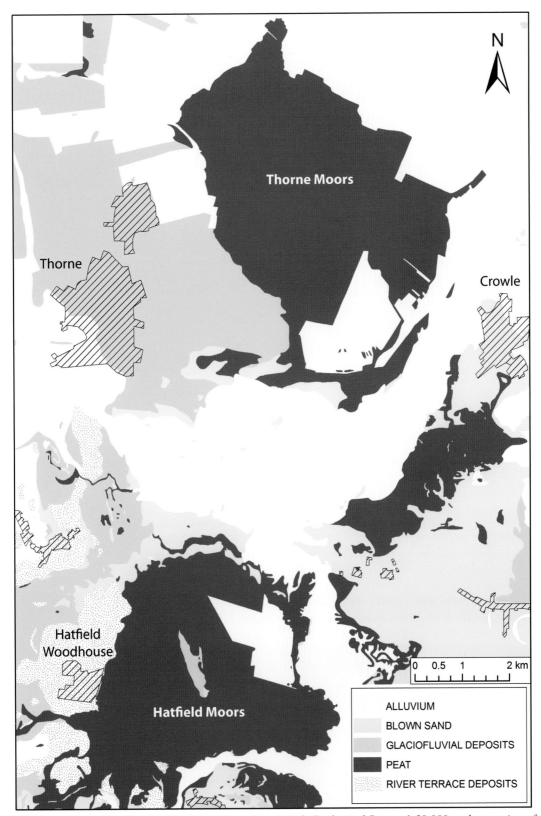

Figure 2.14 Drift geology of the Humber Peatlands (based on British Geological Survey 1:50,000 scale mapping of the area)

from the rivers Trent, Idle, Torne, Don, Aire, Ouse, Derwent and Ancholme (Bateman and Buckland, 2001, Gaunt 1994). The 'highstand' of Lake Humber has recently been dated to *c.* 16,000 years before present (Bateman *et al.* 2008).

The Hatfield Moors area is underlain by coversands, which were probably deposited during the Loch Lomond stadial, although some re-working during the early Holocene is possible (Bateman 1998, Gaunt 1994, Whitehouse,

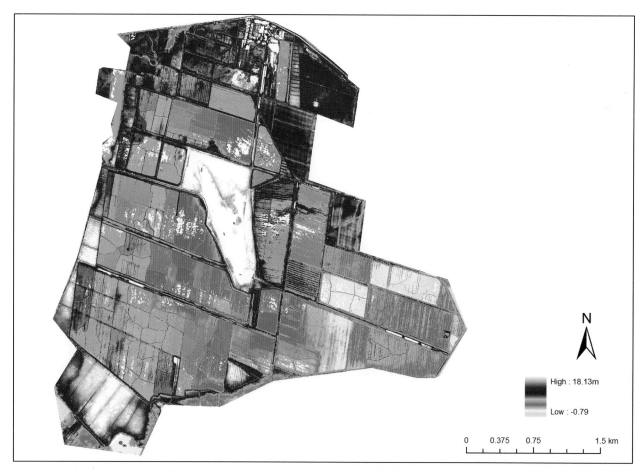

Figure 2.15 The LIDAR data for Hatfield Moors (provided by the Environment Agency)

Buckland, Boswijk *et al.* 2001). Lindholme Island is formed of sands and gravels including Permian Magnesian limestone pebbles, deposited by the Late Devensian terminal moraine within Lake Humber (Gaunt 1976). Whilst the geological maps provide evidence of the nature and extents of these drift deposits, they lack sufficient detail on peat depth in three dimensions to permit an understanding of the morphology of the pre-peat landscape.

Three-dimensional surface data

The Ordnance Survey provides two resolutions of three-dimensional surface data within the British Isles: Landform Profile at 1:10,000 and Landform Panorama at 1:50,000. These datasets can be used to generate Digital Elevation Models (DEMs), within the GIS representing the landscape at cell sizes of 10 × 10m and 50 × 50m respectively. For Hatfield Moors, the higher resolution, 1:10,000 scale DEM displayed a range in elevation values between 0m and 4.20m OD, with a mean value of 2.51m OD. For Thorne Moors, the DEM ranged between 0.60m and 4.00m OD, with a mean value of 2.60m OD.

Additional land-surface information was available for Hatfield Moors from the Environment Agency. A LIDAR (Light Detection and Ranging) dataset covering the area provided higher resolution information regarding the contemporary land-surface of Hatfield Moors. LIDAR is an airborne remote sensing approach whereby the landscape is scanned using laser technology to create high resolution surface data corrected using Global Positioning System (GPS) equipment (English Heritage 2007). The data for Hatfield Moors were collected using the Environment Agency ALTM 3100 LIDAR instrument on the 28 February and 1 March 2006 (Environment Agency 2006) and processed at a surface resolution of 2m. Two different datasets were provided for the study area. The first was a Digital Surface Model (DSM) which included the heights of modern features including buildings, treetops and other vegetation, thereby obscuring the land-surface of much of the Moors, and hence was less useful when considering the earlier landscape. The second dataset was a Digital Terrain Model (DTM) which represents a 'bare-earth' elevation model, generated by passing the DSM through a classification and filtering routine that attempts to strip out vegetation and buildings providing a model of the current surface of Hatfield Moors (Figure 2.15).

2.6 The archaeological record from Hatfield and Thorne Moors

The archaeological record for both Hatfield and Thorne Moors is dominated by the finds made by antiquarians during the 17th, 18th and 19th centuries, which were mostly discovered during drainage activities, peat cutting and the excavation of 'bog oaks' for firewood, fence staves and ships' masts (Stovin 1747). These activities have since been replaced by a mechanised process of peat milling which does not offer the same opportunities for archaeological discovery compared with fieldwalking (Coles and Coles 1996). Peat extraction itself has been described as having one of the most damaging impacts on the archaeology of wetlands (Van de Noort et al. 2002) although ironically it is also the reason why mires have yielded so many archaeological discoveries (Buckland 1993). With the exception of a site on Thorne Moors (see below), no archaeological finds have been reported during the period of mechanised milling of the Moors. The uncertainties relating to the provenance of the majority of past discoveries mean that it is hard to draw firm conclusions regarding the nature or extent of past human activity. For example, a number of antiquarian sources relating to archaeological material refer to the 'Hatfield Chase'; an area taken to include what is currently named the Hatfield Chase, both High and Low Levels, Hatfield Moors and sometimes even Thorne Moors, Goole Moor, Marshland and North Moor.

Hatfield Moors

The earliest references to the archaeological record of Hatfield Moors was by Abraham De la Pryme (1671–1704) who recorded bog oaks and Roman coins and postulated a number of interpretations regarding the landscape, as well as recording a bog body from the region (De la Pryme Ms, Turner and Scaife 1995). George Stovin (c. 1695–1780) also discussed the archaeology of the region alongside anecdotal stories and comments on the drainage (e.g. Stovin 1747). Joseph Hunter (1783–1861) described finds from the Hatfield area, including a second bog body (Hunter 1828, Turner and Scaife 1995), whilst John Tomlinson focussed on the drainage and history of the region, but recorded finds and also discussed the 15th century hermitage on Lindholme Island at the centre of Hatfield Moors (Tomlinson 1882, Eversham et al. 1995).

Lindholme Island has remained an area of dryland within the peatland. Its isolated location recalls Irish early monastic sites and hermitages (Nóra Bermingham, pers. comm.) as well as prehistoric sites on the Somerset Levels, such as Meare (Bulleid and Gray 1948, Coles 1987). Despite this, there is no unambiguous evidence for prehistoric activity on Lindholme. Eversham et al. (1995) recorded concentrations of Neolithic and Bronze Age finds, but Van de Noort et al. (1997) suggested this might relate to the description of 'a flint in the shape of a battle axe' made whilst cutting drains for warping on the eastern side of Lindholme Island in the 19th century (Tomlinson 1882). This indicates some form of prehistoric activity in the area of the island, but the lack of detailed contextual information makes any further comment speculative. The hermitage on the island (see above) was occupied in the 15th century by one William of Lindholme. William's remains were reported in the *Gentleman's Magazine* in January 1747 to have been discovered during excavation in the 18th century. This may also relate to the discovery in 1868 of human bones from a brick-built vault (Tomlinson 1882).

Finds of 'worked timber', in conjunction with Roman coins, objects including 'wedges' for wood working, 'horses heads' and 'possible evidence of early cultivation' are all mentioned (De la Pryme 1701, Stovin 1747). Van de Noort et al. (1997) regarded these findings as possibly relating to areas outside of the current limits of the Moors. Three bog bodies have been recorded from the vicinity of Hatfield Chase. The primary source for the discovery of the first of these finds is a letter from De la Pryme to the Royal Society (de la Pryme 1701) and the second is recorded by Hunter (1828). Both are listed as bog bodies 58/1 and 58/2 within Turner and Scaife's (1995) gazetteer, along with the following oft-repeated quotes:

> '... about 50 years ago, at the very bottom of a turf-pit, there was found a man lying at his length, with his head upon his arm, as in a common posture of sleep, whose skin being tanned as it were by the moor-water, preserved his shape entirely, but within, his flesh and most of his bones were consumed.' (De la Pryme 1701). 'A pair of similar sandals to those from the Amcotts body; (Turner and Scaife 1995, 30/1) taken from the feet of a body found in the reign of Elizabeth, formerly hung up in the hall of the Triggots at South Kirkby, which ... were said to be taken from the feet of an antediluvian.' (Hunter 1828, 154).

The similarity between the sandals on the latter body and those on the Amcotts body has suggested a date in the late 3rd or 4th century AD for body 58/2 (Turner and Rhodes 1992). The third body, Turner and Scaife's (1995) bog body 58/3, was a male dressed in 'Saxon' clothes, discovered at some time before 1720 (Bakewell 1833).

References to charred material within the peats were commonly made. The discovery of 'bog oaks' displaying evidence of charring led De la Pryme (1671–1704) to conclude in a letter to the Dean of York in 1699 that the forest of Hatfield Levels had been destroyed by the Romans. This idea appeared to be confirmed by the discovery of eight or nine Roman coins that were found near a tree root within the Hatfield area (De la Pryme 1701). This interpretation of the Romans being responsible for the destruction of the previous forest cover of Hatfield Levels is one that remained popular with subsequent antiquarians, including Stovin (1747), Hunter (1828) and Tomlinson (1882). A summary of previous finds from the Hatfield Moors area is provided in Table 2.2.

The previous discussion therefore demonstrates that a number of archaeological sites and artefacts have been found on, or close to, the peatland in the past. However, little archaeology has been discovered in recent times. Archaeological survey including fieldwalking was carried out as part of the English Heritage funded Humber Wetlands Project (Van de Noort and Ellis 1997), but no additional finds were made. More recently, Hatfield Moors was examined as part of archaeological mitigation work (Lillie and Gearey 2001) although the combination of ditch

Table 2.2 Archaeological finds from Hatfield Moors

Area	OS grid reference	Site type	Date	Reference
Hatfield Moors	Unknown	Bog oaks and 8 Roman coins	Unknown, Roman	de la Pryme 1701, Turner and Scaife 1995
Hatfield Moors	Unknown	Flint battle axe	Unknown	Tomlinson 1882
Lindholme Island	Unknown	Hermitage	Unknown	Tomlinson 1882
Lindholme Island	Unknown	'Finds'	Neolithic, Bronze Age	Eversham *et al.* 1995
Hatfield Chase	Unknown	Roman coins, battle axes, wedges & horses' heads	Roman, unknown	de la Pryme 1701, Stovin 1747
Hatfield Chase	Unknown	Bog body (58/1)	Unknown	de la Pryme 1701, Turner and Scaife 1995
Hatfield Chase	Unknown	Bog body (58/2)	?Roman	Hunter 1828, Turner and Scaife 1995
Hatfield Chase	Unknown	Bog body (58/3)	?Saxon	Bakewell 1833, Turner and Scaife 1995

survey and fieldwalking also failed to reveal any further sites or artefacts.

A collection of material which allegedly includes artefacts recovered by peat cutters on Hatfield Moors has recently come to light within the possession of a local landowner. This collection remains to be analysed in detail, although it includes a number of Neolithic polished axes (including Langdale Group IV series), Bronze Age axe hammers and polished stone balls (Peter Robinson pers. comm.). No further archaeological features are recorded within the Sites and Monuments Record for South Yorkshire. There was a report by a local landowner of a possible 'wooden trackway' observed some years ago within the section of a field drain cutting arable land on the eastern fringes of the Moors, although limited dyke cleaning of this location in 2002 by the Wetland Archaeology and Environments Research Centre, University of Hull, failed to locate any archaeology. However, in the winter of 2004, a cluster of worked wooden poles were discovered on the cut-over surface of the peatland to the north of Lindholme Island. The excavation of this site revealed a wooden trackway structure and platform dating to the end of the Neolithic period, and provided one of the locations for further fieldwork (see Chapter 3). The results of these archaeological excavations are detailed in Chapter 7.

Thorne Moors

Antiquarian discoveries from Thorne Moors include two bog bodies, although each of these was reported indirectly and at a much later date than their discovery. Both were recorded in a letter by Stovin to the Royal Society (Stovin 1747) and included in Turner and Scaife's (1995) gazetteer as bog bodies 59/1 and 60/1. Several prehistoric finds were made including a Neolithic stone axe, a Mesolithic tranchet axe and flint flakes from Nun Moor, the area to the southwest of Thorne Moors (Magilton 1977). The landscape context of each of these finds may indicate a focus on sand ridges beneath the peat, subsequently exposed through peat wastage (Van de Noort *et al.* 1997). In addition, the *Yorkshire Philosophical Society Annual Report* of 1862 mentions the discovery of an undated sword from the Hatfield Chase region, although a more specific location was not recorded. Middle Bronze Age

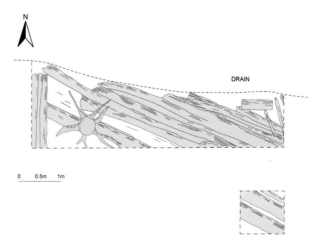

Figure 2.16 Plan of the Thorne Moors trackway excavation trenches (re-drawn after Buckland 1979)

finds, allegedly from the region of Crowle in 1747 (Dudley 1949), appear to have actually come from Burringham, to the east of the River Trent (Buckland 1979).

The most significant archaeological site on Thorne Moors, and indeed the only one to have been discovered and investigated in modern times until the excavation of the Hatfield trackway and platform (see Chapter 7), was made by William Bunting in October 1971 close to the southern edge of the Moors. Subsequent investigations were carried out by Buckland (1979) who reported that the site was considered to be below likely levels of peat extraction. It was thus left *in situ* but limited excavation and detailed palaeoenvironmental analyses were carried out (see below). Two trenches were excavated; the first was 6m long cut back from the ditch side to a width of between 1m and 2m and to a depth of 1m. The second was a 1 × 1m test pit excavated to the south of the first trench (Figure 2.16).

These excavations revealed 'a short stretch of a rough trackway, constructed of timbers of various sizes, orientated approximately southeast to northwest' (Buckland 1979, 10–11). Seven split timbers of *Betula* (birch) and *Pinus sylvestris* (Scots' pine), with a maximum length of 3m, were identified. However, the ends of the timbers had been truncated by the ditch cutting and were probably originally somewhat longer. A radiocarbon determination from the bark of one of the *Pinus* timbers provided a date of 1450–990 cal. BC (Birm-358; 2983±110 BP; see Chapter 5). Several of the timbers appeared to show oblique cut-marks, although no associated artefacts were discovered. Buckland and Dinnin (1997) speculated that the trackway connected Pony Bridge Marsh and Pighill Moor, two areas of raised basal topography that may have remained relatively dry at the time. However, this interpretation is somewhat tentative and it is possible that it might have been a platform or other structure of unclear function (Dinnin 1997b). The location of this site is now flooded (Figure 2.17).

Finds outside the boundaries of the present areas of peatland have been similarly limited. In addition to those mentioned above, North Lincolnshire Sites and Monuments Record lists a concentration of Roman period finds, Bronze Age pottery and later Neolithic/Bronze Age flint remains found directly to the southeast of Thorne Moors adjacent to the Old River Don. Survey and limited excavation by the Archaeological Research and Consultancy Unit (ARCUS, formerly University of Sheffield) produced only a single flint of possible Neolithic date (Dinnin 1994). A summary of previous finds from the Thorne Moors area is provided in Table 2.3.

Figure 2.17 The location of the Thorne Moors Bronze Age trackway following re-wetting, looking east

2.7 Previous palaeoenvironmental research on Hatfield and Thorne Moors

The Moors have been foci for palaeoenvironmental study for a considerable time and the results from most of this work have been published in journal or monograph form. Hence, the aim of this section is not to provide detailed summaries of these studies but to briefly describe the history of palaeoenvironmental research on Hatfield and Thorne Moors, to outline the nature and scope of the work that has been carried out and to identify gaps in knowledge. In the case of unpublished 'grey literature' reports, the results of which are not otherwise easily accessed, more detailed summaries will be provided.

Palynological and coleopteran analyses account for the greater bulk of the work on both Moors to date, with plant macrofossil, peat humification and testate amoebae studies also being significant. Dendrochronological studies have been carried out concentrating on the northern part of Thorne and the western edge of Hatfield. Some limited micromorphological analysis of the interface between the pre-peat land-surfaces and the base of the peat on both Moors was undertaken as part of the work of the Humber Wetlands Project survey (Dinnin 1997b). This work primarily aimed at identifying evidence for disturbance caused by forest clearance, ploughing and other human activity, and to compare results between the two landscapes. Whilst no direct evidence of these processes was discovered within either, on Hatfield Moors cryoturbation structures indicated the freezing and thawing of groundwater during the Devensian. It also identified well developed soil horizons beneath the peat on the northern part of the Lindholme ridge, although evidence for soil faunal activity was limited. On Thorne Moors, this work indicated the presence of soil faunal activity prior to peat formation, and that the pre-peat land-surface was probably around pH neutral, most likely with a brown earth soil which would typically have supported deciduous woodland prior to the onset of peat growth.

In general, more palynological sequences have been analysed from Thorne Moors but Hatfield Moors has seen more detailed work on Coleoptera in the form of complete sequences supported by radiocarbon dates. In contrast, more radiocarbon dates are available for the pollen (and humification, plant macrofossil and testate amoebae) sequences from Thorne Moors. The spatial distribution of study sites (Figure 2.18 and 2.19) shows that Thorne

Table 2.3 Archaeological finds from Thorne Moors

Area	OS grid reference	Site type	Date	Reference
Thorne Moors	Unknown	Bog body (59/1)	Unknown	Turner and Scaife 1995
Thorne Moors	Unknown	Bog body (60/1)	Unknown	Turner and Scaife 1995
Thorne Moors	SE718151	Possible trackway	1450–990 cal. BC*	Buckland 1979
Nun Moor	SE725123	Neolithic axe, Mesolithic tranchet axe & flint flakes	Mesolithic–Neolithic	Magilton 1977
Hatfield Chase	Unknown	Sword	Unknown	Yorkshire Philosophical Society Annual Report 1862

*This is a modelled date: see Chapter 5

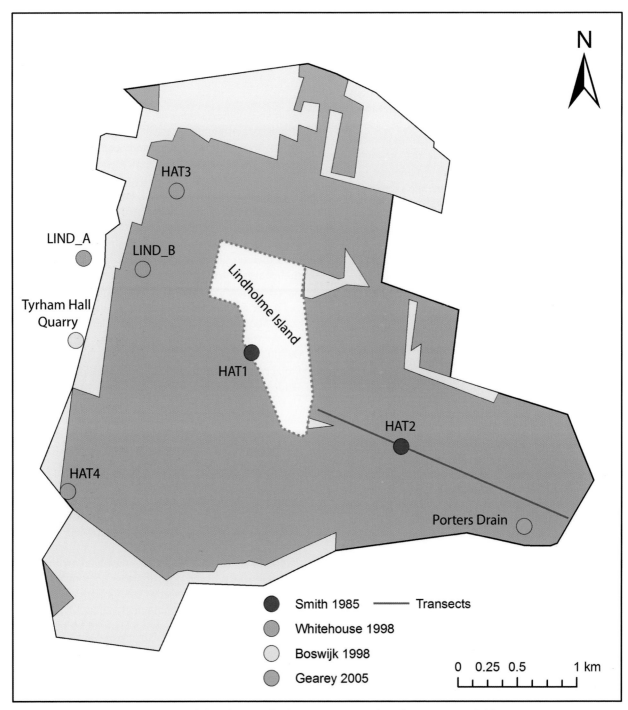

Figure 2.18 Spatial distribution of palaeoenvironmental study sites on Hatfield Moors

has also seen a comparatively wider spread than Hatfield, with the majority of the more detailed work on Hatfield concentrating in the southern end of the peatland.

Erdtman (1928) carried out palynological studies and concluded that peat formation had commenced in the 'Atlantic' period (Pollen zone VIIa in Godwin's (1940) scheme). Clapham (in Piggott 1956) also investigated the age and depth of the peat deposits on Thorne concluding that peat formation had begun in Sub-Boreal and Sub-Atlantic periods (PZ VIIb and VIII). Smith (1958) carried out stratigraphic recording on Hatfield and produced an undated pollen diagram. The basal deposits were suggested to be of 'Atlantic' age with evidence for the Neolithic Elm Decline (*c.* 3800 cal. BC) associated with a layer of macroscopic charcoal. Later clearance episodes in the diagram were attributed to the Iron Age and Romano-British periods. Wetland formation processes were regarded as similar to those of 'blanket mire', implying that paludification of the pre-peat land-surface led to peat accumulation. Changes in humification of the *Sphagnum* deposits indicated four wet shifts, which were attributed to the effects of climatic deteriorations.

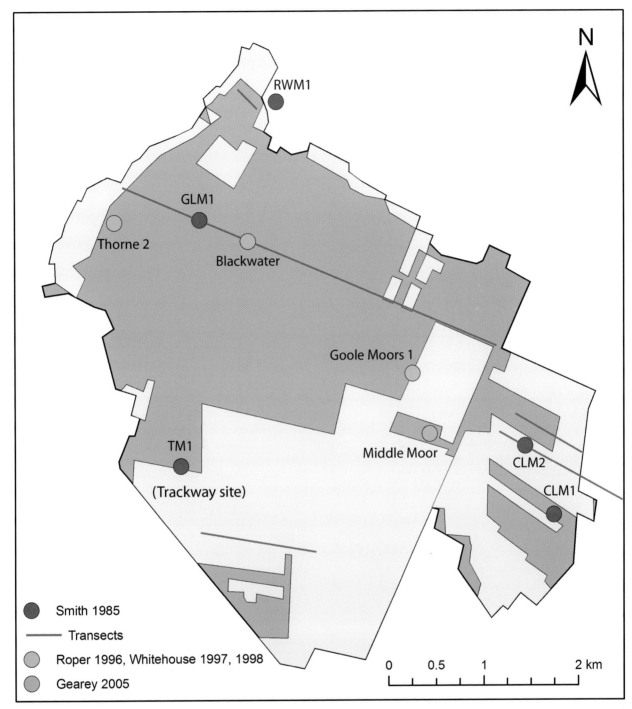

Figure 2.19 Spatial distribution of palaeoenvironmental study sites on Thorne Moors (CLM = Crowle Moor, GLM = Goole Moor, RWM = Rawcliffe Moor, TM = Thorne Moors)

Turner (1962, 1970) followed this work up by producing the first radiocarbon dated pollen diagrams from Thorne Moors, concentrating on the evidence for the *Tilia* (lime) decline which she hypothesised was a result of anthropogenic clearance activity on the dryland around the Moors during the Bronze and Iron Ages. The first coleopteran analyses and also the first modern archaeological investigations were carried out by Paul Buckland and Harry Kenward, following the discovery close to the southwestern edge of the Moors of a structure identified as a short trackway dating to the Bronze Age (described above; Buckland 1979,

Buckland and Kenward 1973). It was also proposed that the pre-peat *Quercus* woodland had been of a relatively open structure that had been pollarded and managed as a form of 'wood pasture' (Buckland and Kenward 1973). The evidence for burnt *Quercus* and *Pinus* timbers close to the trackway site was also interpreted as an indication of deliberate anthropogenic firing of the woodland. However, subsequent investigations of the *Quercus* woodland suggested that timbers had straight trunks with few side branches, typical of closed canopy woodland (Buckland and Dinnin 1997). The issue of the anthropogenic use of fire versus the natural flammability of *Pinus* heathland and mixed deciduous woodland has been much debated (see Whitehouse 1993 and Dinnin 1997b for a comprehensive review of the evidence from both Hatfield and Thorne).

This work was followed by the postgraduate studies of Brian Smith (1985) which was published in monograph form more recently (Smith 2002). The work focussed on the ontogeny and character of wetland development on both peatlands and comprised stratigraphic investigation alongside palynological, plant macrofossil, peat humification and testate amoebae analyses of two peat sequences from Hatfield Moors and five from Thorne Moors. A comprehensive radiocarbon dating programme resulted in 23 radiocarbon dates from the Hatfield Moors sequences and 54 from Thorne Moors. Five 'recurrence' surfaces (*sensu* Granlund 1932) were identified (Table 2.4), indicative of increased BSW that appeared to correlate across both peatlands and with data from mires elsewhere in both Britain and Europe. This was regarded as providing proxy evidence for climatic changes and hence support for the 'phasic' theory of raised mire development (*cf.* Barber 1981).

The palynological data from both peatlands were correlated to define five regional pollen assemblage zones (the 'Humberhead Levels' zones – HHL/A–E; Table 2.5) reflecting a series of phases of woodland development and anthropogenic activity episodes from the Bronze Age through to the Medieval period. Smith (2002) regarded human activity as being of a low level during the Neolithic (HHL/A), although sufficient to have modified the mixed *Quercus–Corylus–Tilia* (oak–hazel–lime) forest to permit the spread of some secondary *Fraxinus* (ash) woodland. The opening of HHL/B was defined by a decline in *Pinus* and is suggested as representing a period of early–middle Bronze Age clearance activity, perhaps including the use of fire to open up the woodland canopy. *Alnus* (alder) also expanded at the opening of this zone, reflecting the development of fen carr related to wetland expansion.

Subsequent rises in taxa indicating open grassy habitats such as *Plantago lanceolata* (ribwort plantain), *Urtica* (nettle) and *Pteridium aquilinum* (bracken) and a pronounced decline in trees including *Tilia* indicated the impact of human activity on the vegetation continuing into the Iron Age (HHL/C) with a predominantly 'pastoral' economy. A recovery in woodland in HHL/E was regarded as representing a 'collapse in agriculture' dating to the end of the Romano-British period and related to the possible impact of a sea level rise and/or the effects of large scale tree clearance on the fluvial regime which might have resulted in increased flooding. The final zone HHL/E reflected renewed woodland clearance during the Medieval period with evidence for the expansion of arable cultivation of a variety of crops including *Cannabis*-type (hemp, hops), *Secale* (rye) and *Triticum/Avena* (oats, wheat).

The early 1990s saw the beginning of a concerted phase of palaeoenvironmental work on both Hatfield and Thorne Moors with postgraduate studies by Nicki Whitehouse (1993, 1998), Tessa Roper (1993) and Gretel Boswijk (1998). Roper (1993) worked on Coleoptera samples from Thorne Moors, concentrating on the transition from fen to raised mire. Whitehouse's (1993, 1998)

Table 2.4 Summary of the Humberhead Levels 'Recurrence Zones'. After Smith (1985, 2002)

HHL recurrence surface	Date
I	AD 1300–1400
II	AD 630–740
III	AD 300–400
IV	350–250 BC
V	650–850 BC

Table 2.5 Summary of the Humberhead Levels 'Regional Pollen Assemblage Zones' (based on Smith 1985, 2002)

Regional zone HHL	Date	Vegetation change and human impact
E	?	Norman and later cultivation, emphasis on arable cultivation (*Cannabis*, *Triticum* and *Secale*)
D	AD 1000–1050	Anglo-Saxon clearance. Arable and pastoral farming, some woodland regneration
C	AD 500–300	Pre-Roman/Iron Age and Roman clearance, pastoral and arable farming
		Tilia decline
B	150–350 BC	Bronze Age and Early Iron Age clearance, pastoral farming, limited woodland recovery
		Pinus decline
A	2350 BC	Post-Neolithic modified *Quercus* woodland, *Pinus* dominant on mire margins and sandy soils
	2650 BC	
		Mixed deciduous woodland

coleopteran analyses on Hatfield Moors, produced the first radiocarbon dated, continuous coleopteran sequence from the southwestern corner of this peatland, as well as analysing a series of discrete samples from other locations across the Moors (Whitehouse 1997a). This work illustrates mire development from *c.* 3000 cal. BC in central areas and from *c.* 1500 cal. BC on the southern margins of Hatfield and provides a record of the processes and timing of wetland development and raised mire growth from a sub-fossil beetle perspective. The results are available in a series of journal papers and other publications (Roper 1996, Whitehouse 1997a and b, 2000, 2004, Boswijk and Whitehouse 2002). Concise summaries of much of the work carried out during the 1990s are available in Bateman *et al.* (2001).

Dendrochronological studies were also undertaken at this time, with Boswijk's (1998) work on Thorne Moors and Hatfield Moors producing both *Pinus* and *Quercus* chronologies from the extensive remains of these trees preserved in the basal peat. The *Pinus* chronology from Tyrham Hall Quarry on the western edge of Hatfield was dated to 2921–2445 BC and a single sample of *Quercus* to 3618–3418 BC (Boswijk 1998). *Quercus* and *Pinus* chronologies from Thorne were also constructed; the former produced a chronology of 3777–3017 BC and the latter showed three distinct phases of woodland growth between 2916–2475 BC. These data were subsequently linked with *Pinus* chronologies from Cheshire and Ireland (Chambers *et al.* 1997). Integrated study of associated sub-fossil insect assemblages from Tyrham also provided considerable information on the date, composition, age structure and entomological biodiversity of the mid-Holocene woodland (Boswijk and Whitehouse 2002). More recently, multi-proxy analyses (pollen and coleopteran analyses supported by radiocarbon dating) have been carried out on sequences from a palaeochannel of the River Torne and its floodplain between Hatfield Moors and Wroot (Mansell 2011a, 2011b).

The study of Coleoptera from both sites has been important in other ways, specifically from the biogeographic and conservation perspectives (e.g. Whitehouse *et al.* 2008). These data are significant parts of the record of changing biodiversity for the area and illustrate the wider importance of palaeoenvironmental data. The list of sub-fossil beetles recovered from Thorne, for example: '… contains 45% of the species known to have been extirpated from Britain during the last 3000 years …' (Buckland and Dinnin 1997, 11, see also Whitehouse 2006).

Further archaeological work was carried out on Thorne Moors during the 1990s which also included excavation and palaeoenvironmental assessment of samples collected during assessment excavations by ARCUS (Dinnin 1994). Subsequently, the Humber Wetlands Project (Van de Noort and Ellis 1997) undertook fieldwalking as well as stratigraphic description, palynological assessment of 'spot' samples and micromorphological study of buried pre-peat soils from both Moors. This work produced data regarding the depth and approximate temporal span of the remaining peat deposits on both Moors and illustrated: '… the great potential for further research into the nature of landscape development at the two sites' (Dinnin 1997b, 188) and highlighting the difficulties of locating archaeological sites in such landscapes.

Other work carried out on both Moors has included stratigraphic survey, palynological study and analyses of testate amoebae and plant macrofossils, supported by the radiocarbon dating of two sequences from Porters Drain on Hatfield Moors and Middle Moor on Thorne Moors (Gearey 2005).

The analyses at Porters Drain on Hatfield Moors focussed on a peat sequence some 2.5m deep. The sequence reflected the initial accumulation of peat in relatively dry conditions with fen woodland including *Quercus*, *Corylus* and *Alnus* growing locally (Table 2.6). The local environment subsequently seems to have favoured the development of heath communities in which *Pinus* initially expanded. However, *Calluna vulgaris* rapidly spread across the landscape *c.* 2085 cal. BC, indicating that such conditions might have developed in perhaps as little as 200 years following peat inception. Continuing sediment accumulation seems to have resulted in the development of a community characterised by *Calluna*, *Eriophorum* and other monocotyledons with *Sphagnum* as a subordinate component. Increased concentrations of testate amoebae in the sequence some 300 years later *c.* 1700 cal. BC provided evidence of a transition to more acidic and wetter conditions, although the presence of charcoal and charred *Calluna* leaves reflected the persistence of dry conditions and occasional burning of the peatland surface.

Such a phase of a relatively dry 'heath-like' or 'psuedo-raised bog' community (also apparent at Middle Moor, see below) has been identified at the base of other ombrotrophic mires in England (Hughes *et al.* 2000, Hughes 2000). Raised mire developed at Porters Drain around *c.* 1130 cal. BC and the subsequent record illustrates a series of fluctuations in BSW (Table 2.7), with wetter phases at 650 cal. BC, 220 cal. BC, 90 cal. BC, cal. AD 180 and cal. AD 700. Evidence for human impact on the dryland woodland around Hatfield appears to have been of a low level during the Neolithic but increased into the Bronze Age (*c.* 1530 cal. BC) prior to marked impacts on the vegetation during the Iron Age (*c.* 400–200 cal. BC) and into the Romano-British periods.

The Middle Moor sequence from Thorne Moors was some 4.14m deep and sampled from a pronounced depression or possible palaeochannel feature. The results demonstrated that peat accumulation began *c.* 3140 cal. BC, with the deposition of silt-rich woody sediment in an *Alnus* carr system in either a slow moving or still waterbody. The infilling of the palaeochannel with silt rich wood peats was followed by the local spread of *Betula* and *Pinus* by *c.* 1600 cal. BC with the growth of a fern rich understory including *Osmunda regalis*. The decline in *Alnus* at this time was interpreted as demonstrating that the local soil pH had fallen below the optimal range of 4–7 for this tree (Bennett and Birks 1990).

Betula continued to expand at the expense of *Alnus*

Table 2.6 Summary of vegetation changes at Porters Drain and comparison with other palaeoenvironmental studies on Hatfield Moors

Date BC/AD	Porters Drain Mire vegetation	Interpretation	HAT2 (Smith 2002)	HAT4, LINDB, Whitehouse (2004)
2300 BC	Betula-Quercus-Alnus woodland	Rheotrophic woodland	2385 BC Betula-Alnus woodland	3350–3030 BC Mesotrophic Pinus-Calluna heath with Quercus & Betula (HAT3)
			2305 BC Calluna-Eriophorum	2900-2350BC Acid heath but with ombrotrophic elements (LIND_B)
2170 BC	Pinus and Calluna expanding	Acid heath		
	Pinus decline			
1760 BC	Calluna dominant Calluna-Eriophorum Sphagnum			1520–1300 BC: Eutrophic fen-heath community HAT4)
1130 BC	Sphagnum-Section cuspidata	Ombrotrophic raised mire	1290 BC Sphagnum cuspidatum (850–650 BC: HHL V)	1130–840 BC: Ombrotrophic raised mire (LIND_B) 1000 BC: Ombrotrophic raised mire (HAT4)
815 BC			725 BC S.imbricatum, sect.Acutifolia	
	Peak in Calluna vulgaris			
c. 400 BC	Sphagnum imbricatum section Acutifolia			
350–300 & 212–237 BC	Section Acutifolia-S.imbricatum-Calluna		(350–550 BC: HHL IV) c. 305 BC S.imbricatum, sect. Cuspidata	
156 BC–AD 81	Monocots-S.imbricatum-Calluna			
AD 180	S.imbricatum-Calluna		AD 175 Eriophorum-Tricophorum	
AD 250–433	S.imbricatum		AD 250 S.imbricatum, S.cuspidatum (AD 300–400: HHL III)	
AD 720	Monocots-sect.Acutifolia		(AD 650–740: HHL II)	

and *Pinus* (TMM3) and by *c.* 1070 cal. BC, the fern understorey was replaced by *Menyanthes* and Cyperaceae, indicating a wetter environment. The decline in *Alnus* was also closely associated with the beginning of the *Sphagnum* curve indicating the demise of this tree was associated with the local spread of increasingly oligotrophic conditions.

Another change in the local vegetation *c.* 1070 cal. BC reflected acidic communities, with a decline in *Betula* fen and a local expansion in *Calluna* and other Ericaceous plants, and perhaps also *Myrica gale* (Sweet gale).

The pollen and macrofossil data therefore portray further isolation from base-rich ground waters during

Table 2.7: Summary of palaeohydrological changes at the Porters Drain (Hatfield Moors) and Middle Moor (Thorne Moors) study sites (from Gearey 2005). Proxy records: Tests= testate amoebae derived watertable reconstruction; Macros=Sphagnum leaf counts; Pollen=evidence for hydrological changes inferred from palynological data (e.g. relative fluctuations in Calluna and Sphagnum percentages)

Site	Calibrated date	Proxies	Comments
Thorne Middle Moor	AD 450	Tests/Macros/?Pollen	Section *Cuspidata* dominated peats, rising watertables
	220 BC	Tests/Macros/Pollen	Peak in Section *Cuspidata* and shallow watertable, *Calluna* declines
	750 BC	Macrofossils	Shift to full ombrotrophic mire, Section *Cuspidata* replacing monocot/ericaceous peat
	860 BC	Tests/?Macros	Appearance of *Sphagnum* macros, rise in watertable following brief dry shift
	1070 BC	Tests	Rising test concentrations, watertable curve begins
	3140 BC		Paludification, peat accumulation begins
Hatfield Porters Drain	AD 450–720	Tests	Fluctuating watertables, some evidence for wetting up
	AD 180	Tests/?Macros/Pollen	High watertables following pronounced dry shift
	After 90 BC	Tests/Pollen	Rising watertables following pronounced dry shift
	220 BC	Tests/Macros/Pollen	Slight rise in watertables, peak in Section *Cuspidata*, aquatic taxa recorded in pollen record
	650 BC	Tests/Pollen	Slight rise in watertables following dry shift with *Calluna* peak
	1130 BC	Tests/Macros/Pollen	Shift to full ombrotrophic mire, rising watertables, Section *Cuspidata* dominated peat (after *c*. 890 BC)
	1640 BC	Tests/?Macros	Rising watertables & appearance of *Sphagnum* macrofossils
	(1760 BC)	?Tests/Pollen	Rising test concentrations, watertable curve begins, *Calluna* decline
	2300 BC		Paludification, peat accumulation begins

resulting in a transition from birch fen to what may be described as a dry, acid heath, which persisted for a *c.* 350 years, prior to a final shift to true ombrotrophic raised mire after *c.* 720 cal. BC. By this point, conditions would have become inimical to the growth of trees in the near vicinity of the sampling site, with *Alnus, Betula* and possibly in part the *Quercus* curves are likely to reflect local woodland, with *Alnus, Betula* and *Quercus* in the 'lagg' areas around the expanding mire and possibly also on any higher and drier areas of the landscape. Accumulation of sediment within an *Alnus* fen system thus continued for some 1200 years, suggesting that fen carr was a longer lived stage in the succession to raised mire at Thorne Moors than previously identified, with previous palaeoecological data (Smith 2002) indicating *Alnus* and *Betula* communities growing on the central area of Thorne Moors for an estimated maximum of 300 years. The testate amoebae and plant macrofossil record indicated a series of shifts in BSW at dates of 860 cal. BC, 750 cal. BC, 220 cal. BC and cal. AD 450 (Table 2.7).

The pollen record from Middle Moor was interpreted as reflecting low levels of human disturbance to the woodland during the later Neolithic–early Bronze Age. Tentative evidence for a reduction or shift in anthropogenic activity during the later Bronze Age (*c.* 1440 cal. BC) was followed by significant woodland clearance during the Iron Age (*c.* 730 cal. BC) and into the Romano-British periods. The later Romano-British into the early Medieval periods (between cal. AD 350–540 and *c.* AD 680), by contrast, were interpreted as indicating woodland regeneration but with evidence for continuing arable activity, perhaps representing a shift in the character or location of settlement and farming activity, rather than actual contraction in agriculture.

In addition to previous work undertaken on Hatfield and Thorne Moors, a range of stratigraphic and palaeoenvironmental studies have also been carried out on the Rivers Torne, Idle, Don and Went, which are closely associated with the Humber peatlands. Prior to drainage and canalisation in the 17th century, the rivers formed meandering courses around Thorne and Hatfield Moors, draining into the Humber Estuary, although subsequent drainage and has altered the configuration of the river network. Gaunt (1976, 1987, 1994) summarised the known extent of the channel and floodplain deposits and proposed an outline of the timing and character of Late Devensian and Holocene fluvial development. The Humber Wetlands Project (Dinnin and Weir 1997b, 1997c) investigated the depth and extent of the fluvial and floodplain deposits of these rivers in the vicinity of the peatlands, although the chronology was based on relative 'pollen dates' with no radiocarbon determinations obtained. The floodplain and channel of the River Idle to the east of Hatfield Moors was around 1.5km wide, significantly larger in comparison to the other locations investigated along its course (Dinnin and Weir 1997c).

Other studies by Buckland (1979) of the River Don at Thorne Waterside and on the River Idle, Buckland and

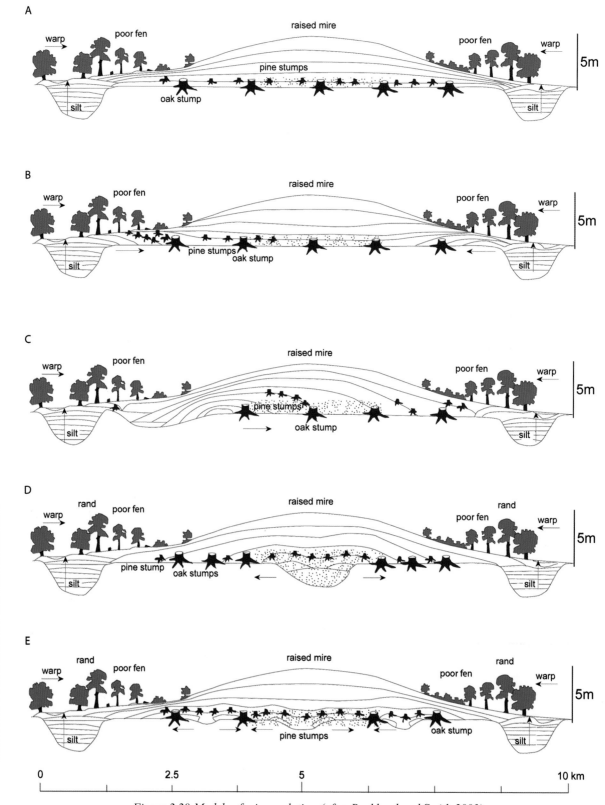

Figure 2.20 Models of mire evolution (after Buckland and Smith 2003)

Dolby (1973) at Misterton Carr (see also Whitehouse 1998) have produced additional information including radiocarbon dates. The very general pattern of development (see also Van de Noort 2004 for a summary) indicates river incision due to low base (sea) levels during the Late Devensian/early Holocene, with channel aggradation from c. 7000–8000 BC. River channels appear to have infilled with associated floodplain aggradation by c. 3000–4000 BC with associated overbank flooding and alluviation during the later Holocene. It should be stressed that significant local variation in this general sequence is implied and the chronology is based on relatively few radiocarbon dates. However, recent study by Mansell (2011a, 2011b) has produced closely dated multi-proxy sequences from the floodplain of the river Torne.

2.8 Models of peatland development

The raised mire deposits of both peatlands in fact represent only the final stage of the successional processes of peat accumulation that began on both areas during the mid-Holocene. The environmental changes represented by the growth and spread of peat will have had implications for human activity in these landscapes through time and, potentially, for the archaeological record. The various palaeoenvironmental studies outlined above indicate that the actual timing and character of the wetland vegetation development preceding the growth of ombrotrophic peat varied significantly between each area. It has been hypothesised that a number of factors may account for these differences, which include the contrasting character of the pre-peat deposits (see above) and hence the composition of the vegetation communities growing on the pre-peat landscapes (Dinnin 1997b). In addition, the actual processes that led initially to paludification and peat inception, and hence to raised mire development, have been the subject of much discussion (Buckland 1979, Smith 1985, Whitehouse 2001a, 2004). Smith (2002) hypothesised that peat formation was linked to climatic deterioration and the effects on local watertables of rising sea levels. Buckland and Smith (2003) subsequently proposed five falsifiable models of peatland development (Figure 2.20, models A–E):

- Wetland spread as a result of rising watertables leading to the widespread paludification of the landscape (Figure 2.20; model A). This may lead to either a multi-focal or single focal development, but lateral spread is regarded as slight. In this case the earliest peat spread should be from the topographically lowest parts of the landscape, since these would be those most prone to rising watertables, and hence the earliest dates for peat growth should be located in these areas.
- Wetland spread from river floodplains (Figure 2.20; models B and C). In this case the areas nearest the river channels should reflect the earliest peat development with dates progressively younger towards the 'core' of the mire. This model stresses the link between the local rivers, rising base levels and impeded drainage.
- Wetland spread as a result of the 'classic' processes of hydroseral succession from an infilling lake basin (Figure 2.20; models D and E). Model E is a 'polyfocal' version of Model D, with restricted run-off causing paludification and the development of numerous *foci* of wetland development. In this model surface drainage may be restricted by local topographic variations. Given the free-draining nature of the basal sandy substrate of Hatfield Moors, it seems likely that any local variation would have been determined in part by base levels which would control local watertables.

All of these models are underpinned by the concept of the accumulation of peat in a fen environment, resulting in progressive acidification and the growth of oligotrophic plant communities. Once such communities were established, the continued accumulation of peat results in the raising of the sediment mass above the influence of groundwater. The mire then expands laterally until a barrier such as high ground or a river course is reached. Other factors such as climate are implicated in these models, especially since the interplay between rising base levels under the influence of sea level change and a wetter/colder climate would be expected to exacerbate and accelerate processes of paludification.

2.9 Summary

Despite the considerable contribution to a range of issues in Holocene palaeoecology made by the various studies of Hatfield and Thorne Moors carried out over the years, the full potential of some of these data are currently somewhat restricted due to an incomplete understanding of the landscape context of the various sampling sites. The archaeological record of both peatlands is strongly biased towards antiquarian finds which provide tantalising evidence for human activity on and around the wetlands but a frustrating lack of surviving, securely provenanced material. It is unclear whether the lack of finds during recent times is a result of the mode of peat extraction which does not lend itself to archaeological discoveries or whether this reflects low concentrations of human activity in the past. Despite the effects of drainage and cutting, these areas still hold the potential for the preservation of organic material, including archaeological features and artefacts (*cf.* Coles and Coles 1996), but the depth and chronological span of these remaining peat deposits is unknown. The next chapter will consider the specific challenges of integrating spatial and temporal datasets in order to map the 'hidden landscapes' of the peatlands for the contextualisation, interpretation and management of the archaeo-environmental resource.

3. Building from the bottom up: aims and approaches

3.1 The challenge: space, time and raised mires

Chapter 1 outlined some of the challenges associated with exploring the spatial and temporal dimensions of past landscape change and human activity, arguing that the investigation of raised mires presents an excellent opportunity for integrating spatial and temporal datasets to enable study rooted within a broad tradition of landscape archaeology. Clearly, whilst wetlands provide the potential for the preservation of an integrated archaeo-environmental record, they have not generally been regarded as representing contexts which can generate information with direct relevance for understanding aspects of past human activity beyond their immediate confines. However, investigating some of the processes that are defining features of wetlands and peatlands, such as wetland inception and peat spread (Chapter 2), may allow an insight into the theoretical and methodological aspects of linking other landscape scale processes of change to past human activity and the archaeological record.

In addition to practical issues (Figure 3.1), a landscape archaeology approach to peatlands presents significant methodological challenges, as aspects of what may be regarded as 'traditional' approaches to landscape archaeology are of limited value in wetland environments (see Chapter 1). Furthermore, few pristine peatlands survive and, in the case of those that have been drained and cut for peat, the landscape has already been effectively already been 'stripped backwards' in time, destroying a significant part of the archaeo-environmental record (e.g. the archaeological and palaeoenvironmental resource that *in situ* peat deposits represent) in the process. Since the established methods of prospection and investigation of dryland landscapes cannot be used 'off the shelf'

Figure 3.1 Landscape archaeology approaches can be challenging within peatland environments

for peatlands, the landscapes below and within the peat cannot be observed directly and data must be collected largely through indirect methods including borehole excavation and palaeoenvironmental study, in conjunction with scientific dating provided by radiocarbon. The complexity and variability of the various datasets means that the application of computer technologies for data management, manipulation and visualisation is very well placed (discussed below).

This chapter presents the approaches used in the study of the 'hidden' landscape archaeology of Hatfield and Thorne Moors. It begins with an outline of the various challenges of studying peatlands as archaeological landscapes, presenting the four specific objectives which relate to the modelling of these dynamic environments. The first three objectives comprise the generation of models of the pre-peat landscapes, the analysis of the factors underlying wetland inception and peat spread and the generation of models of environmental change through the interpretation and chronological modelling of the palaeoenvironmental records. The fourth objective concerns cultural resource management issues, focussed on the quantification of the surviving peatland resource. Finally, the approaches and methodologies used to achieve these objectives are presented.

3.2 Key research objectives

The Holocene history of the landscapes of Hatfield and Thorne Moors begins with the dryland pre-peat environments and continues through to their final form as extensive raised mires (see Chapter 2). Thus, in order to reconstruct the landscape evolution of Hatfield and Thorne Moors, the form and character of the hidden land-surfaces beneath the peat had to be established, and it was necessary to formuate a method for this. The modelling of these landscapes had four principal objectives. Firstly, establishing the morphology of the pre-peat, dryland landscape was required to provide a framework for interpretation and as a foundation for further analyses. Secondly, modelling the inception of wetland and the chronological and spatial spread of peat across the landscapes was necessary to assess implications for past human activity. Thirdly, exploring the patterns and processes of wider landscape and environmental change was needed in order to provide a broader context for the archaeological record.

The subsidiary aim was to explore practical methods that might assist in the cultural resource management of peatlands. This aim was addressed by the fourth objective, to create outputs that could contribute to the future management of these landscapes within the context of current policies relating to peatland restoration. This was based on outputs generated from the first two of the objectives outlined above – the modelling of the pre-peat landscape, and of wetland inception and peat spread – in addition to the production of new models. Together, these provided a foundation for the quantification of the contemporary peatland resource in terms of both the depth of surviving peat and the chronological range represented by these deposits across both landscapes.

Objective 1: the morphology of the pre-peat landscapes

Previous work (see Chapter 2) had demonstrated that peat growth on the Hatfield and Thorne Moors did not commence until the mid-Holocene and that, prior to this, these landscapes were dryland. The areas preserved beneath the peat therefore represent surviving fragments of the wider prehistoric landscape of the region, and it has also been noted that the pre-peat landscapes of both Hatfield and Thorne Moors may be representative of the wider prehistoric landscape of the Humber lowlands (Van de Noort and Davies 1993, Van de Noort 2004). Therefore, despite the loss of peat through cutting, the sealed landscape beneath preserves valuable information regarding the context of past human activity (see Chapter 7).

However, beyond their intrinsic value as preserved prehistoric landscapes, these pre-peat contexts have additional worth. Without some knowledge of the morphology of this basal topography, it is not possible to test the different models of mire development (described in Chapter 2; Buckland and Smith 2003), to assess the significance of the radiocarbon dates for peat spread and development or to fully contextualise the range of dendrochronological, palynological, plant macrofossil, coleopteran and testate amoebae derived data (Chapter 2). These gaps in knowledge have implications for investigating the different potential 'drivers' in peatland development, such as changes in climate, relative sea level and human activity. The problem of direct access to 'hidden' land-surfaces generally dictates that any investigation requires the use of remote methods: techniques such as trench and test pit excavation or restricted intervention through the excavation of boreholes.

Objective 2: wetland inception and peat spread

As discussed above, the timing and nature of wetland inception and spread has significant implications for both the interpretation and management of these peatland landscapes. Previous palaeoenvironmental research carried out on both Hatfield and Thorne Moors suggests the spatial and temporal patterns in wetland development and associated evidence for human impact on the wider landscape. However, a series of outstanding questions remain regarding the timing and pattern of wetland inception and subsequent peat spread. The principal hypothesis for Hatfield holds that there was a rapid transition from mesotrophic (of moderate biological productivity) to ombrotrophic peat, with the existing suite of basal radiocarbon dates suggesting this happened '… more or less simultaneously …' around 3000 cal. BC (Whitehouse, Buckland, Boswijk *et al.* 2001, 196). However, this is based on relatively few radiocarbon dates and the spatial and temporal spread of peat remains poorly understood.

It has been proposed that peat formation commenced on Thorne Moors in the north at Rawcliffe Moor around 3370–3100 cal. BC and spread southwards (Boswijk et al. 2001). The sub-fossil remains of Quercus preserved in the basal peats suggests that mature woodland was growing around this time but died back as Alnus–Betula fen developed. Buckland and Dinnin (1997) proposed that the extant area of Thorne Moors represents not the original core of the mire, but the remnant of a complex which had spread south from around Goole Fields. Three phases of dendrochronologically analysed Pinus woodland (2921–2445 BC, 2227–1810 BC and 1690–1489 BC) have been regarded as supporting a southeasterly spread of mire (Whitehouse, Buckland, Boswijk et al. 2001).

The lack of detailed knowledge of the morphology of the pre-peat landscape hampers attempts to contextualise the results of these various detailed analyses and therefore understand the spatial processes of peat growth and spread (Buckland and Smith 2003; see Chapter 2). As outlined above, from an archaeological perspective, wetland development on Hatfield and Thorne Moors would also have altered the character of the landscapes, with implications for past human activity. Establishing the morphology of the pre-peat landscapes (see objective 1 above) also provided a context for the interpretation of archaeological sites (Chapters 2 and 7). In addition, determining the spread of peat in four dimensions can generate data which might be used in a 'predictive' capacity to identify locations where archaeological remains might survive. Little previous research elsewhere has attempted to investigate peat inception and spread in four dimensions and it was intended that this objective would explore methods and data requirements for future integrated studies of peatland development.

The reconstruction of the pre-peat landscapes of Hatfield and Thorne Moors thus formed the foundation to assess the pattern of peat growth in three dimensions. Firstly, it permitted the results of previous palaeoenvironmental study to be more accurately assessed in terms of the character and spatial patterning of change. However, in order to investigate the spread of wetland in *four* dimensions (spatial and temporal), a suite of radiocarbon dates from the base of the peat at locations from across the Moors was required with associated accurate contextual information including elevation. Such data could then be used to test the models of wetland inception and peat spread outlined in Chapter 2. Chronological control was available from various palaeoenvironmental studies carried out on the Moors, but the value of such legacy datasets for further analysis was dependent on their accuracy, both in terms of the georeferencing of individual studies as well as the robustness and precision of the associated chronologies. Hence, these spatial and temporal aspects of legacy data required further testing and validation (see Chapter 5).

The spatial distribution of basal dates from Hatfield and Thorne Moors from past work was relatively limited, especially for the former, and so additional radiocarbon dates were needed to investigate peat inception. The integration of new datasets alongside the legacy datasets from previous research therefore provided the basis, in conjunction with the model of the pre-peat landscape, to generate GIS models of peat inception (Chapter 6) to test hypotheses regarding the growth and development of the peatlands. Thus, the second objective focussed on the integration of data relating to the timing of wetland formation from across both Moors obtained from both the legacy and the new data from fieldwork and subsequent analysis in order to test the hypotheses relating to peatland inception spread (Buckland and Smith 2002; see Chapter 2). This process led to the creation of models of wetland inception and spread across both Hatfield and Thorne Moors.

Objective 3: modelling and interpreting palaeoenvironmental records: evidence for mire development, landscape change, human impact and the archaeological record

As Van de Noort and O'Sullivan (2006) have observed 'umbrella' terms such as 'wetland' have only been apparent in English, French and Dutch and Danish languages from relatively recently. The nutrient poor, ombrotrophic environments which represent only the 'end point' of the development of the Humber peatlands, would have been very different compared with the various earlier stages of development (see Chapter 2 for details) and would have provided different 'affordances' or opportunities to human communities in the past for perception or action (Gibson 1977, 1979, Kaufman and Clement 2007). For example, from a functional or economic perspective, fen woodlands would have presented a different set of affordances, compared to an acidic mire environment. Understanding the different implications of these changes through time for human perception and action, in part necessitates a quantitative and qualitative approach to the palaeoenvironmental record. Such datasets also reflect the context and impact of human activities on the wider landscape, providing another perspective on the regional archaeological record.

Investigating 'events' such as changes in vegetation and climate, and human perception and possible responses to such processes (as outlined in Chapter 1), requires palaeoenvironmental sequences with robust chronological control. There are a large number of radiocarbon dates from the previous palaeoenvironmental studies on Hatfield and Thorne Moors summarised in Chapter 2. These data provide critical information regarding the character, pattern and process of environmental change during the later Holocene. Formal methods are now available (discussed below) which permit the modelling of sequences of radiocarbon dates in order to establish the chronological relationship between different archaeological and palaeoenvironmental 'events'. Hence, the third objective was to explore the patterns, processes and timing of environmental changes through the integration of past and bespoke datasets using chronological modelling.

Objective 4: assessing and quantifying the surviving peat resource

Peat on Hatfield and Thorne Moors has been extracted for several hundred years, resulting in the transformation of large areas of the mires into post-industrial landscapes and the potential loss of archaeological sites and artefacts (see above and Chapter 2). The heritage value of the surviving peat is twofold: firstly, it represents the direct potential for the preservation of archaeological and palaeoenvironmental source material within its matrix; and secondly, the peat blankets the underlying landscape (as discussed above) and effectively protects any archaeological remains associated with it. Hence, quantifying the peat resource in terms of the depth of deposits and the temporal range these represent is critical for providing a foundation for both interpreting the potential of these deposits as an archaeo-environmental resource and hence informing the future management of these landscapes.

The chronological span of the surviving peat resource has never been established on a landscape scale although, as Dinnin (1997b, 186) observed: 'The elucidation of the temporal variation in the spread of mire has important implications for the age of the landscape buried and preserved by the peat and the maximum age of the heritage resource preserved within it.' The failure of past projects to identify any archaeology on the peatlands is clear, but it was not certain if this represented an actual dearth of surviving archaeology or if other factors were at play. Whilst the peatlands clearly have exceptional potential for the preservation of organic archaeological remains, identifying or locating sites was highly problematic. Traditional approaches to archaeological prospection such as aerial photography and geophysical survey are of limited value in wetlands (Chapter 2). Historically, the majority of finds have been made during peat cutting, but the mechanisation of this process has made such discoveries considerably less likely.

The first stage in the process of understanding the surviving peatland resource was to establish peat depth across each landscape using a combination of the results of previous studies and bespoke fieldwork. The second stage was to define the chronological span represented within the resource for any location, and this was achieved through a programme of radiocarbon dating of basal and surface deposits of the surviving peat (see objective 2 above). The extent of cutting has varied across the study areas and hence the cut-over surface will not be the same age on an intra or inter site basis.

Peat depth for specific locations had been determined by previous investigations such as borehole excavation. However, for the modelling of peat depth on a landscape scale, it was necessary to generate models of the topography of both the pre-peat land-surface (see objective 1 above) and the current cut-over surface; the difference between the two models equating to the depth of the surviving peat. Establishing the chronological span represented within this body of peat required scientific dates for both the base and top of the peat. Tens, or perhaps hundreds, of such dated samples would generally be needed to determine this precisely across the entire extracted area, with all the subsequent resource implications of such an extensive dating programme. The aim was to establish whether an alternative approach could be formulated, using modelled basal dates and interpolation between more limited numbers of radiocarbon dated samples from the cut-over peat surface to permit an estimation of the extent of the loss of the peat resource and the associated archaeo-environmental resource (Chapter 6).

3.3 Approach and methods

The objectives of the project required the integration of a wide variety of legacy and bespoke datasets in order to generate models of environmental change and the peat resource. The legacy datasets (see Chapter 2) and those resulting from fieldwork (see below) consisted of different formats which varied in resolution and the accuracy of their georeferencing. Some datasets were 'continuous surfaces' (such as the Environment Agency LIDAR data or the Ordnance Survey mapping) whereas others were effectively single points in space (such as borehole locations). In other words, the datasets were far from systematic, with some areas of the peatlands explored in greater detail and resolution than others. The challenge therefore lay in combining different sources of data at varying resolutions to create continuous spatial surfaces and models, whilst chronological control underpinned the modelling of spatial patterns of environmental change. GIS provided the platform for manipulating the complex of data and for generating the different models.

Geographical Information Systems: a platform for modelling past landscapes

The use of GIS technology is well attested within studies of landscape archaeology, as it provides the ability to integrate and manage spatial data of varying resolutions, and to interpolate and model surfaces from such diverse datasets (Aldenderfer 1996). It is also capable of analysing models to generate additional information such as slope, aspect, hydrology or visibility and, from such approaches, to analyse archaeological site locations for interpretation and for heritage management (e.g. Kohler and Parker 1986, Altschul 1990, Marozas and Zack 1990, Warren 1990, Kuna and Adelsbergerová 1995, Westcott and Brandon 2000, Kamermans *et al.* 2002, Van Leusen and Kamermans 2005). This analytical capability of GIS means that it is possible to rapidly test varying hypotheses, leading it to be described as 'a place to think' as much as 'a simple data management and mapping tool' (Gillings and Goodrick 1996; *cf.* Gillings and Wise 1999).

The key spatial outputs include three-dimensional deposit models (*cf.* Chapman and Gearey 2003). These 'Digital Elevation Models' (DEMs) provide continuous surfaces that represent the three-dimensional survey data stemming from legacy data and fieldwork. These are

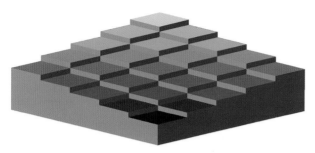

Figure 3.2 How three-dimensional space is represented within a Digital Elevation Model (DEM; after Chapman 2006)

generated in the GIS by mathematically interpolating a surface between measured positions, with the resulting surfaces representing simulations of the physical three-dimensional landscapes, but within the GIS they are expressed as two-dimensional regular grids of square 'cells'. Each cell represents a given physical area (e.g. 1 × 1m) and so the size of the cells within the DEM reflects its resolution (Figure 3.2). Each cell contains a value representing its elevation, and hence these DEMs are commonly referred to as being 2.5D rather than strictly 3D (Burrough 1986).

The accuracy of a DEM is determined by the quality and arrangement of the input data (for example, the three-dimensional position of the base of peat derived from borehole data) and the method of interpolation used to effectively 'fill the gaps' between this data. Factors that can influence the process of interpolation include the regularity of the input data (*cf.* Chapman and Van de Noort 2001, Fletcher and Spicer 1988) and the specific mathematical model used within the software (*cf.* Burrough 1986, Chapman 2006, Conolly and Lake 2006, Robinson and Zubrow 1999, Wheatley and Gillings 2002). For the purposes of generating multiple deposit models of a single landscape (for example, the base and the surface of peat) it is important that the same methods of interpolation and the same 'cell' resolutions are used throughout to ensure comparability and to avoid the addition of error (e.g. Chapman 2001, Chapman and Cheetham 2002).

Methods for modelling pre-peat landscapes (objective 1)

As discussed above, the reconstruction of the pre-peat landscapes of Hatfield and Thorne Moors provided both the context for early Holocene human activity and the literal 'foundations' with which to contextualise palaeoenvironmental evidence for peatland development. Furthermore, determining the form of the pre-peat landscape provided a critical basis for the quantification of the surviving peat resource. As summarised in the previous chapter, in addition to the outputs from previous archaeological and palaeoenvironmental research, a range of other datasets were available for the study areas. The geological maps indicated the general spatial extent and character of the peatlands, and Ordnance Survey and LIDAR (for Hatfield Moors) data provided accurate information regarding the current, cut-over surface of the peatland. Geophysical data (GPR) were available for the southern portion of Hatfield Moors, and contour plots were available from Natural England for the shape of the pre-peat landscapes based on previous borehole surveys carried out in 1996.

In order to generate DEMs of the pre-peat landscapes of Hatfield and Thorne Moors, the suitability and coverage of available legacy data had first to be assessed. In order to achieve this, three main stages of research were undertaken. In the first stage, all pre-existing archaeological, palaeoenvironmental and non-archaeological datasets (including all mapping, LIDAR data and geophysical data, where available) for both peatlands were collated within the GIS and assessed in order to identify gaps in spatial coverage and areas where validation through additional fieldwork was required, due to factors such as uncertainty in georeferencing or where the resolution of the data was relatively low.

Data regarding the morphology of the pre-peat land-surfaces of both Hatfield and Thorne Moors were available in the form of contour plots of peat depths generated from previous borehole surveys carried out in 1996. These plots were digitised within the GIS and processed to create continuous DEM surfaces at 20m resolution (*cf.* Chapman 2006). The resulting models provided initial representations of the morphology of the pre-peat landscape for both study areas (Figures 3.3 and 3.4).

Terrestrial remote sensing data were obtained from the Scotts Company who had undertaken Ground Penetrating Radar (GPR) surveys of peat depths across areas of both Hatfield and Thorne Moors. Whilst the utility of geophysical approaches within peatland environments is relatively limited (see Chapter 1), GPR is effective at identifying transitions between organic and inorganic deposits (e.g. Buteux and Chapman 2009). However, the only data available in digital format covered two areas in the south of Hatfield Moors comprising 2000 × 1400m (western section) and 2000 × 1200m (eastern section). Transects had been surveyed every 130–200m, with the data exported to provide three-dimensional positions of peat surface and base every 50m along individual transects. The resulting data were interpolated within the GIS to generate two separate models of the pre-peat basal topography (Figure 3.5). The topography of the exposed pre-peat land-surface of Lindholme Island at the centre of the Moors meant that, for this area at least, surface datasets were also relevant, including both the Ordnance Survey 1:10,000 scale Landform Profile data and the LIDAR data provided by the Environment Agency.

In addition to these datasets, the results of previous palaeoenvironmental studies were incorporated into the GIS (see Chapter 2). Borehole surveys from both Moors had the potential for providing additional information regarding the elevation of the pre-peat landscape. On Hatfield Moors, these studies were focussed on the western edge of the

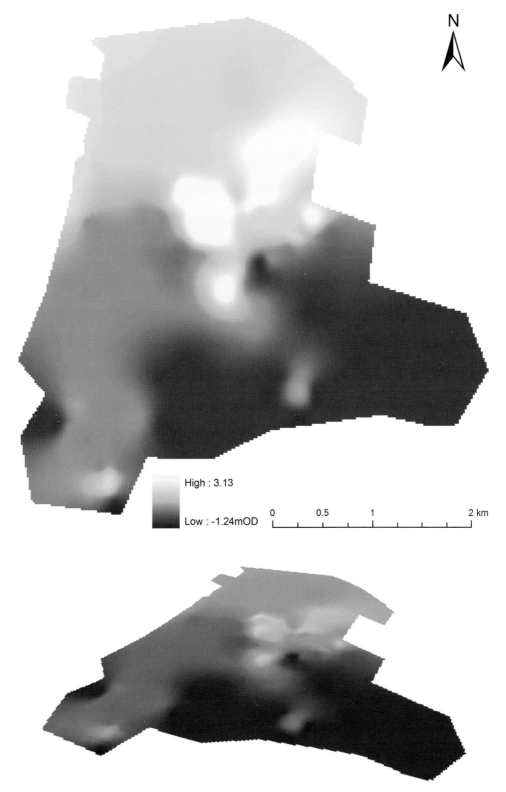

Figure 3.3 Pre-peat landscape of Hatfield Moors modelled from Natural England contours derived from peat-depth data

Moors (Whitehouse 1998, Whitehouse, Buckland, Boswijk *et al.* 2001) with some additional work on the western edge of Lindholme Island and a single transect across the southeastern area of the Moors (Smith 1985, 2002) (see Figure 2.18). On Thorne Moors, the spatial coverage of previous studies was focussed on the peripheral areas of the peatland on Rawcliffe Moor, Goole Moor and Crowle Moor, in addition to some work on the southwestern edge of the Moors (see Figure 2.19). However, the georeferencing of much of this work was problematic, with the majority

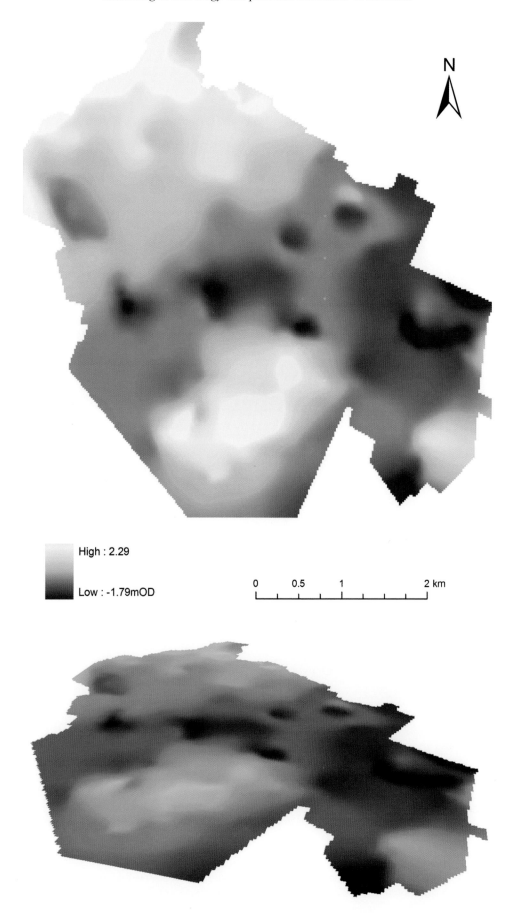

Figure 3.4 Pre-peat landscape of Thorne Moors modelled from Natural England contours derived from peat-depth data

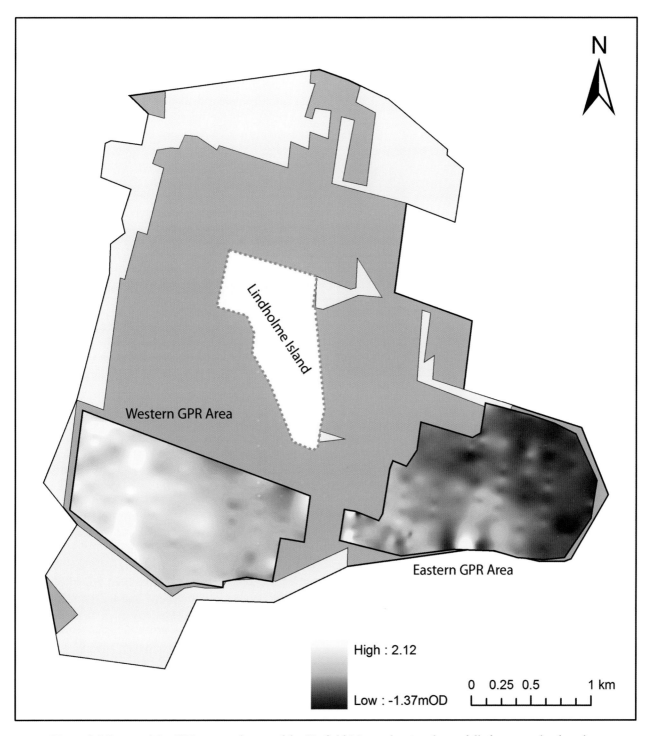

Figure 3.5 Extent of the GPR survey data used for Hatfield Moors showing the modelled pre-peat land-surface

of sites supported by six-figure grid references accurate to just 100m, meaning that they were of limited value for the purposes of integration and collation of subsequent models.

Following the collation of legacy datasets into the GIS, the second stage consisted of a programme of fieldwork consisting primarily of borehole excavation (Figure 3.6). The selection of locations for further fieldwork was partially restricted by problems of accessibility, due to both vegetation coverage and flooding as a result of the ongoing restoration programme by Natural England. As it was not feasible to investigate the entirety of the pre-peat landscape in extensive detail, a sampling approach was adopted, aimed at validating the legacy datasets and obtaining data for areas which had seen little previous study. The strategy consisted of gridded borehole survey (Chapman and Gearey 2003) which produced data to generate deposit models of discrete areas. This also enabled the comparison of peat depth information with that from the Natural England contour data and the GPR data provided by the Scotts Company (see above). Additional

Figure 3.6 Excavating boreholes on Hatfield Moors

borehole transects were also excavated in specific locations across both peatlands to provide basal elevations and peat stratigraphic information over wider areas. All fieldwork sites were positioned relative to Ordnance Datum (OD) using survey grade differential Global Positioning System (GPS) equipment, corrected three-dimensionally to the National Grid. For the majority of fieldwork, a Trimble R8 GPS was used.

All borehole data were collated within the GIS, including the DEMs generated from the areas of gridded borehole survey. The DEMs generated from the gridded borehole surveys were compared statistically with those of the various legacy datasets including the DEMs derived from Natural England contour data and from the GPR data for the southern portion of Hatfield Moors. Statistical comparison for each area was achieved by calculating minimum and maximum values, in addition to mean values, and enabling areas of disjuncture to be identified. This provided a validation and if necessary a rejection of the legacy datasets, whilst generating additional empirical data. Hence, at the end of phase 2, the distribution and quality of the various datasets were established prior to the generation of new DEMs of the pre-peat landscapes.

In the final stage, the combination of validated legacy and new datasets derived from fieldwork were collated in order to generate a single dataset that was used to interpolate DEMs of the pre-peat landscapes. Where appropriate, all data were reduced to their original format which consisted of three-dimensional points, and the datasets were combined into a single file. For Hatfield Moors, the datasets displayed a very wide range of spatial resolutions, including the high (2m) resolution LIDAR dataset and the low resolution contour datasets from Natural England. In order to standardise data for the intended resolution of the output DEM, such high resolution datasets were sub-sampled (see Chapter 4). The modelling of the pre-peat landscape of Thorne Moors was problematic due to the relatively limited range of legacy datasets (see Chapter 4). For the interpolation of continuous surfaces representing both landscapes a consistent tension spline method of interpolation was used for the generation of the resulting pre-peat surface for Hatfield Moors due to the spatial irregularity in the resolution of the input data points (*cf.* Burrough 1986, Chapman 2006). This method provided an output DEM which retains the values for all of the input data whilst restricting the number of interpolation 'artefacts' (Cebecauer *et al.* 2002).

Methods for modelling wetland inception and peat spread (objective 2)

As discussed above, the peat conceals and preserves earlier dryland landscapes, and the date of the basal peat deposits across both areas therefore provides a *terminus ante quem* for the buried landscape. Similarly, the presence of peat equates to the *potential* for organic preservation of the archaeo-environmental resource; all peat will have some palaeoenvironmental potential irrespective of the presence of an archaeological site or artefact. Understanding the timing and patterns of wetland inception and peat spread is thus significant both for the archaeological interpretation of these landscapes and for their management as cultural resources (see objective 4).

However, in order to model peat spread, it is necessary to explore the relationship between spatial and temporal patterns of peat inception. The causes of peat growth and the subsequent raised mire formation have been the subject of some debate and have been presented as five testable hypotheses (outlined in Chapter 2; see Figure 2.20, Buckland and Smith 2003).

The first stage of this process focussed on testing these five hypotheses using a combination of legacy and bespoke data. Ideally, for such large areas, tens or perhaps hundreds of radiocarbon determinations would be required to reconstruct patterns of peat inception. Part of the project rationale was to investigate whether targeted dating of samples from specific locations, alongside good contextual control (e.g. basal topography and stratigraphic data) could produce coherent datasets to test the various models of peat inception. A selection of radiocarbon dates from the base of peat from across both Hatfield and Thorne Moors was available from previous research projects. However, for Hatfield Moors, due to low resolution georeferencing of the sample sites, these dates were insufficient for incorporation into the initial modelling, although they did provide independent data against which the final model could be tested (see Chapter 5).

The new radiocarbon dates were collected from locations identified during the gridded borehole surveys and boreholes transects (see objective 1 above), and were sampled from open peat faces via the excavation of test-pits. For Hatfield Moors, these samples provided both a lateral spread and a range of elevations from across the landscape. This phase of work also included the assessment of selected samples to determine the macrofossil content of the peat samples in advance of radiocarbon dating (Chapter 5; Appendix 1).

As discussed earlier, access problems, due to the flooding as part of the peatland restoration works, meant that the radiocarbon dating programme focussed on Hatfield Moors with no new samples obtained from Thorne Moors, although previous palaeoenvironmental work on the latter had produced a greater number of basal dates (Chapter 2). This phase of the project also included assessment of the chronological robustness (*sensu* Bronk Ramsey 2008) of previous palaeoenvironmental analyses (Chapter 2). The radiocarbon dating and Bayesian modelling programme is discussed in greater detail in Chapter 5.

The second stage of the process focussed on the generation of models testing the hypotheses of wetland inception and peat spread across the Moors, outlined above (and detailed in Chapter 2), using the DEMs of the pre-peat landscapes (see objective 1) as a foundation. This process comprised the testing of models relating to ground water rise and elevation (Figure 2.20; model A), proximity to water courses (Figure 2.20; models B and C), and surface hydrology relating to processes of hydroseral succession (Figure 2.20; models D and E). These models were generated within the GIS using a range of analytical tools, and validated where appropriate against legacy datasets (Chapter 6).

The modelling of wetland inception resulted in the generation of chronozones of peat inception across both Hatfield and Thorne Moors. The outputs consisted of models that can be used to move towards a landscape archaeology of the peatlands (Chapter 8) and most importantly, these are testable models which future work can seek to develop and refine. The results of the modelling of wetland inception and peat spread for both landscapes are described in Chapter 6.

Methods for interpreting palaeoenvironmental records: evidence for mire development, landscape change, human impact and the archaeological record (objective 3)

Robust chronological control is essential in establishing the rate and tempo of environmental changes in the past from proxy records, whether vegetation from pollen, Bog Surface Wetness from testate amoebae or relative sea level changes from diatoms. Previous palaeoenvironmental research on Hatfield and Thorne Moors had generated a series of hypotheses regarding mire development, palynological evidence for human activity across the mid to late Holocene and for changes in BSW attributed to climatic deteriorations (Chapter 2). In order to test these hypotheses, it was first necessary to assess the robustness of the radiocarbon chronologies associated with the different palaeorecords.

Clearly, reliable interpretation requires accurate and precise chronological control. Understanding the temporal relationship between different 'events' interpreted from these records permitted an understanding of the spatial component of different processes. For example, the identification of the HHL 'recurrence surfaces' as evidence of climatic deterioration is in effect a spatial interpretation derived from apparent temporal synchronicity in multiple records. The spatial understanding of palaeorecords is thus reliant first and foremost on chronology. The Bayesian chronological modelling (Chapter 5, see also Appendix 1 for further discussion of the methdology) therefore allowed an inter- and intra-site comparison of palaeoenvironmental evidence for mire development, vegetation change and human activity, feeding into subsequent understanding of the archaeological record.

Methods for assessing and quantifying the surviving peat resource (objective 4)

A subsidiary aim of the project was to provide a quantification of the peatland resource in terms of peat depth and associated chronology, to establish where wet-preserved archaeological remains might be present on the basis of a chronological 'preservation envelope' for the surviving peatland resource. For this quantification of resource, a series of different models were required. A model of peat depth across the Moors was required and this was achieved through the comparison between the DEM of the pre-peat landscape (objective 1) and data relating to the surface of the peat. For Hatfield Moors, the availability of LIDAR data provided a suitable foundation, such that, by subtracting the DEM of the pre-peat landscape from the LIDAR DEM provided a model of peat depth across the Moors. For Thorne Moors, this was more problematic due to the lack of LIDAR data. Instead, a Natural England contour dataset of the modern land-surface of Thorne Moors was used to establish a model relating to the surface of the peat.

The second model that was required for the quantification of the surviving peatland resource was of the chronological range represented by the deposits. The date of the bottom of the peat was provided through the modelling of wetland inception and peat spread (see objective 2 above). The dating of the top of the peat required the collection of radiocarbon samples and a similar methodology to interpolate these data across the study area. On Hatfield Moors, samples from close to the peatland surface were collected from the same sequences as the basal samples (see objective 2 above) and these dates were used in conjunction with estimated dates from previous work (Dinnin 1997) to provide a greater spatial coverage. These dates were then statistically compared with the elevation values for peat surfaces represented within the LIDAR data to generate a correlation algorithm that could be applied to the LIDAR data in order to transform elevation values into calendar years (further detail of this process is provided in Chapter 6). The limited availability of appropriate datasets meant that this process was not possible for Thorne Moors.

3.4 Summary

Understanding spatial and temporal patterns of change is both a theoretical as well as a methodological challenge. Any approach is ultimately determined by the availability of robust datasets. Using the case studies of Hatfield and Thorne Moors, this aim was addressed through a series of specific objectives. Firstly, models of the pre-peat landscapes were generated using a combination of pre-existing and bespoke datasets, and the resulting models were used for subsequent analyses, and provided a foundation for 'visualisations' of the data. The second objective centred on the modelling of wetland inception and peat spread, whilst the third focussed on the interpretation of the palaeoenvironmental records in terms of chronology, evidence for mire development, landscape change and past human activity. The fourth objective centred on the quantification of the peat resources from both Hatfield and Thorne Moors. This was achieved through establishing peat depths, and chronological modelling to interpret the likely age-span represented at locations within these landscapes. In terms of the management of raised mires and other wetland landscapes, this approach to the quantification of the peatland resource provides further methods for linking spatial and chronological datasets for applied purposes. GIS provided an excellent platform for integrating datasets, for interpolation and for the spatial mapping of hypothetical scenarios. The results are presented in Chapters 4–6, and integrated in Chapters 7–8.

4. Laying the foundations: modelling pre-peat landscapes

4.1 Introduction

The contemporary landscapes of Hatfield and Thorne Moors, like those of many other lowland peatlands in the UK, have been heavily modified through drainage, peat cutting activities and also by recent conservation measures. The current land-surfaces are mosaics of bare, cut-over peat, standing water and areas of shrub and woodland vegetation, which conceal the earlier pre-peat landscapes that would once have been dryland and which form the foundation for the environmental evolution of the Moors (Figure 4.1). This chapter discusses the approaches adopted for the generation of GIS deposit models of these pre-peat landscapes.

Figure 4.1 Pre-peat sands forming a low dune revealed in a drain section on Hatfield Moors

The sources of available data were summarised in Chapter 2 and these included legacy datasets relating to peat depth derived from previous borehole survey and geophysical (GPR) survey. Digital mapping and LIDAR also provided elevation data relating to the contemporary surface of the peatlands, although the latter was only available for Hatfield Moors. The process began with an assessment of legacy datasets, through cross comparison of the various sources of topographic information, enabling a consideration of relative accuracy and an estimation of error. The accuracy of these datasets was further assessed through targeted fieldwork consisting of the excavation of boreholes in grids and transects across selected areas of Hatfield and Thorne Moors, which in turn also provided additional data for the modelling of the pre-peat landscapes. The subsequent generation of GIS models of each of the pre-peat landscapes was based on the robust data highlighted by these analyses. The resulting models were subsequently used as the foundations for the modelling of peat inception and spread (see chapter 6).

4.2 Hatfield Moors pre-peat landscape

Assessing the value and coverage of legacy datasets for modelling the pre-peat landscape

Modelling the pre-peat landscape of Hatfield Moors was complicated by the presence of Lindholme Island, which currently stands above the peat and is occupied by woodland, farmland and buildings. Lindholme Island appears to have survived as an area of dryland within the peat, although the current extent of the feature presumably represents just the higher and drier part of a once larger island (Whitehouse, Buckland, Wagner *et al*. 2001). The complexity for Hatfield Moors, therefore, is that the study area includes both this 'visible' or exposed area of Lindholme Island and the 'hidden' area of the landscape that is concealed beneath the peat.

For Lindholme Island, topographic datasets were available from the Ordnance Survey (*Landform Profile*)

Table 4.1 Comparison of statistical results from the various surface datasets for Hatfield Moors

Dataset	Resolution	Minimum (m OD)	Maximum (m OD)	Mean (m OD)
OS Profile	10m grid	0.00	4.20	2.51
EA LIDAR DTM	2m grid	-0.79	5.78	1.40
NE basal topography	Uncertain	-1.24	3.13	0.13
GPR – western area	c. 50m	-0.57	1.49	0.17
GPR – eastern area	c. 50m	-1.37	2.16	-0.46

and the Environment Agency (EA LIDAR; Figure 2.15). The LIDAR data were supplied in two formats; a Digital Surface Model (DSM) which represents the vegetation canopy, and a Digital Terrain Model (DTM) which represents the 'bare earth' surface. For the purposes of generating a model of the pre-peat landscape the DSM was discounted since it represented the contemporary land surface complete with buildings and other features (see Chapter 2). For the areas beneath the peat, two principal datasets were available resulting from work undertaken by Natural England (NE), in the form of contour data derived from boreholes and interpolated to create a Digital Elevation Model (DEM) (Figures 3.3 and 3.4), and from geophysical work undertaken by the Scotts Company (the peat extractors), using Ground Penetrating Radar (GPR) to assess peat depths across the southern part of Hatfield Moors, calibrated through borehole survey. In practice, the LIDAR and GPR datasets might be regarded as the most robust due to the ability to access the raw or source data. However, it was felt necessary to explore uncertainties in the relative accuracy of these datasets and to define which was the most appropriate for use in the creation of a model of the pre-peat landscape. Metrical comparisons of these data are provided in Table 4.1.

A comparison of the various datasets revealed maximum values that range between 4.20m OD (10m resolution OS Profile) to 5.78m OD (2m resolution EA LIDAR DTM), both representing the area of Lindholme Island. The higher definition of the LIDAR data (Environment Agency 2006, English Heritage 2007) compared with the more generalised, 10m resolution OS Profile provided a more reliable dataset. This is confirmed by an Ordnance Survey benchmark on the footings of a building on Lindholme Island which has an elevation value of 5.25m OD. Hence, the OS Profile data were rejected for purposes of modelling the topographically higher extent of Lindholme Island. Similarly, the LIDAR data were rejected for areas away from Lindholme Island since these data correspond to the contemporary, cut-over surface of the peatland.

For the pre-peat landscape, a relatively low resolution DEM generated from Natural England peat depth data provided a range in elevation between -1.24m OD and 3.13m OD. Whilst the GPR data had a more restricted spatial coverage, it provided the lowest elevation (-1.37m OD) recorded in any of the datasets. The variability of this dataset compared with the generalised contour dataset provided some indication of the likely variability in the topography of the pre-peat land-surface. The systematic sampling of the areas surveyed by GPR (sampled at approximately 50m intervals) indicated greater confidence although the identification of localised variability indicated that a process of 'ground-truthing' using borehole survey would be required at a higher resolution than 50m. This was necessary in order to, firstly, establish the accuracy of the GPR data, and secondly, to determine the degree of localised topographic variation in this southern area. In addition, the spatial coverage of the GPR was limited to the southern parts of Hatfield Moors leaving areas to the north requiring additional data capture.

Ground-truthing and fieldwork locations

The assessment of the existing datasets relating to the pre-peat landscape indicated that additional data were required that could refine the basal topography at higher resolutions. Previous studies of the Moors have generated additional borehole data (e.g. Smith 1985, see Chapter 2) and, initially it was considered that these datasets could assist in this process of validation described above. However, limitations in the spatial positioning of these previous sampling sites meant that the six-figure grid references were insufficient to incorporate into the GIS accurately and so these datasets were of limited value. Hence a programme of further ground-truthing fieldwork consisting of borehole excavation was developed to assist in this process.

Three areas were chosen for the gridded borehole survey (Figure 4.2) to validate the data generated by the GPR survey. These borehole surveys were aimed at providing a grid of data points in order to explore the pre-peat surface and peat depths above it three-dimensionally. The surveys focussed on the areas of Packards South, to the north and south of Porters Drain. Packards South (Grid 1) lies within the eastern section of the western GPR survey. Fieldwalking had indicated that the sands beneath the peat appeared to form a series of pronounced dunes in this area, but these were not clearly reflected in the GPR implying that the geophysical data was limited in terms of resolution of localised variation. Boreholes were excavated in this location over an area of approximately 70 × 230m in a regular grid with intervals of approximately 20m between boreholes. A second area (Grid 2), measuring 350 × 250m with a 50m resolution was located in the northern part of the eastern GPR survey area. The third area (Grid 3), to the south of Porter's Drain was 60 × 180m with boreholes excavated at 20m intervals.

Five additional borehole grids were positioned else-

4. Laying the foundations: modelling pre-peat landscapes

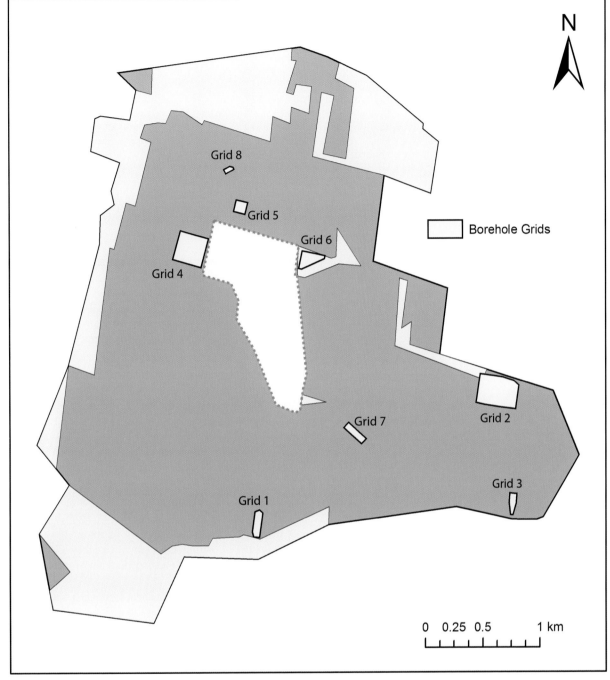

Figure 4.2 Locations of gridded borehole surveys on Hatfield Moors

where on Hatfield Moors (Figure 4.2): three of these grids were located in the area around Lindholme Island to explore the topography of the edges of this feature beneath the peat. On the northwestern side (Grid 4) an area measuring 250 × 250m was investigated at a resolution of 50m intervals. To the north of Lindholme Island the Neolithic trackway and platform site (see Chapter 7) an area measuring 100 × 100m was investigated at a higher resolution of 10m. In the third area adjacent to Lindholme Island, in the region of Spinney Corner (Grid 6), the grid took the form of a triangular area measuring 200 × 150m with boreholes excavated at a spacing of 20–25m.

Grid 7 was positioned to the southeast of Lindholme Island, across a possible sub-peat ridge identified on the model derived from the Natural England data, and perhaps also reflected by the position of a pathway on the 19th century mapping of the area (see Chapter 2). This grid covered an area of 200 × 60m at intervals between boreholes of 20m. The final grid of boreholes was excavated further to the north of Lindholme Island to explore an area of the northern part of the Moors at Kilham West (Grid 8), where there had been no previous investigations. Here, an area measuring 80×40m was investigated with boreholes positioned at 20m intervals.

Table 4.2 Borehole transects on Hatfield Moors

Transect no.	Length (m)	Borehole spacing (m)	Comments
T1	800	100	To assess any topographic evidence for extension of Lindholme Island beneath peat
T2	440	100	Angled extension of T1
T3	300	50	30m north of Grid 5. Borehole 8 off alignment due to access
T4	90	10	Adjacent to drain at northern edge of Lindholme, 80m south of Grid 5
T5	600	200	To the southwest of Lindholme Island
T6	350	50	Positioned along altitudinal gradient of pre–peat topography as defined by Grid 2
T7	980	40–120	Along Porters Drain
T8	980	60–120	Parallel to T7
T9	750	50–100	Perpendicular to T7 and T8, extending across Grid 3
T10	320	20–40	To assess pre–peat topography and stratigraphy towards current edge of Hatfield Moors

Table 4.3 Basal topography statistics from the values of boreholes within the grids for Hatfield Moors

	Minimum (m OD)	Maximum (m OD)	Range (m)	Mean (m OD)
Grid 1	-1.60	0.78	2.38	-0.48
Grid 2	-1.21	-0.23	0.98	-0.77
Grid 3	-1.35	-0.92	0.43	-1.11
Grid 4	-0.05	0.98	1.03	0.27
Grid 5	-0.38	1.72	2.10	0.59
Grid 6	-0.40	0.14	0.54	-0.19
Grid 7	-0.29	0.36	0.65	0.05
Grid 8	0.25	0.62	0.37	0.76

To supplement the data gathered by the gridded survey, a programme of borehole transects was undertaken in order to provide a wider coverage with the aims of firstly generating data for areas that had seen less investigation, and secondly to validate the results of previous borehole studies. Hence, four transects of boreholes were positioned in the area to the north of Lindholme Island (Figure 4.3; T1–T4) and a fifth transect (Figure 4.3; T5) was excavated in the area to the southwest of Lindholme Island at intervals of 200m, where four boreholes were excavated over an overall distance of 600m from the centre of the peatland up to the edge of the island. An additional five transects totalling 3.5km were excavated across the eastern GPR survey area along and to the north and south of Porters Drain, with boreholes at intervals of between 20m and 100m (Figure 4.3; T6–T10), comprising a total of 51 boreholes. Transect details are provided in Table 4.2.

Results from the gridded borehole surveys

The statistical results from the eight areas investigated through gridded borehole survey are provided in Table 4.3 and the resulting surfaces (DEMs) from each of these areas are provided in Figures 4.4–4.11. The results show an overall maximum variation in basal elevation of between -1.35m and 1.72m OD: a range of 3.07m across the surveyed areas of Hatfield Moors. Whilst the lowest recorded borehole base was in the area of Packards South (Grid 1), the mean values indicate that the lowest area overall was at Porters Drain (Grid 3) in the southeast corner of the Moors.

Taking the results of the gridded borehole surveys overall, in the northern segment of the study area, the greatest local variation in the basal topography was recorded directly to the north of Lindholme Island (Grid 5). The DEM derived from this data (Figure 4.4) demonstrates that this range reflects the edge of the Island, with the basal topography rising up relatively sharply to the south. The survey displaying the second greatest local variation was northwest of Lindolme Island (Grid 4; Figure 4.5), although this appears to reflect low ridges, most probably sand dunes, running across the centre of an otherwise relatively flat area, rather than a fall in elevation from the edge of the Island. In this area the maximum value (0.98m OD) is considerably lower than that for Grid 5 (1.72m OD) indicating that it was well away from higher relief at the edge of the Island. Very little topographic variation was identified within the area of Spinney Corner (Grid 6; Figure 4.6), or in the area of Kilham West to the north (Grid 8; Figure 4.7).

The elevation of the pre-peat landscape was generally lower in the south of Hatfield compared with areas to the north, broadly reflecting the relief apparent in the data provided by Natural England. On this southern side, the grid closest to Lindholme Island (Grid 7; Figure 4.8) revealed only slight variations in the topography of the pre-peat land-surface across what was otherwise higher ground; perhaps a ridge which was reflected in both the Natural England data and by the presence of a pathway on the 19th century Ordnance Survey mapping (see Chapter 2; Figure 2.12). The area of Packards South (Grid 1; Figure 4.9) displayed considerable local variation with an overall range of 2.38m, the greatest recorded on Hatfield Moors and coinciding with pronounced sand dunes beneath the peat. The areas to the north and south of Porter's Drain within the southeastern part of the Moors (Grid 2 and Grid 3; Figures 4.10 and 4.11), identified as the lowest lying areas on the Moors, displayed relatively slight ranges in elevation compared with the size of the areas that were investigated (a range of less than 1m across a combined area of 8.75ha).

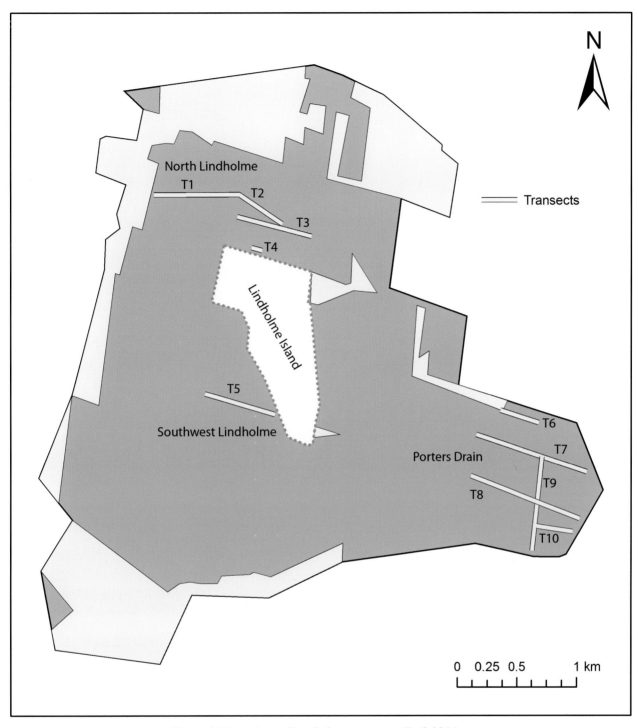

Figure 4.3 Locations of borehole transects on Hatfield Moors

Comparing legacy datasets with data from the gridded borehole surveys

The resulting DEMs generated from the borehole grids provided a relatively high-resolution foundation against which to assess the accuracy of the legacy datasets. For each of the gridded areas, the legacy datasets were sub-sampled so that values could be calculated in terms of minimum, maximum and mean elevation. For this comparison, all available datasets were included (see Table 4.4).

The overall trends in the pre-peat land-surface derived from the Natural England data were very similar to those apparent in the other datasets, indicating generally lower areas of topography in the southern area compared with the north, and some similarities were implied with the general morphology of the pre-peat landscape in the data derived from the GPR survey. However, comparison between the model generated from the Natural England contour data and the LIDAR data indicates that the value of the former

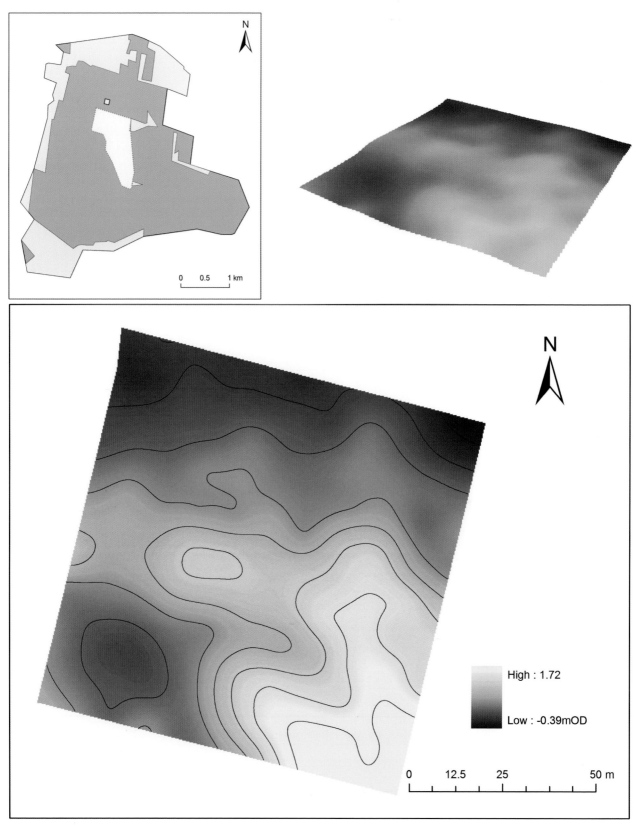

Figure 4.4 DEM derived from borehole Grid 5 to the north of Lindholme Island over the area of excavations of the late Neolithic trackway and platform (see Chapter 7). Contour lines on this and subsequent figures show 25m intervals

is probably somewhat restricted. Since the LIDAR data records the current land-surface, it would be expected that its values should always be higher than those derived from models pertaining to the underlying, pre-peat landscape. Whilst this is the case for the majority of investigated areas, at Kilham West (Grid 8) and North Lindholme (Grid 5) the height values for the peat surface derived from the LIDAR data are lower than those for the pre-peat landscape

4. Laying the foundations: modelling pre-peat landscapes

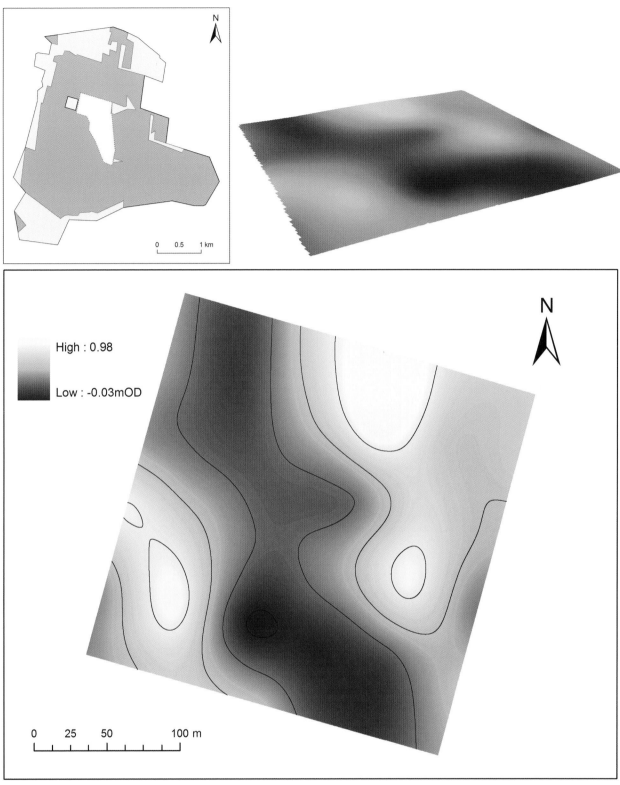

Figure 4.5 DEM derived from borehole Grid 4 on the northwestern side of Lindholme Island

derived from the Natural England data. Similarly, the area of Spinney Corner, to the northeast of Lindholme (Grid 6) has values of just 0.01m difference. Variations in the apparent accuracy of the Natural England data mean that it was not possible to reliably determine or compensate for the degree of error, and so this dataset was therefore discounted from further modelling and analyses.

A comparison between the GPR data and the results from the borehole grids provided broad similarities indicating the overall quality of this dataset. For the three areas of overlap the GPR data was consistently contained within the overall range of elevations provided by the higher-resolution borehole data. At Packards South (Grid 1) and to the north of Porters Drain (Grid 2), the maximum (highest)

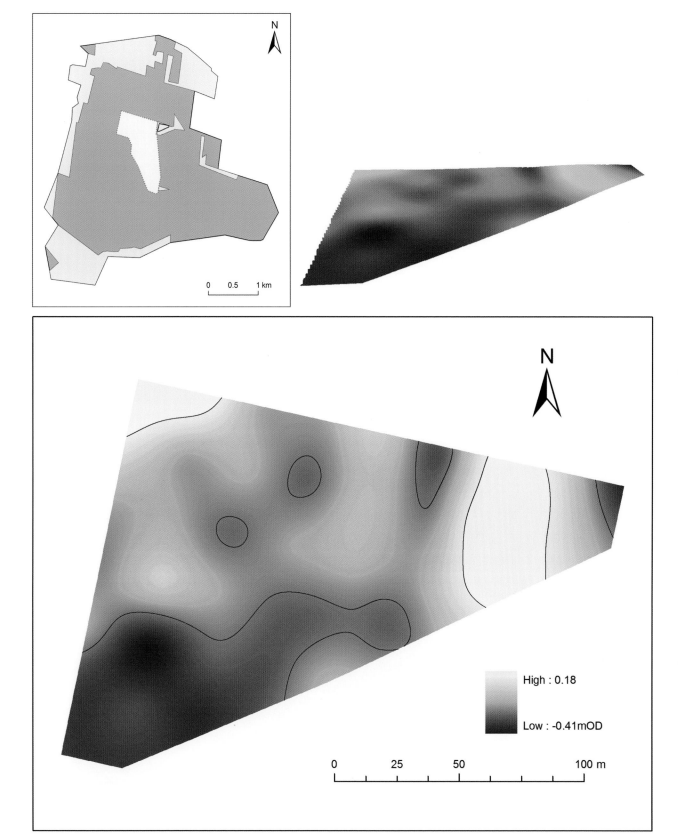

Figure 4.6 DEM derived from borehole Grid 6 on the northeastern side of Lindholme Island within the area of Spinney Corner

values relating to the pre-peat landscape were comparable, whereas in all cases greater variation was apparent with the minimum (lowest) values. This reflects the lower resolution of the GPR data and the resulting shortcomings in terms of identifying more subtle variations in the basal topography and, in particular, the inability to identify localised features between the 50m sampling resolution. Despite these somewhat local differences, it was broadly

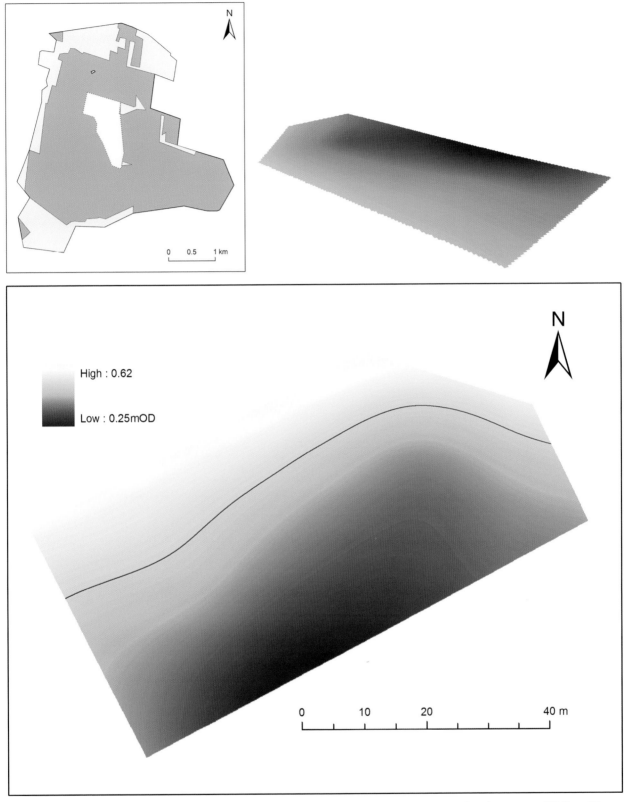

Figure 4.7 DEM derived from borehole Grid 8 in the northern part of Hatfield Moors within the region of Kilham West

consistent with the elevations derived from the borehole grids. Hence, the GPR data were regarded as sufficiently robust to be incorporated into the final modelling of the pre-peat landscape.

Results from the borehole transects

The results of the ten borehole transects are outlined and discussed below in relation to the three areas of Hatfield Moors that were investigated: to the north of Lindholme

58 *Modelling archaeology and paleoenvironments in wetlands*

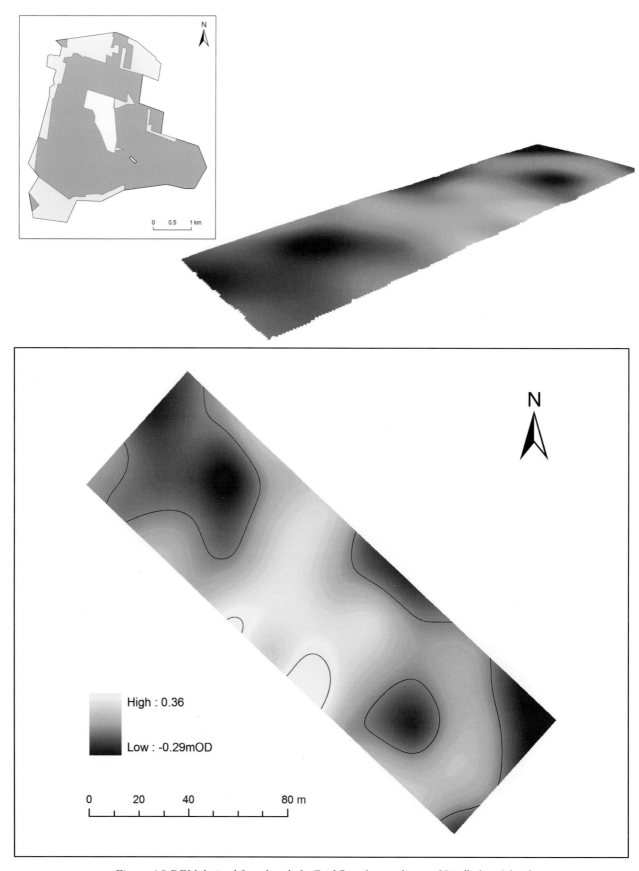

Figure 4.8 DEM derived from borehole Grid 7 to the southeast of Lindholme Island

4. Laying the foundations: modelling pre-peat landscapes

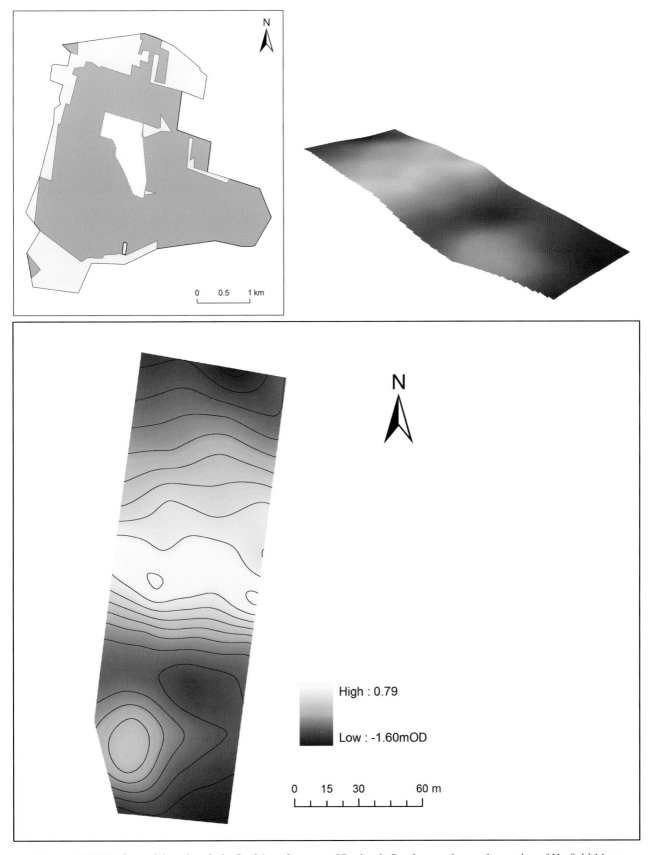

Figure 4.9 DEM derived from borehole Grid 1 in the area of Packards South near the southern edge of Hatfield Moors

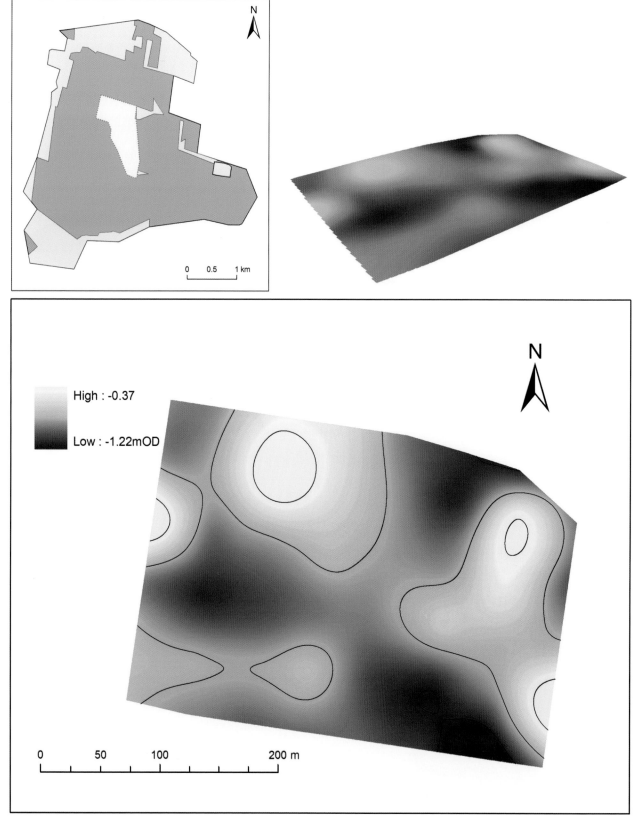

Figure 4.10 DEM derived from borehole Grid 2 in the area to the north of Porters Drain

4. Laying the foundations: modelling pre-peat landscapes

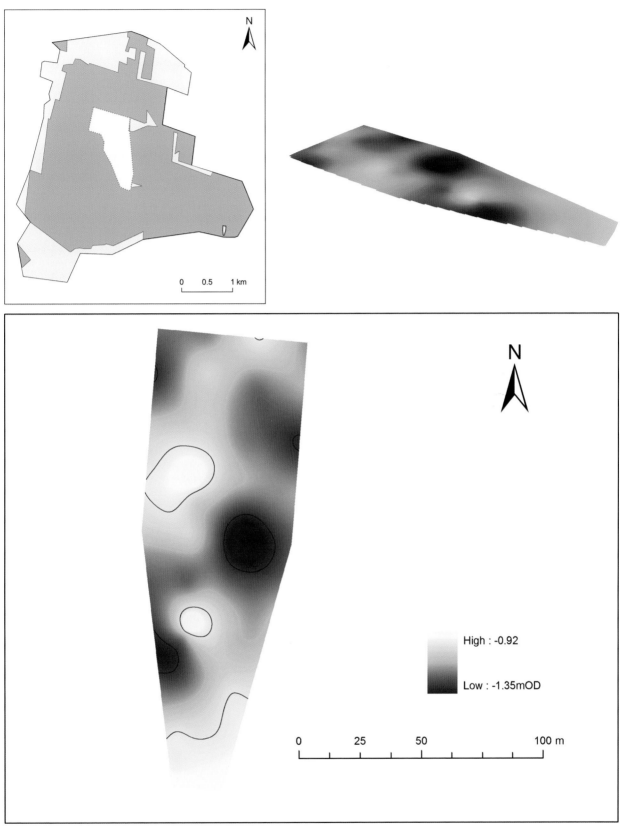

Figure 4.11 DEM derived from borehole Grid 3 in the area to the south of Porters Drain

Table 4.4 Values from the GIS models for each of the gridded borehole survey areas in comparison with the results from other datasets for the same areas on Hatfield Moors

	Dataset	Minimum (OD)(m)	Maximum (OD)(m)	Mean (OD)(m)
Grid 1	Borehole DEM	-1.60	0.79	-0.41
	Natural England data	-0.91	-0.89	-0.90
	GPR data	-0.44	0.61	0.09
	LIDAR DTM	0.57	1.32	0.81
Grid 2	Borehole DEM	-1.22	-0.37	-0.76
	Natural England data	-0.90	-0.86	-0.89
	GPR data	-0.90	-0.36	-0.75
	LIDAR DTM	-0.10	2.00	0.21
Grid 3	Borehole DEM	-1.35	-0.92	-1.11
	Natural England data	-0.91	-0.90	-0.91
	GPR data	-0.80	-0.57	-0.67
	LIDAR DTM	0.46	1.39	0.85
Grid 4	Borehole DEM	-0.03	0.98	0.42
	Natural England data	0.17	0.72	0.46
	LIDAR DTM	0.95	1.88	1.13
Grid 5	Borehole DEM	-0.40	1.72	0.61
	Natural England data	0.87	0.90	0.89
	LIDAR DTM	0.74	1.87	1.06
Grid 6	Borehole DEM	-0.41	0.18	-0.18
	Natural England data	1.14	1.83	1.70
	LIDAR DTM	1.15	1.64	1.28
Grid 7	Borehole DEM	-0.29	0.36	0.05
	Natural England data	-0.53	-0.33	-0.47
	LIDAR DTM	0.43	1.67	0.88
Grid 8	Borehole DEM	0.25	0.62	0.43
	Natural England data	0.90	0.90	0.90
	LIDAR DTM	0.76	1.82	0.96

Table 4.5 Statistical results from the borehole transects on Hatfield Moors

Transect	Minimum peat depth (m)	Maximum peat depth (m)	Lowest level pre-peat landscape (m OD)	Highest level pre-peat landscape (m OD)
N of Lindholme				
T1	0.18	0.75	-0.06	0.84
T2	0.08	0.62	-0.06	0.43
T3	0.33	0.92	-0.16	0.33
T4	0.00	0.30	1.41	2.06
SW of Lindholme				
T5	0.30	0.40	0.31	0.59
Porters Drain				
T6	0.40	0.75	-0.89	-0.38
T7	1.60	2.15	-1.24	-0.71
T8	1.40	2.04	-1.03	-0.60
T9	0.80	2.40	-1.10	-0.59
T10	1.15	2.33	-1.46	-0.96

Island (transects T1–T4), southwest of Lindholme Island (transect T5) and around Porters Drain in the southeastern corner of Hatfield Moors (transects T6–T10). Statistical results from these transects are provided in Table 4.5.

North Lindholme (T1–T4)
The results from the borehole transects excavated in the area to the north of Lindholme are presented in Figure 4.12. Peat depth on transect T1 ranged between 0.18m and 0.75m with associated variations in basal elevation of between -0.06m and 0.84m OD. For transect T2 peat depths ranged between 0.08m and 0.62m, providing a variation in basal elevation between -0.06m and 0.43m OD. These transects together provide a 1km section across the northern part of Hatfield Moors. This shows the basal topography falling over a distance of 400m from 0.84m OD towards the western edge of Hatfield Moors to a level of 0.13m OD in the east (an overall drop of 0.71m). For the next 500m there is a slight rise of up to 0.30m before a fall to -0.06m OD at the eastern end of the transect.

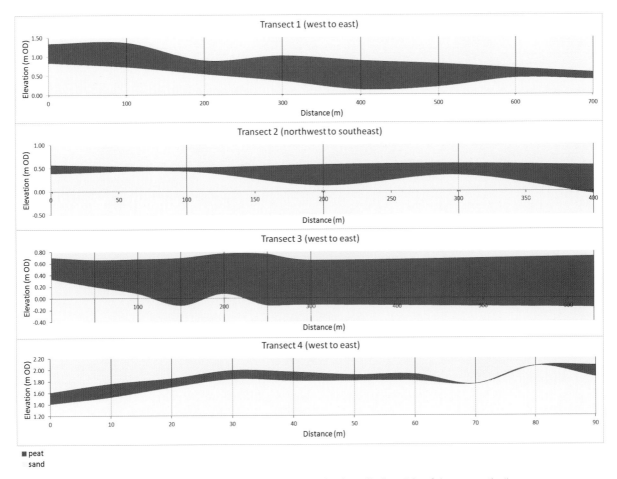

Figure 4.12 Borehole transects to the north of Lindholme Island (transects 1–4)

Peat depths on transect T3 ranged between 0.33m and 0.92m, with basal values of between -0.16m and 0.33m OD. The basal topography along this transect rose from 0.15m to 0.33m OD over the westernmost 50m before very gradually dropping off to 0.07m OD over 100m. Over the next 100m the elevation drops sharply to -0.54m OD and rises again to -0.16m OD, before levelling off for the next 380m and reaching -0.07m OD to the east. The ridge towards the western end of Transect 3 corresponds with the slight 0.30m high ridge to the north, with a similar fall in elevation on its eastern side. It is possible that this very slight ridge reflects a northern extension of the Lindholme moraine, which is also apparent in the gridded borehole survey Grid 5. Depths of peat on transect T4 ranged from 0m to 0.30m, with absolute elevations ranging between 1.41m and 2.06m OD.

Southwest Lindholme (T5)

The results from the borehole transect excavated to the southeast of Lindholme Island are presented in Figure 4.13. The basal topography ranged from between 0.31m and 0.37m OD within the western three boreholes, rising to 0.59m OD towards the edge of Lindholme Island, indicating a relatively sharp gradient between the Island and the wider pre-peat landscape. The peat surface followed a similar trend, from between 0.61m and 0.76m OD within the westernmost three boreholes, rising to 0.99m OD within the eastern borehole, demonstrating that the contemporary landsurface reflects variations in the underlying topography. Depth of peat across this area was relatively consistent, ranging between 0.30m and 0.40m over the entire distance. Overall, the results demonstrate that the pre-peat landscape to the west of Lindholme Island was relatively high, particularly compared with those areas to the south (see below).

Porters Drain (T6–T10)

In the northernmost part of the Porters Drain area in the southeastern portion of Hatfield Moors, transect T6 revealed an overall variation in the topography of the pre-peat landscape between -0.38m OD and -0.89m OD, with the level of this surface dropping towards the east (Figure 4.14). The overall trend in the basal topography is in part reflected by that of the surviving peat surface which ranges between -0.06m OD at its western end to -0.14m OD at its eastern end, with peat depths increasing from 0.40m to 0.75m.

Transect 7 was located approximately 300m to the south of transect 6, but following the same overall alignment, and reflected a relatively consistent basal topography of between -0.71m and -1.24m OD, a maximum range of 0.53m. However, there was no overall clear trend in the basal topography. The surviving surface peat was recorded

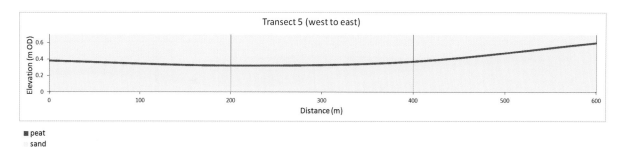

Figure 4.13 Borehole transect to the southwest of Lindholme Island (transect 5)

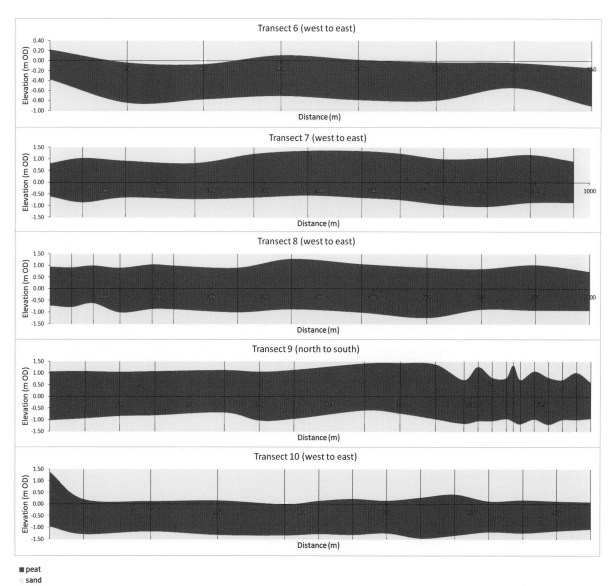

Figure 4.14 Borehole transects in the Porters Drain area of Hatfield Moors (transects 6–10)

at a maximum of 1.28m OD and a minimum of 0.74m OD, with depths of peat ranging between 1.60m and 2.15m.

Transect T8 was excavated just over 300m to the south of transect T7 following a parallel west-northwest by east-southeast alignment. As with transect T7, the basal topography reflected by transect T8 was relatively uniform, ranging between -1.03m and -0.60m OD, a total range of 0.43m. The basal topography therefore proved to be slightly higher than that to the north, although there was again no clear trend associated with proximity to the current edge of the Moors or the adjacent river floodplain. The top of surviving peat along this transect ranged between a maximum of 1.39m OD and a minimum of 0.80m OD.

Transect T9 was aligned broadly north–south, from the east of the centre of transect T7 southwards, across transect T8 and past the westernmost edge of transect T10, continuing to the current southern edge of the peatland. The basal topography along transect T9 demonstrates no clear trend in the level of the pre-peat landscape along this line, with local variations between a minimum of -1.10m OD, and a maximum of -0.59m OD. The surviving surface peat was recorded at elevations of between 0.95m and 1.43m OD, with peat depths of between 0.80m and 2.40m.

Transect T10 extended eastwards from the southern part of transect T9. There were no distinct trends in the topography of the pre-peat landscape along this transect, with elevations ranging between -1.46m and -0.96m OD, a total variation of 0.5m over a distance of approximately 320m. Levels of peat extraction in this area have led to a significant variation in the elevation of the surface peat along this transect, ranging between -0.01m and 1.37m OD. Consequently, peat depths decreased along the eastern edge of Hatfield Moors, with maximum values of 2.33m, compared with as little as 1.15m of surviving peat towards the central and eastern end of the transect.

Overall, the results from the transects across this southeastern portion of Hatfield Moors show the low elevations of the pre-peat landscape compared with other areas of the Moors, confirming the results from the gridded borehole surveys. There appears to be relatively little pronounced localised variation in the surface of the pre-peat landscape.

Summary of gridded borehole surveys and transects

In total, the gridded survey consisted of 392 boreholes excavated within eight discrete locations, covering a total area of nearly 22ha, approximately 1.6% of the total land area of Hatfield Moors, including all effectively inaccessible areas, such as those under dense woodland. In addition to these boreholes, additional data were generated from over 9km of borehole transects. Together, the results provide a foundation for assessing the validity of the legacy data from Hatfield Moors. For example, they show that the overall relief of the pre-peat landscape indicated by the Natural England contour data compares favourably to the results from the ground-truthing. However, it appears that local detail is lacking in certain areas, such as in the area of Packards South (Grid 1). The gridded borehole survey at this location revealed the highest range of variation in the topography of the pre-peat landscape (2.38m), which presumably reflects the topographic variation associated with the dune system (*cf.* Buckland 1982).

Similar levels of topographic variation were identified in the pre-peat land-surface in the area proximal to the northern part of Lindholme Island (2.10m) marking an extension of the topographic feature beneath the peat which appears to extend to both the north and the south. There was no evidence of similar ridges extending to either the east or west of the Island. Elsewhere, the variation in the pre-peat topography was shown to be very slight whilst the depth of the surviving peat across the study area is on average less than 1m.

Peat stratigraphy on Hatfield Moors

Very little stratigraphic variation in the surviving peat was recorded and the deposits consist generally of black, highly humified organics with few identifiable macrofossil remains. Only occasional *Sphagnum* peat was recorded, indicating that the majority of the true raised mire deposits have been removed by peat cutting.

Four main stratigraphic units were recognised during the grid and transect surveys (base upwards):

1 Yellow–brown coarse sands.
2 Black, mineral rich well humified, amorphous sediment with few recognisable plant remains but with wood fragments locally abundant.
3 Well humified *Sphagnum* with *Eriophorum* and ericaceous remains locally abundant.
4 Moderately humified *Sphagnum* with *Eriophorum*.

There was little variation in this general sequence evident. Sands underlie the organic sequences in all the boreholes, and were overlain by the black, well humified sediment unit, up to c.1.0m thick. In places, this unit was overlain by well humified *Sphagnum* peat, with the remains of *Eriophorum* and ericaceous plants evident, with occasionally an uppermost deposit of less well humified *Sphagnum*. However, *Sphagnum* peat was rarely recorded in significant thicknesses, indicating that the majority of the true raised mire deposits have been removed by peat cutting. Limited laboratory assessment of the macrofossil content of the samples confirmed that the peat consisted mainly of unidentifiable organic matter, with occasional ericaceous roots and stems, *Eriophorum* and in some samples charred twigs and other material (see Chapter 5). Only one sample contained *Sphagnum* macrofossils at its base (Grid 4, basal sample HM 5, also discussed further in chapter 5). This relatively shallow depth of extant peat restricted the potential for detailed straigraphic recording or reconstruction of the mire surface through time (*cf.* Derryville; see chapter 2), or any concerted palaeoenvironmental analysis and further sampling of the deposits was carried out for radiocarbon dating purposes only (see Chapter 5). Hence, for Hatfield Moors, the palaeoenvironmental data were limited to that derived from previous research (Chapter 2) informed and contextualised by the modelling of the pre-peat land-surface (see Chapter 5).

Generation of a new Digital Elevation Model of the pre-peat landscape of Hatfield Moors

The various borehole surveys provided empirical data on the morphology of the pre-peat landscape and the depth of surviving peat which in turn provided data for the creation of a new model of the pre-peat landscape of Hatfield Moors, and also allowed some estimation of the accuracy of the other datasets. Whilst the DEM derived from the Natural England contour data broadly corresponded with results of the ground-truthing, it appeared to be less accurate in terms of local detail. However, the GPR data were shown to be broadly accurate covering a considerable area of the

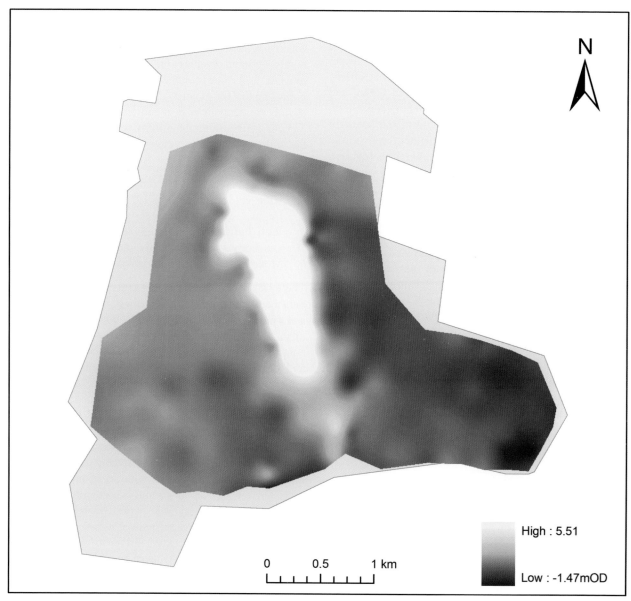

Figure 4.15 DEM of the pre-peat landscape of Hatfield Moors

southern part of Hatfield Moors. Furthermore, the bare-earth DTM model derived from the Environment Agency LIDAR data was considered 'fit for purpose' of modelling the exposed area of Lindholme Island. The eight borehole grids also provided further topographic data, which were augmented by that from the borehole transects to provide data for those parts of Hatfield Moors for which little other data were available.

The total area of Hatfield Moors as managed by Natural England comprises approximately 1360ha, but significant portions of the Moors were effectively inaccessible due to dense woodland, such as in the north and south, along the western fringes, or by standing water due to ongoing restoration works. Hence, the accessible area of Hatfield Moors comprised approximately 883ha: about 65% of the total managed landscape. Within this redefined study area, the 22ha assessed through the gridded borehole surveys amount to nearly 2.5%. The GPR DEMs covered a total area of 352ha (*c.* 40% of the total study area), of which 10ha were replicated by the borehole grids. Furthermore, the LIDAR coverage of Lindholme Island accounted for 79ha. Hence, taking this overlap into account, the total area of surface data regarded as 'fit for purpose' of modelling the morphology of the pre-peat landscape of Hatfield Moors was 443ha, or just over 50% of the study area.

In addition to this surface data, the results from the borehole transect surveys provided additional data points from which to interpolate a model of the pre-peat landscape. Overall, the spatial coverage was uneven, with the area to the south of Lindholme Island having greater coverage compared with those to the north which were more reliant on the results of the transect borehole surveys. This inevitably resulted in a greater reliance on interpolation and hence a higher level of uncertainty in terms of the deposit modelling.

Four datasets were combined for the generation of the model of the pre-peat landscape of Hatfield Moors: 2m resolution LIDAR data for the area of Lindholme Island covering a total area of 79ha, the GPR dataset covering a total area of 352ha, the eight borehole grids covering a total area of 22ha and the series of borehole transects from across the Moors. Where overlap existed between the datasets (approximately 10ha), the higher resolution borehole data took priority.

All datasets were processed into point data and combined into a single file. The variability in spacing between points, compared with strictly gridded data, limits the range of appropriate interpolation techniques (*cf.* Burrough 1986). Given the challenges of the variable resolution of the input data, a number of approaches were tested to explore the potential for the generation of anomalous artefacts in the resulting GIS surface. Based on the results of these experiments, and following processes used elsewhere with similarly irregular datasets (e.g. Chapman 2006), a spline interpolation method was used, maintaining tension in the model to reduce the number of interpolation artefacts (see also Chapter 3). The resulting model (Figure 4.15) was generated at a cell resolution of 5m reflecting the variability in the resolution of the input data. The higher resolution LIDAR data was thus sub-sampled to a 5m resolution from its original 2m point density. The combination and interpolation of legacy and bespoke datasets produced a continuous DEM, the first representation of the three dimensional morphology of the pre-peat landscape of Hatfield Moors.

The resulting DEM shows the higher area of Lindholme Island as a pronounced feature, although it demonstrates that the current extent of the Island represents only the higher portion of a more extensive morainic ridge that extended to the south and partially to the north and which subsequently became partially subsumed by peat growth. The DEM indicates that the lower lying areas of the pre-peat landscape of Hatfield Moors are relatively flat and level, dropping slightly within the southeastern area. The dune systems in the south of the Moors are apparent within the DEM, although local detail is somewhat generalised due to the variable resolution of the input data. Whilst the overall variation in elevation for the landscape ranged between -1.47m OD and 5.51m OD, away from the pronounced topographic high of Lindholme Island, elevations varied by less than 2m. This final model provided the basis for all further deposit modelling of Hatfield Moors (see Chapter 6).

4.3 Thorne Moors pre-peat landscape

Assessing the value and coverage of legacy datasets for modelling the pre-peat landscape

The range of legacy datasets for Thorne Moors was restricted compared to those available for Hatfield Moors. For example, whilst previous GPR surveys had been conducted on Thorne Moors, the resulting data were not in a digital format that could be readily re-used and no LIDAR data were available. The relatively low resolution georeferencing of previous palaeoenvironmental studies, were also apparent for Thorne Moors, meaning that the associated data could not be incorporated into the pre-peat landscape DEM. Thus, of the various legacy datasets, only the DEM derived from Natural England peat depth data was available (see Chapters 2 and 3, Figure 3.4), indicating a range in elevation of between -1.79m and 2.29m OD, with a mean elevation of 0.27m OD. Without additional topographic information it is difficult to assess the accuracy of this DEM.

Testing the legacy data and determining fieldwork locations

As for Hatfield Moors, fieldwork was aimed at obtaining data that could be used for the validation of the legacy data, as well as providing new data that might feed into the model of the pre-peat landscape, especially in areas that had not been investigated previously. In practice, the locations for the gridded borehole surveys were restricted due to the flooding of large areas of Thorne as part of the restoration works (see Chapter 1). The targeting of fieldwork was therefore based upon the DEM derived from the Natural England contour data and, in particular, upon the areas of pronounced pre-peat topographic variation in the areas of Middle Moor and Rawcliffe Moor (Figure 4.16). Work also focussed on the highest areas of Thorne Moors which lie in two areas: the first within the southern part of Thorne Moors, around Birtwistle, and the second within the northern area of the Moors to the south of Rawcliffe Moor.

A comparison with the Natural England habitat information collected in 2002 (see Chapter 2) showed a correlation between the low lying central areas and the bare, cut-over peat, compared with the higher areas which were largely wooded. A large proportion of Thorne Moors, particularly covering the southern areas and periphery, were dominated by *Betula* dominated woodland (29%) and scrub and regenerating grassland (nearly 8%). Areas under restoration when fieldwork commenced were relatively high (14%) and only 46% of the landscape was accessible.

The first two areas investigated focussed on Middle Moor, directly south of the Limestone Road (see Figure 2.11), depicted within the DEM derived from the Natural England data as an area of considerable topographic variation in the pre-peat land-surface, as indicated by previous stratigraphic and palaeoenvironmental investigation (Gearey 2005, see Chapter 2). These two areas were assessed using gridded borehole surveys over areas of 170 × 200m (Grid 1) and 240 × 200m (Grid 2) respectively, positioned 300m apart. Grid 1 consisted of 90 boreholes positioned at 20m intervals, and Grid 2 consisted of 115 boreholes also positioned at 20m intervals. A third location (Grid 3) in the area to the north of the Limestone road, approximately 750m to the north of Grids 1 and 2, focussed on a shallow depression in the pre-peat landscapes revealed by the DEM derived from the Natural England data. It consisted of a total of 103 boreholes positioned at 20m intervals over a total area of 210 × 120m. A fourth grid (Grid 4) was positioned over the area of Rawcliffe Moors at the northern edge of Thorne

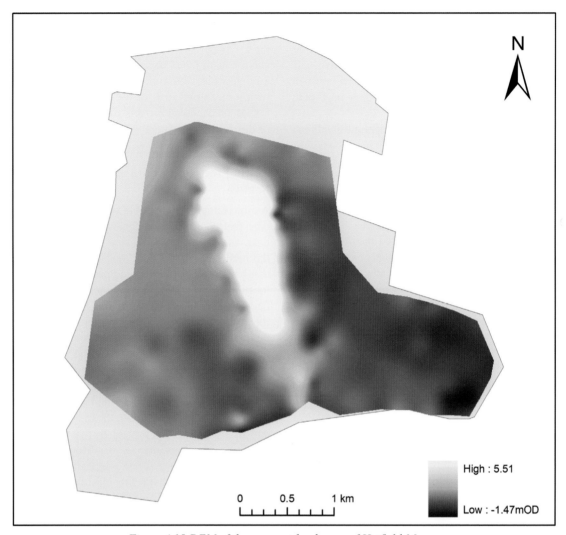

Figure 4.15 DEM of the pre-peat landscape of Hatfield Moors

Moors, an area previously identified as having the deepest known surviving peat (Smith 2002). Grid 4 consisted of 58 boreholes excavated at intervals of between 40m and 50m over a total area of approximately 300 × 600m.

In addition to the gridded borehole surveys, two borehole transects were excavated across Thorne Moors (Figure 4.17). Transect 1 was positioned in the area of relatively low elevation pre-peat land-surface identified within the Natural England data in the area of Cottage Dyke, parallel to the southern side of the Limestone Road. This transect extended for 460m at intervals of between 10m and 120m (in response to variations in the pre-peat topography) within an area of scrub and *Betula* woodland. Transect 2 was 250m long with boreholes excavated at intervals of between 40m and 60m and was positioned on the southeastern side of Birtwistle, also in an area of scrub and woodland, to investigate an area of relatively high pre-peat topography shown in the DEM derived from the Natural England contour data.

Table 4.6 Basal topography statistics from the values of boreholes within grids for Thorne Moors

	Minimum (m OD)	Maximum (m OD)	Range (m)	Mean (m OD)
Grid 1	-2.69	1.36	4.05	-0.36
Grid 2	-2.86	1.62	4.48	-0.34
Grid 3	-0.49	0.83	1.32	0.21
Grid 4	-0.04	1.83	1.87	1.07

Results from the gridded borehole surveys

A total of 366 boreholes were excavated within the four grids (Table 4.6) covering a total area of nearly 18ha, just over 0.93% of the total area of Thorne Moors (including all inaccessible areas such as those under dense woodland and flooded areas). The resulting DEMs interpolated from the borehole data within Grid 1 (Figure 4.18) and Grid 2 (Figure 4.19) both revealed a pronounced sinuous

4. Laying the foundations: modelling pre-peat landscapes

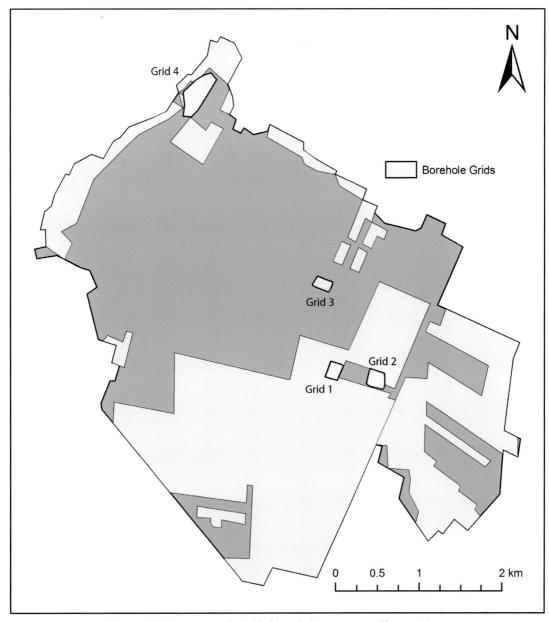

Figure 4.16 Locations of gridded borehole surveys on Thorne Moors

depression of similar form and depth following a similar north-northwest to east-southeast alignment. The lower elevation values related to the base of this feature and were relatively consistent (-2.69m OD for Grid 1 to the west, and -2.86m OD for Grid 2 to the east). This may imply that these grids record part of the same feature, possibly a palaeochannel related to a similar feature identified to the east and referred to as the 'Crowle Depression' (Smith 2002, Gearey 2005). To the north, within the area of Grid 3, the variation in the pre-peat topography (Figure 4.20) was relatively slight, with an overall range of just 1.32m relating to a linear hollow aligned east–west. The consistency of alignment between the features identified in all three of these areas might reflect the principal direction of drainage across this central part of Thorne Moors prior to the formation of peat.

Grid 4 (Figure 4.21), centred on Rawcliffe Moor at the northernmost part of Thorne Moors, and demonstrated a pronounced depression in the pre-peat topography on the northwestern side of the gridded area where the topography drops off rapidly towards the northwestern edge of the current peatland. The deepest peat deposits and the earliest known peat inception on Thorne Moors have been recorded in this area (Smith 2002), although this area is far from being the topographically lowest part of the Moors, as there are considerably lower areas within the Crowle Moors area.

Comparison of legacy datasets with the data from the gridded borehole surveys

The total area covered by the gridded borehole surveys on Thorne Moors represent just less than 1% of the total land

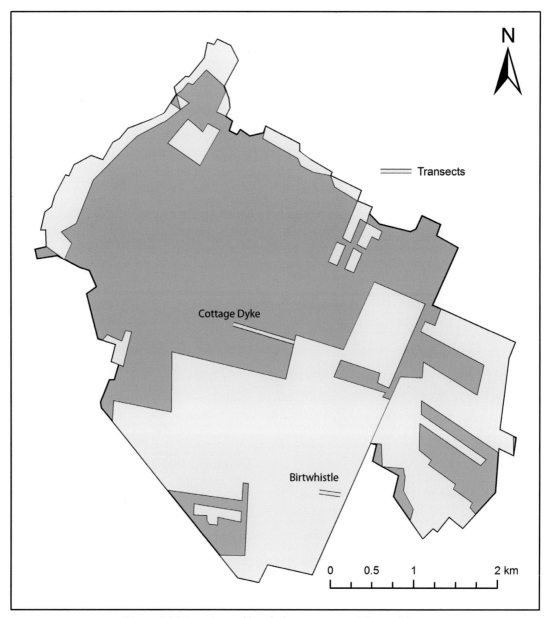

Figure 4.17 Locations of borehole transects on Thorne Moors

area of the Moors. Whilst clearly insufficient for generating a complete model of the pre-peat landscape on their own, they do provide a basis for partially validating the legacy data for this purpose. As for Hatfield Moors, in order to facilitate comparison between datasets, localised DEMs were extracted from the DEM based on Natural England data using the spatial limits of the higher-resolution borehole surveys. Statistics were then generated for each of the areas (Table 4.7).

Within the areas of Grid 1, Grid 2 and Grid 4, the values from the DEM based on Natural England data were contained within the range of elevations in the borehole DEMs, indicating a broad match but with generalised local detail, as is reflected by the variation in mean values. The area covered by Grid 3 provided less of an overlap, with slightly lower minimum values demonstrated by the DEM based on the Natural England dataset. The overall comparison between the two datasets indicated that the model generated from the initial DEM was relatively robust, although lacking in the subtleties of local variations and detail.

Results from the borehole transects

The two transects (Figure 4.22) on Thorne Moors amounted to over a kilometre in length, collected at a variety of intervals between boreholes, and focussed on both the highest (Birtwhistle) and lowest (Cottage Dyke) areas as indicated by the DEM derived from the Natural England contour data and the results of previous study (e.g. Smith 2002, see Figure 4.17).

4. Laying the foundations: modelling pre-peat landscapes

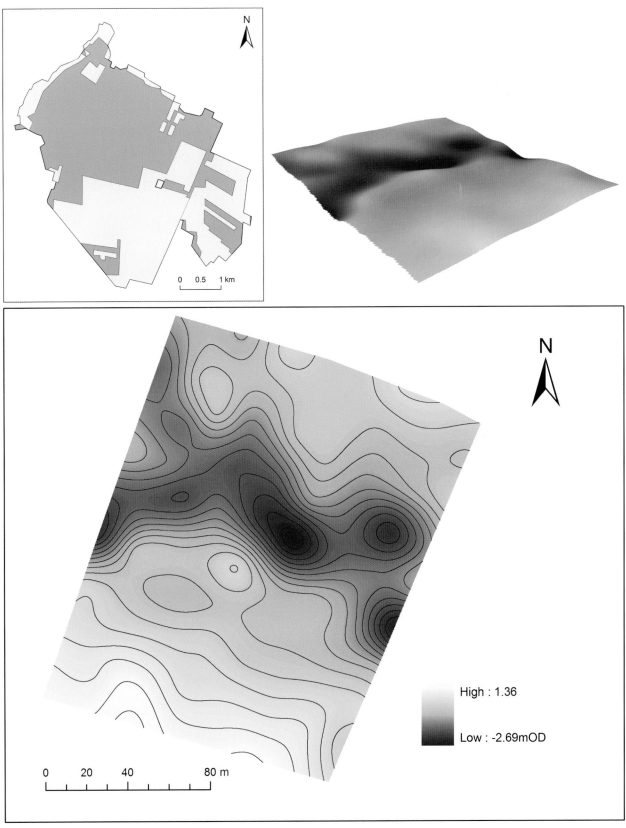

Figure 4.18 DEM derived from borehole Grid 1 in the area of Middle Moor. Contour lines on this and subsequent figures show 25m intervals

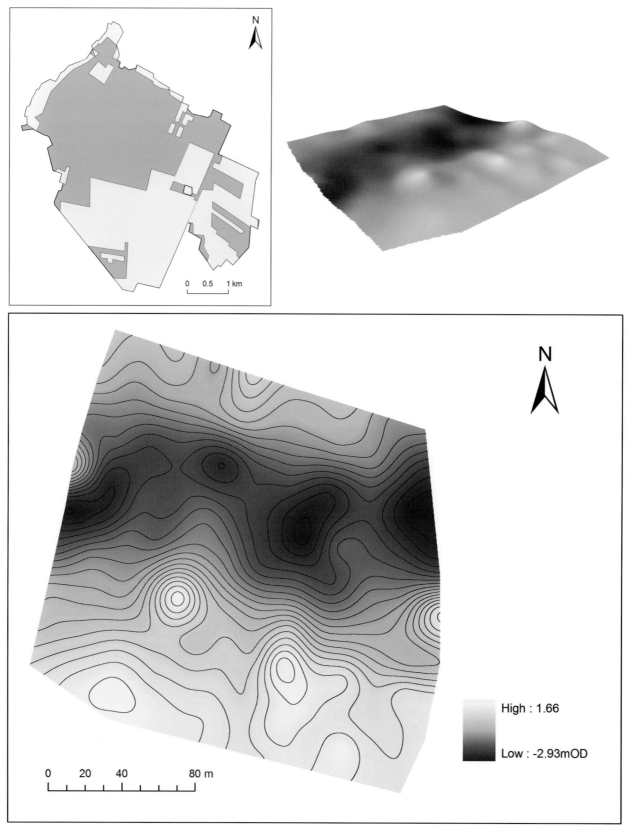

Figure 4.19 DEM derived from borehole Grid 2 in the area of Middle Moor to the east of Grid 1

4. Laying the foundations: modelling pre-peat landscapes

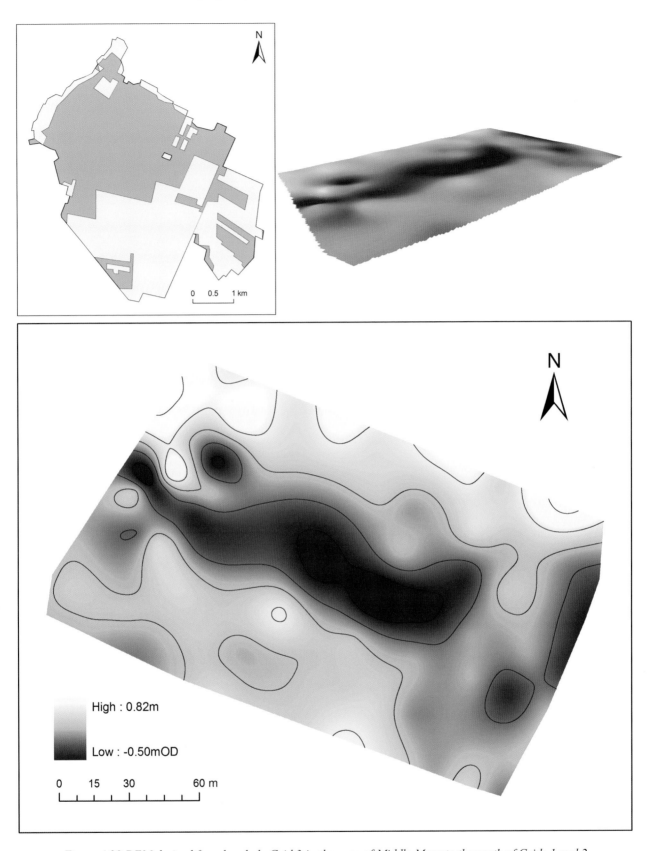

Figure 4.20 DEM derived from borehole Grid 3 in the area of Middle Moor to the north of Grids 1 and 2

74 *Modelling archaeology and paleoenvironments in wetlands*

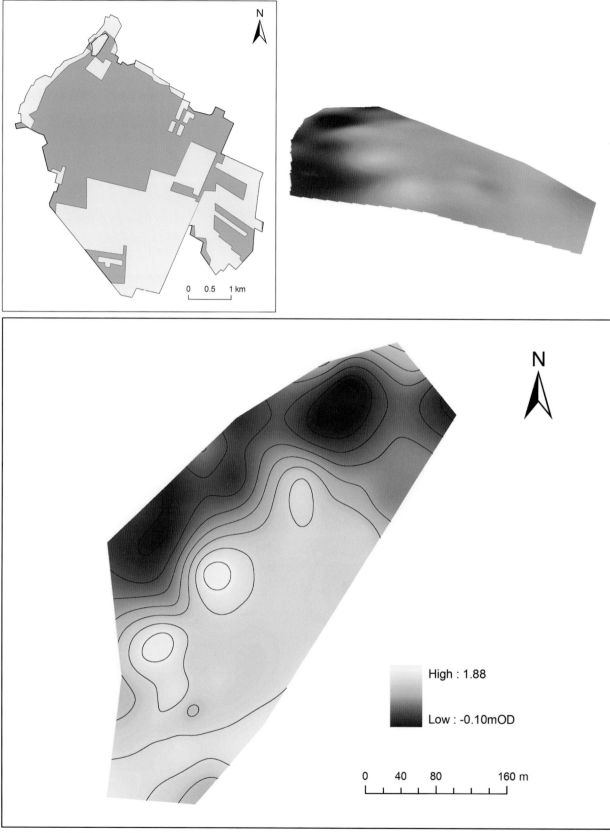

Figure 4.21 DEM derived from borehole Grid 4 in the area of Rawcliffe Moor towards the northern edge of Thorne Moors

4. Laying the foundations: modelling pre-peat landscapes

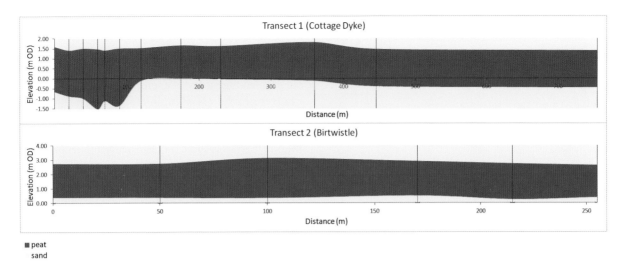

Figure 4.22 Borehole transects in the areas of Cottage Dyke and Birtwistle on Thorne Moors

Table 4.7 Comparative statistics from the GIS models of each of the gridded areas

	Dataset	Minimum (m OD)	Maximum (m OD)	Mean (m OD)
Grid 1	Borehole DEM	-2.69	1.36	-0.45
	Natural England data	-1.01	0.27	-0.55
Grid 2	Borehole DEM	-2.93	1.66	-0.60
	Natural England data	-0.28	0.07	-0.09
Grid 3	Borehole DEM	-0.50	0.82	0.19
	Natural England data	-0.58	-0.24	-0.47
Grid 4	Borehole DEM	-0.10	1.87	1.07
	Natural England data	0.43	1.77	0.34

Transect 1 (Cottage Dyke)

In the area of Cottage Dyke, transect 1 extended over a total distance of 750m parallel with and to the south of the Limestone Road. A total of 12 boreholes were excavated at intervals of between 15m and 320m, in response to the inaccessibility due to the density of scrub and woodland and to the local variations in the basal, pre-peat topography. Basal elevations along the transect ranged between -0.02m and -1.50m OD, with the lowest areas identified within the western-most quarter of the transect. The depth of peat along this transect in part reflected the elevation of the pre-peat landscape, with the deepest peat towards its western end where a maximum depth of 2.97m was recorded. At its shallowest the peat was 1.63m deep towards the centre of the transect, and slightly deeper to the east, up to 1.85m.

Transect 2 (Birtwistle)

The area of Birtwistle was targeted due to the fact that this was evidently a higher area of the basal topography of Thorne Moors (see figures 2.11 and 3.4). It was hypothesized that this area might have remained a higher, drier area of the Moors during early peatland development, similar to Lindholme Island on Hatfield Moors. However, this area could not be easily accessed due to flooding, although six boreholes were excavated at the southern end of the feature at approximately 50m intervals over a distance of 250m. Basal levels ranged in this area between 0.25m and 0.52m OD, a range of 0.27m, with no significant trend along its length and peat depths were relatively consistent, ranging between 2.22m and 2.77m.

Peat stratigraphy on Thorne Moors

Detailed stratigraphic recording was carried out in a transect across the Middle Moor area (Grids 1 and 2). Five sediment units were recognised (base upwards):

1. Pinkish–grey clay silts (all boreholes)
2. Grey–brown silt rich humified peat with abundant wood macrofossils. Monocotyledonous remains locally present.
3. Saturated brown detrital peat with fibrous material and wood fragments
4. Humified brown peat with *Phragmites* (reeds) and *Carex* stem and root fragments. *Menyanthes* (bog-bean) and ericaceous remains locally present.
5. *Sphagnum* peat with ericaceous remains and *Eriophorum* locally present (all boreholes).

Some local variations in both the composition and thickness of these units were recognised. The thickest deposit was that of unit 2, which thickened to a maximum depth of 3.05m in the centre of the possible palaeochannel feature. The profile was a roughly compressed 'v' shape, with the clay/organic sediment contact present at just over OD at each end of the transect and dropping to nearly -3m OD towards the centre.

Detailed stratigraphic recording was also carried out along a transect of six cores across Rawcliffe Moors (Grid 4). Here, the organic deposits were identified to a maximum depth of *c.* 1.5m and three main units were recognised (base upwards):

1. Dense pinky–grey sandy clay
2. Wood rich fen peat with occasional monocot remains
3. Moderate to highly humified *Sphagnum* peat, with occasional *Eriophorum,* ericaceous rootlets and monocot remains

The thickest deposits corresponded to the central area of the gridded borehole survey area where peat accumulation appears to have commenced in a pronounced depression in the basal pre-peat landscape (see above). The thickest surviving deposits (maximum 1.1m) of *Sphagnum* peat were also recorded in this part of the grid, The peat thinned to *c.* 0.50m at the ends of the transect, reflecting the edges of the depression and the effects of peat cutting. Thicker peat deposits were present in the southwestern extent of the grid, where the depression is deepest, with a maximum recorded depth of 2.75m, consisting of 2.2m of *Sphagnum* and 0.55m of wood rich fen peat.

Generation of a new Digital Elevation Model of the pre-peat landscape of Thorne Moors

Accessible areas of Thorne Moors, defined as those free of dense scrub and woodland, amounted to approximately 46% of the total area of the Moors, although this reduced throughout the fieldwork phase of the project due to the increased flooding as part of the peatland restoration processes. Of the available 547ha, a total of 18ha was investigated through gridded borehole survey. Although this figure accounts for less than 1% of the total area of Thorne Moors, it amounts to approximately 3.3% of the available area. Furthermore, these datasets were supplemented by over 1km of borehole transects.

The quantity of existing datasets for Thorne Moors was thus slightly less than those available for Hatfield Moors, limiting the potential for generating a robust GIS model of the pre-peat basal topography. Moreover, there was considerable variation in the spatial resolution of the two principal datasets. From the assessment of the gridded borehole data in relation to the model derived from the Natural England contour data it appeared that the latter was generally robust but lacking in local detail, such as the apparent channel feature running west–east across the centre of the Moors identified in the gridded borehole surveys (Grid 4). The lack of other datasets such as GPR meant that the potential for integrating the data from the pre-existing DEM, the gridded borehole surveys and the boreholes transects into a single model was restricted compared with the situation for Hatfield Moors. Thus, for the purpose of further analyses, the DEM derived from the Natural England data (Figure 3.4) was used for broader analyses, but was calibrated locally by the DEMs from the gridded boreholes as appropriate.

Overall, the DEM indicated significantly greater variation in the pre-peat topography in comparison to Hatfield Moors. Within the southern part of Thorne Moors, there was to be a consistently higher area around Birtwistle. Within the northern part, the topography was also relatively consistent except for localised areas such as at Rawcliffe. A possible palaeochannel has been identified running approximately west–east through the Middle Moor area in the central part of Thorne Moors (Grids 1 and 2).

4.4 Summary

The assessments of legacy datasets in addition to the acquisition of additional stratigraphic and topographic information through gridded borehole survey and transects, models have been generated of these pre-peat hidden landscapes. These models can perhaps best be considered as *data visualisations*, rather than as *representative visualisations* (see McCoy and Ladefoged 2009) of the 'true' morphology of the pre-peat landscape. The generation of these DEMs of the basal topography of Hatfield and Thorne Moors formed the critical initial stage of understanding both the morphology of the pre-peat landscape. For Hatfield Moors, the wider availability of datasets enabled the generation of a refined model of basal topography using a hierarchical choice of datasets for different parts of the Moors. Hence, the resulting DEM combined data from LIDAR survey, GPR survey, gridded borehole survey and a series of borehole transects, but excluded all other forms of legacy data.

In contrast, the DEM derived from the Natural England contour data for Thorne Moors was demonstrated to be more robust than that for Hatfield Moors, perhaps due to the relatively high level of localised variation in the pre-peat topography and the subsequent increased resolution of the contour data itself. Hence, despite the lack of multiple datasets for Thorne Moors, it was possible to use this overall model for the generation of further models, supplemented by the additional results from the borehole surveys. These models provide a foundation for the analysis and reconstruction of the evolution of the landscapes of Hatfield and Thorne Moors, from peat inception through to mire formation, and the quantification of the archaeo-environmental resource represented by the surviving organic deposits. In the following chapter, the chronology and character of wetland inception for both peatlands is discussed.

5. Modelling, dating and contextualising palaeoenvironmental records

with Pete Marshall (Sections 5.2, 5.3, 5.4 and 5.10) and Nicki Whitehouse (Section 5.5)

5.1 Introduction

The previous chapter detailed the generation of GIS deposit models of the pre-peat landscapes of Hatfield and Thorne Moors. These models provided the basis for the subsequent stage of study: the establishment of a chronology for wetland inception and peat spread, and also for estimating the chronological 'envelope' of the surviving peat resource (see Chapter 3). The first step in this process was to determine the robustness of the radiocarbon chronologies of previous palaeoenvironmental studies. Subsequently, a programme of radiocarbon dating of additional samples from contexts from the areas investigated through the fieldwork on Hatfield Moors, described in Chapter 4, was undertaken. Following this process, the basal models (Chapter 4) were then used to contextualise information provided by previous palaeoenvironmental work and the new radiocarbon dates (Chapter 2), to identify the location and timing of wetland inception and investigate pattern and process of mire development. The use of these data for the construction of GIS deposit models of wetland inception and peat spread in four dimensions, and for the quantification of the surviving archaeo-environmental resource, is discussed in the following chapter (Chapter 6).

As discussed in Chapter 2, a significant number of palaeoenvironmental and archaeological studies have been carried out on Hatfield and Thorne Moors over the last few decades, generating a range of interpretations regarding specific 'events', including the development of the mire systems as well as changes in the wider environment around the peatlands. Hence, the available radiocarbon dates for peat inception were used to test the different spatial models of wetland development and spread. Bog Surface Wetness (BSW, see Chapter 2) records were used to identify a series of 'synchronous' events that have been attributed to climatic forcing of the peatlands. Palynological evidence indicated variations in the spatial and temporal patterns of vegetation change and prehistoric human impact on the vegetation. These and related interpretations could only be assessed and modelled if the chronological control associated with these events could be demonstrated. Any sequences identified as having poor or suspect chronological control could not be meaningfully incorporated into any subsequent narratives or models of landscape development.

A total of 171 radiocarbon measurements were associated with these previous palaeoenvironmental studies (chapter 2; Figures 5.1 and 5.2). These included 39 AMS (Accelerator Mass Spectrometry) and a single radiometric measurement obtained from the Scottish Universities Environmental Research Centre (SUERC), East Kilbride, together with 16 AMS measurements from the Oxford Radiocarbon Accelerator Unit described below (see Chapter 6). A further 109 measurements have been obtained from the following laboratories: Beta Analytic Inc, Birmingham University, Cambridge University, Cardiff University, Rafter, and NERC Scottish Universities Research and Reactor Centre (Glasgow).

5.2 Radiocarbon dating peat and modelling chronologies

As discussed in Chapter 1, robust chronologies are essential to establish the rate and tempo of change in the past from proxy records, whether vegetation from pollen, Bog Surface Wetness from testate amoebae or relative sea level changes from diatoms. Peat deposits can present particular problems for radiocarbon dating given that they are composed of a heterogeneous mix of organic matter, potentially of different ages and origins, and in different states of biological decay and humification (Brock *et al.* 2011). There has been much debate concerning what type of samples, including different carbon/organic fractions, may yield the most reliable age estimation of a given horizon (Bartley and Chambers 1992, Brock *et al.* 2011, Cook *et al.* 1998, Lowe and Walker 2000, Piotrowska *et al.* 2011, Shore *et al.* 1995; Walker *et al.* 2001).

The most reliable material are short-lived terrestrial plant remains that synthesised atmospheric CO_2 and therefore do not incorporate a reservoir or an 'old-wood

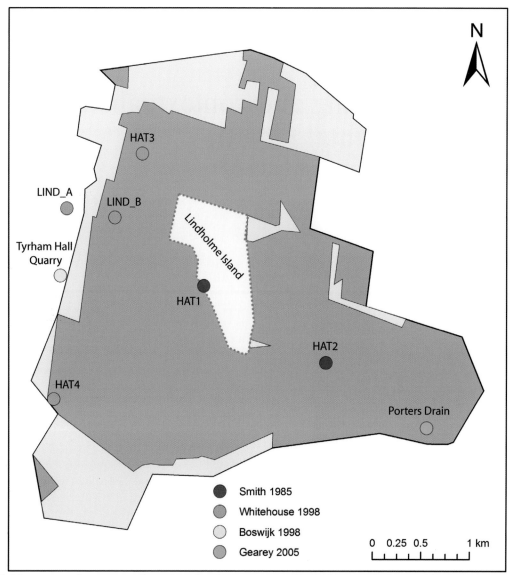

Figure 5.1 Locations of previous radiocarbon dated sequences on Hatfield Moors (referenced by first publication date)

effect' (Bowman 1990). *Sphagnum* macrofossils provide excellent samples, as do the above-ground leaves and stems of dwarf shrubs (e.g. *Calluna vulgaris* or *Empetrum nigrum*), when the former are absent (Piotrowska *et al.* 2011). However, for highly humified peats where it is not possible to select individual macrofossils, either the whole (bulk) peat fraction or different chemical or physical fractions can be dated (Dresser 1970).

The two most commonly defined chemical fractions are the humins (i.e. alkali and acid insoluble organic detritus) and the humic acids (i.e. alkali soluble and acid insoluble matter), with a third fraction that is occasionally dated consisting of the fulvic acids (i.e. the acid soluble fraction). Finally, if there is not enough available material to date the separate fractions then the bulk sediment can be dated (i.e. humin and humic fractions combined). As the humin fraction is composed of the actual organic detritus, the resultant date from measuring this fraction is subject to many of the processes that affect the dating of macrofossils in the same type of environment. Firstly, organic material that forms all or part of the humin fraction could be in-washed, which would result in a date that is too old, although this would not be anticipated to be an issue for raised mire deposits.

The contamination of this organic material by geological age carbon (e.g. hard-water error) would have the same effect. The humin fraction might also be too young if, for example, the environment was prone to wet–dry episodes or bioturbation, allowing intrusive material to migrate down the sediment column. Therefore, the humin fraction is not necessarily homogeneous and it might be best to avoid dating this fraction by AMS as a small amount of contamination would greatly affect the resultant measurement (Nilsson *et al.* 2001). It might, therefore, be more appropriate to date humins using conventional radiocarbon dating techniques, as it is unlikely that a sufficient volume of such contamination would be present to bias the results significantly.

The humin fraction may also be divided into the coarse

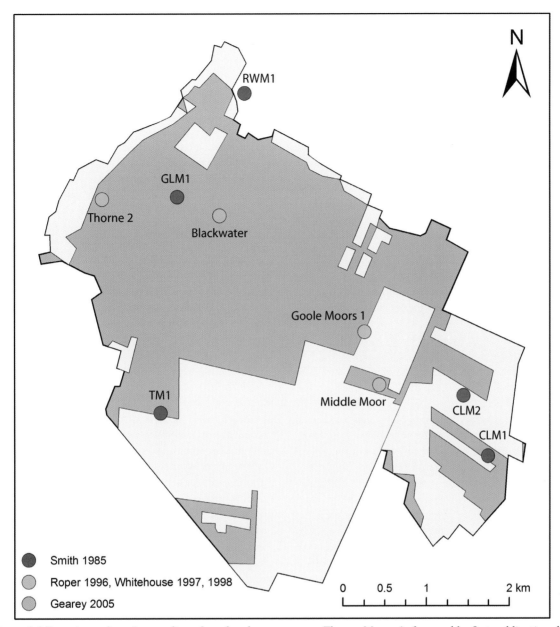

Figure 5.2 Locations of previous radiocarbon dated sequences on Thorne Moors (referenced by first publication date)

(>250μm) and fine (<250μm) fraction and Dresser (1971) concluded that only the fine humin fraction from blanket and reedswamp peats consistently produced reliable dating. However, Bartley and Chambers (1992) argued that the entire humin fraction should be used. It is important to stress once more that, because of the various scientific opinions regarding the fractions of peat that can be reliably dated and site-specific factors, consistency is sought within the radiocarbon dating programme.

The second fraction that may be dated is the humic acids, which are the *in situ* products of plant decay, although it has been shown that they can be mobile in groundwater, both vertically and horizontally (Shore *et al.* 1995). Therefore, humic acids cannot always be relied upon to date the level from which they were collected accurately. However, unlike the humin fraction, humic acids are homogeneous, as they are alkali soluble and therefore can be more reliably dated through AMS.

Fulvic acids are acid soluble and therefore can be suspended in a homogeneous solution. Shore *et al.* (1995) have shown that fulvic acids are nearly always the youngest fraction within a sediment profile and in a few cases they are by far the oldest of the three dated fractions. It is suggested that this is likely to be due to the fact that fulvic acids are soluble in water and therefore are highly mobile within peat, moving up and down with the watertable, and laterally within the sub-surface drainage patterns.

In cases where there is insufficient material for dating separate fractions, the humic acid and humin fractions can be bulked together to provide an average date for all the organic material in a sample. As stated earlier however, it is preferable to have the dates of the two fractions as this provides the data necessary for using replication as a measure of consistency. When measurements based on two fractions are obtained, if they are statistically consistent, a weighted average can be taken before calibration takes place following the method outlined in Ward and Wilson (1978). In most cases this produces a date that is more reliable. However, if the two results are not in agreement then the data will need to be re-evaluated in an attempt to determine which sample more reliably relates to the date of peat formation at the level under consideration.

Bayesian chronological modelling: assessing the synchronicity of palaeoenvironmental 'events'

In order to understand the palaeoenvironmental record fully and, in particular, to understand the syncroneity of events in the stratigraphic data, the establishment of the age of an 'event' in each record is required, as well as the establishment of the character of the 'event' itself (Parnell et al. 2008). Smith and Pilcher (1973) proposed the term 'rational limit' for an event reflected by pollen data, and defined this as the point in a sequence marked by the first rapid increase in the abundance of a given pollen taxon. The term 'empirical limit' refers to the point after which a taxon is consistently present in a pollen diagram. But as Watts (1973) highlighted, such a definition is extremely sensitive to the relative abundance of a pollen taxon, for example the number of grains counted and the resolution of the pollen samples used, introducing considerable uncertainty as to where in a pollen diagram the events should be 'located'.

The three main sources of stochastic 'noise' identified by Parnell et al. (2008) within pollen data are the randomness in collecting and counting grains, the inter-annual variations in pollen production, and the agencies of vegetation disturbance that can operate at local as well as at regional/global scales. Thus, when considering an event as a point in space it can take one of two general forms: a transition from one stable palaoenvironmental state to another, such as the change in pollen values, e.g. the *Alnus* rise, or a short-lived change within a longer term more stable situation, such as the deposition of a tephra layer in a peat sequence.

This study has taken the approach that an 'event' can be associated with a specific depth in a core. Assessing whether synchroneity occurs between events in two or more stratigraphic sequences depends on establishing the age of such records. As outlined in Chapter 1, significant progress has been made utilising Bayesian approaches, in the refinement of archaeological chronologies in particular (Bayliss et al. 2007, Buck et al. 1996), and these methods are now frequently used to analyse and model palaeoenvironmental chronologies. Details of Bayesian age-depth modelling have been widely covered elsewhere and will not be discussed in detail here (e.g. Parnell et al. 2011, Bronk Ramsey 2008, Blaauw and Christen 2005).

The development of the Bayesian approach is significant in a methodological sense as it allows the chronological robustness of palaeoenvironmental records to be assessed using all the stratigraphic and chronological information available for a given sequence, along with the uncertainties associated with these data. Moreover, it permits the estimation of the dates of 'events' of interest in a palaeoenvironmental sequence which fall between directly dated horizons. The correlation of archaeological and palaeoenvironmental chronologies can also be undertaken. For example, Gearey et al. (2009) used this approach to estimate the probability that a radiocarbon dated Bronze Age cremation burial was synchronous with palynological evidence for woodland clearance in a record from a nearby sampling site. The ability to determine the relationship between different records in this manner amounts to more than just a methodological advance. Determining the relative rate of change between cultural and environmental records is arguably the only way to test hypotheses regarding the relationship between environmental changes in the past and the possible human perception and response to such changes (discussed further below).

As discussed in Chapter 2, a significant number of palaeoenvironmental and archaeological studies have been carried out on Hatfield and Thorne Moors over the last few decades and various patterns of change in space and time have been interpreted from these records. These events include both the developments of the mire systems themselves as well as the changes in the wider landscape around the sites. Hence, available radiocarbon dates for peat inception have been used to support different spatial models of wetland development and spread. Changes in Bog Surface Wetness identified from macrofossil and humification data have been used to identify a series of 'synchronous' wet shifts that have been attributed to climatic forcing of the peatlands, suggesting local and possible regional teleconnections in processes of change. Palynological evidence indicates variations in the spatial and temporal patterns of vegetation change and the prehistoric human impact on the vegetation. These and related interpretations can only be assessed and modelled if the chronological control associated with these events can be demonstrated.

Chronological modelling: methods, results and discussion

The results of the modelling are summarised in Table 5.3 and individual models are presented in Appendix 1, along with further details of the chronological modelling. The *P-Sequence* models for Hatfield Moors sites HAT1 (Appendix 1 Figure A1) and HAT2 (Appendix 1 Figure A2) showed good overall agreement between the model and the radiocarbon dates, implying that these provide robust chronologies for the associated palaeoenvironmental events recorded in each. However, the *P-Sequence* and *U-Sequence* models from the HAT4 sequence (Whitehouse 2004, see Chapter 2), from the southwestern corner of Hatfield, and the Porters Drain sequence (Gearey 2005, see Chapter 2), from the southeastern edge of the Moors, showed poor overall agreement between the model and the radiocarbon dates ($A_{overall}$ = <60.0%) (Appendix 1

Table 5.1 Summary of the results of Bayesian modelling of palaeoenvironmental sequences (for further details see Appendix 1)

Area	Name/Code	P-Sequence ($A_{overall}$)	Robust	Reference
Hatfield	HAT1	104.3%	Yes	Smith 2002
Hatfield	HAT2	108.3%	Yes	Smith 2002
Hatfield	HAT4	–	No	Whitehouse 2004
	Porters Drain	–	No	Gearey 2005
Thorne	Crowle Moor CLM1	66.9%	Yes	Smith 2002
Thorne	Crowle Moor CLM2	85.6%	Yes	Smith 2002
Thorne	Goole Moor GLM1	62.0%	Yes	Smith 2002
Thorne	Goole Moor GLM2	25.1%	Yes	Smith 2002
Thorne	Goole Moor, GLM3	133.0%	Yes	Smith 2002
Thorne	Rawcliffe Moor, RWM1	104.2%	Yes	Smith 2002
Thorne	Thorne Moors Sites 1 & 2	100.8%	Yes	Smith 2002, Buckland 1979
Thorne	Trackway Site	112.6%	Yes	Buckland 1979
Thorne	Thorne Waste	66.0%	Yes	Smith 2002
Thorne	Middle Moor	–	No	Gearey 2005

Figure A3). With the exception of the Middle Moor sequence (Gearey 2005; see chapter 2), the *P-sequence* models (Table 5.1; see Appendix 1) from Thorne Moors demonstrated good overall agreement between the models and the radiocarbon dates, indicating that these sequences have robust chronological control.

The modelling of the four sequences from Hatfield Moors indicated that HAT1 and HAT2 provided robust chronologies for the associated palaeoenvironmental records (*posterior density estimates*[1] derived from modelling of the sequences are given in Appendix 1). However, the radiocarbon chronologies associated with HAT4 and Porters Drain were less reliable. The sequences from Thorne Moors were all chronologically robust with the exception of that from Middle Moor. The failure of HAT4, Porters Drain and Middle Moor was probably due to the fact that the radiocarbon dates from these sequences included a number of erroneous determinations (outliers). As outlined above, this might be the result of the incorporation of younger carbon (e.g. through root penetration) or older carbon (i.e. through the deposition of re-worked soils or bioturbation; although this may be regarded as less likely in the case of ombrotrophic peat deposits) into the dated sample fraction.

The most robust sequences were those which had been dated using conventional 'bulk' samples and it is these that may thus be regarded as providing the most reliable records of environmental change. The HAT4, Porters Drain and Middle Moor dates were obtained from bulk macrofossil or sediment samples dated using the AMS method. The six radiocarbon dates (SRR-6121-6126) from the middle of HAT4, for example, are statistically consistent (T' = 8.5; v = 5; T' (5%) = 11.1; Ward and Wilson 1978) and could therefore be of the same actual age. These dates were obtained on *Pinus/Betula* twigs (Whitehouse 2004) and the use of such bulk samples has apparently incorporated wood of varying ages that effectively produced an average age. The Porters Drain and Middle Moor samples were all bulk sediment AMS and it is possible that some of these contained intrusive carbon (see above). However, identifying which measurements might have been affected in this way is not straightforward. The problems with the radiocarbon chronologies of these three sequences mean that it was difficult to assimilate the associated palaeoenvironmental records from each into broader chronologies or models of landscape development.

5.3 Sampling strategy: radiocarbon dating of samples from Hatfield Moors

The results from previous studies were augmented by a new programme of sampling for radiocarbon dating from basal contexts across Hatfield Moors. This new set of radiocarbon samples were chosen from locations from across the Moors to test the hypotheses outlined in Chapter 2 relating to wetland inception and peat growth, and for establishing a chronological framework for understanding the surviving peatland resource. All new radiocarbon samples from Hatfield Moors were coded with as 'HM' (Hatfield Moors) to avoid confusion with previous nomenclatures which have been used, including the prefixes 'HAT' and 'LIND'.

The deposit models generated in Chapter 4 provided the ability to relate timing and processes of peat formation to the topography of the pre-peat landscape in three dimensions for the first time. However, it was clear that the acquisition of a suite of radiocarbon dates would be required to model peat inception and spread. Only three basal dates were available from the previous research in the Hatfield Moors study area, although issues with geo-referencing of HAT1 and HAT2 (see Chapter 3) limited the value of these data in terms of further analyses. The sampling strategy focussed on 'sub-sets' for each of the eight areas for which basal topographic grids had been produced and also from samples from the borehole transects (see Chapter 4).

The selection of basal samples within these 'subsets' (Figure 5.3) was refined to reflect an altitudinal gradient (from *c.* -0.30m OD to 1.00m OD), which was, as close as possible, representative to that of the height range of the basal topography of the Moor (Figure 5.4, see chapter 3). One of the areas to the northeast of Lindholme Island

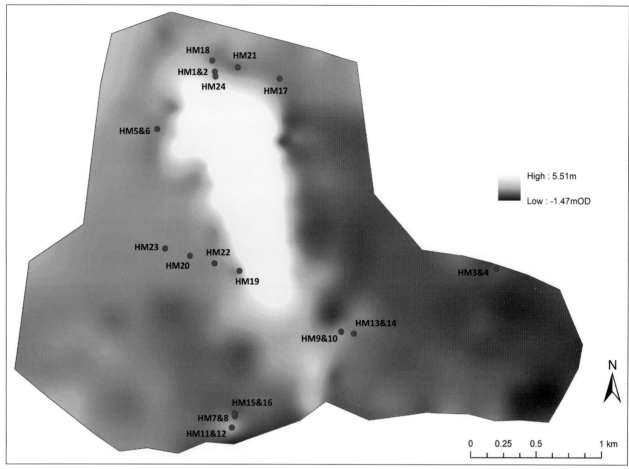

Figure 5.3 Radiocarbon sampling site locations on Hatfield Moors

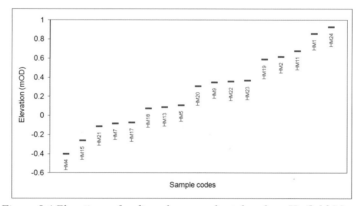

Figure 5.4 Elevations of radiocarbon samples taken from Hatfield Moors

(Grid 6) was inaccessible due to flooding as a result of peatland restoration measures and additional samples from outside the gridded areas to the southwest of Lindholme were therefore selected to complete the altitudinal range. A limited number of samples (eight) were collected from top peat contexts in order to determine the chronological span of the remaining peat deposits on Hatfield Moors (see Table 5.2).

5.4 Results: Hatfield Moors radiocarbon dates

A total of forty measurements (Table 5.2 and Figure 5.5) were obtained from macrofossil and bulk peat samples from Hatfield Moors. The sampling procedure, the sub-sample pre-treatment and the dating procedures are detailed in Appendix 1. Only two of the eight bulk peat samples for which replicate humic and humin fraction measurements were obtained produced statistically consistent results (Table 5.3). Replicate measurements on the humic and

5. Modelling, dating and contextualising palaeoenvironmental records

Table 5.2 Radiocarbon dates from Hatfield Moors

Lab ID	Sample ID	Material	$\delta^{13}C$ (‰)	Radiocarbon Age (BP)	Calibrated date BC (95% confidence)
SUERC-9688	HM2	Peat, humic acid	-27.6	3880±35	2470–2200
SUERC-9689	HM2	Peat, humin	-27.1	4010±35	2620–2460
SUERC-8846	HM2	charred twigs & stems, unident.	-26.0	5430±35	4350–4230
SUERC-8847	HM3	Ericaceae twigs/roots	-27.3	3305±40	1690–1490
SUERC-9636	HM4	Peat, humic acid	-28.6	4425±35	3330–2920
SUERC-9637	HM4	Peat, humin	-27.9	4500±35	2910–2690
SUERC-8848	HM4	charred grass stems	-28.5	4495±35	3360–3020
SUERC-8849	HM4	*Eriphorum* spindle/Ericaceae stems	-27.1	4745±35	3640–3370
SUERC-8850	HM5	twig frag., unident.	-27.9	4630±40	3520–3340
SUERC-8851	HM6	Ericaceae stems	-28.1	4410±35	3320–2910
SUERC-9690	HM7	Peat, humic acid	-27.7	4470±35	3350–3020
SUERC-9691	HM7	Peat, humin	-27.8	4555±35	3490–3100
SUERC-8855	HM8	Ericaceae stems	-28.7	2990±40	1390–1110
SUERC-8856	HM9	Ericaceae stems/roots	-27.2	3470±35	1890–1680
SUERC-8857	HM10	Ericaceae stems/roots	-28.0	3200±40	1530–1400
SUERC-8858	HM11	Ericaceae stems/roots	-27.1	3310±35	1690–1500
SUERC-8859	HM12	Ericaceae stems/roots & charred leaf	-28.1	2960±40	1320–1040
SUERC-9638	HM13	Peat, humic acid	-27.5	4225±35	2910–2690
SUERC-9639	HM13	Peat, humin	-28.0	4395±35	3270–2910
SUERC-8860	HM13	charred twigs, unident.	-26.9	4410±35	3320–2910
SUERC-8861	HM13	charred bark-like material	-24.3	4410±35	3320–2910
SUERC-9692	HM14	Peat, humic acid	-26.9	3205±35	1530–1410
SUERC-9696	HM14	Peat, humin	-26.4	3135±35	1500–1310
SUERC-9640	HM15	Peat, humic acid	-28.3	5350±35	4330–4040
SUERC-9641	HM15	Peat, humin	-28.2	5890±35	4840–4690
SUERC-8947	HM15	charred bark, unident.	-27.5	5985±35	4990–4780
SUERC-8948	HM16	Ericaceae stem/roots	-27.0	3630±35	2130–1890
GU-6365	HM17	Peat, humic acid	-27.2	5650±50	4600–4360
SUERC-9646	HM17	Peat, humin	-26.7	5860±35	4800–4610
SUERC-8949	HM18	charred Ericaceae twig & unident. twig	-26.9	4905±35	3770–3630
SUERC-8950	HM19	Ericaceae stem/roots	-25.3	4655±35	3630–3360
SUERC-8951	HM20	Herbaceous charcoal, small twigs	-25.9	4905±35	3770–3630
SUERC-8952	HM21	Herbaceous charcoal, small twigs	-26.0	6355±35	5470–5220
SUERC-8953	HM22	Herbaceous roots, unident.	-24.6	4605±35	3500–3340
SUERC-8875	HM23	Twigs/Carex nutlets	-27.4	4895±40	3770–3630
SUERC-8876	HM23	*Eriphorum* fibres	-25.4	4730±35	3640–3370
SUERC-9647	HM23	Peat, humic acid	-26.9	4390±35	3270–2900
SUERC-9648	HM23	Peat, humin	-26.3	4400±35	3270–2910
SUERC-8877	HM24	Ericaceae stems/roots	-26.8	4250±35	2920–2760
SUERC-8878	HM24	Ericaceae stems/roots	-26.7	3690±35	2200–1960

Table 5.3 Chi-squared test results of radiocarbon dates from Hatfield Moors

Bulk samples (replicate humic/humin)		
HM2	SUERC-9688 & SUERC-9689	T'=6.9, T'(5%)=3.8, v=1
HM4	SUERC-9636 & SUERC-9637	T'=48.8, T'(5%)=3.8, v=1
HM7	SUERC-9690 & SUERC-9691	T'=955.9, T'(5%)=3.8, v=1
HM13	SUERC-9638 & SUERC-9639	T'=19.7, T'(5%)=3.8, v=1
HM14	SUERC-9692 & SUERC-9696	T'=2.0, T'(5%)=3.8, v=1
HM15	SUERC-9640 & SUERC-9641	T'=118.8, T'(5%)=3.8, v=1
HM17	GU-6365 & SUERC-9646	T'=11.7, T'(5%)=3.8, v=1
HM23	SUERC-9647 & SUERC-9648	T'=0.0, T'(5%)=3.8, v=1
Humic, humin and macrofossils		
HM2	SUERC-9688, 9689 & SUERC-8846	T'=1265.5, T'(5%)=6.0, v=2
HM4	SUERC-9636, 9637 & SUERC-8848, 8849	T'=48.7, T'(5%)=7.8, v=3
HM13	SUERC-9638 & SUERC-9639	T'=19.7, T'(5%)=3.8, v=1
HM23	SUERC-9647, 9648 & SUERC-8875, 8876	T'=0.0, T'(5%)=7.8, v=3

humin fractions of bulk peat samples and the single macrofossils were also statistically inconsistent in all cases.

The basal peats from Hatfield appear to consist of material of a range of different ages (e.g. see Shore *et al.* 1995). In such cases, bulk samples can incorporate material including younger roots and stems from overlying deposits as well as older organic matter formed in the pre-peat mineral soil (Matthews 1993). There is a clear tendency

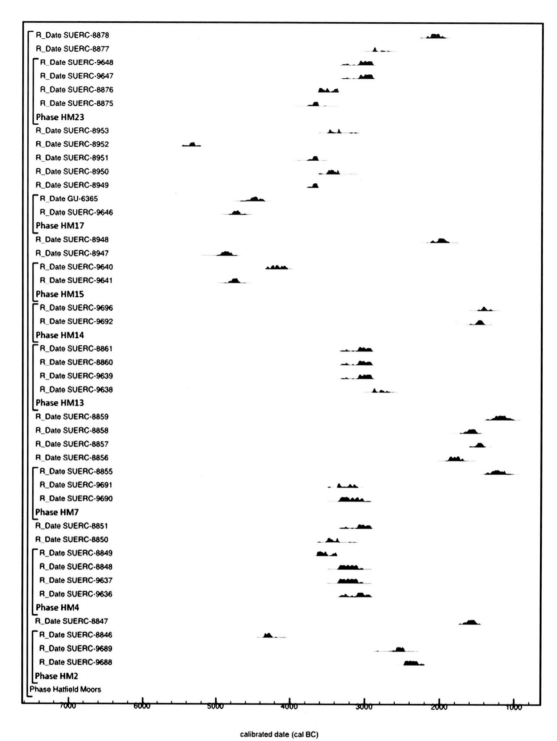

Figure 5.5 Probability distribution of radiocarbon dates from Hatfield Moors. Each distribution represents the relative probability that an event occurred at a particular time and is the result of radiocarbon calibration (Stuiver and Reimer 1993).

for the humic fractions of AMS bulk samples to be younger than the humin fraction (Figure 5.6). In the four cases where the measurements on plant macrofossils from levels where bulk fractions have been also been dated (AMS) (Figure 5.7), the dates from the humic acid fraction and waterlogged plant material are inconsistent. This inconsistency may be related to the formation of these deposits under relatively dry conditions, where a fluctuating watertable might be responsible for mobilising humic acids. This might be expected to be an issue during the earliest stages of peat formation when there is evidence (see below) for fluctuations in hydrological conditions. Identifying which determination might give the 'true' estimate of the age of the deposits with such multiple, statistically inconsistent dates are problematic. The chronological 'envelope' of the surviving peat deposits will be discussed in further detail in Chapter 6.

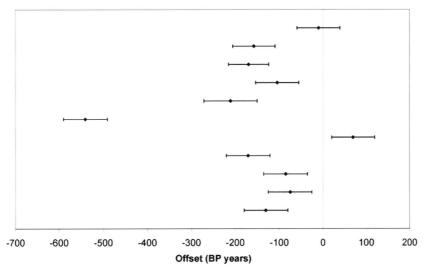

Figure 5.6 Offsets between radiocarbon measurements on the humic acid and humin fractions of bulk sediment samples (error bars are those for 68% confidence)

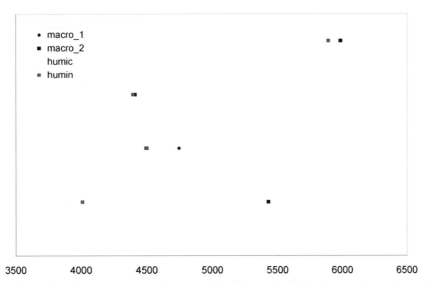

Figure 5.7 Replicate radiocarbon determinations on humic acid and humin bulk fractions and waterlogged plant remains from sediment samples

5.5 Contextualising palaeoenvironmental evidence for wetland inception on Hatfield Moors

This section provides a synthesis and overview of the timing, patterns and processes of wetland inception on Hatfield Moors using the chronological information discussed above and palaeoenvironmental data (Chapter 2) in the context of the topographic information provided by the pre-peat landscape model (Chapter 4). These data also have implications for the possible location and nature of human activity on the peatland in different periods (see Chapter 8). The discussion will concentrate on the timing of peat inception and the nature of the pre-ombrotrophic environment; the transition to ombrotrophy will be discussed separately. As was outlined in Chapter 2, the spatial distribution of the palaeoenvironmental data were uneven. In addition, HAT1 and HAT2 (Smith 2002), LIND_B and HAT4 (Whitehouse 1998, 2004) and Porters Drain (Gearey 2005) are the only sampling sites to have multiple radiocarbon dates, although the robustness of the chronologies from the latter two sites were problematic. To facilitate discussion, the following section will begin with the sampling sites to the north of Lindholme Island and proceed anti-clockwise around the Moors (Figures 5.1 and 5.3).

North of Lindholme Island (northern area of Hatfield Moors)

Five basal radiocarbon dates in total were available from the area to the north of Lindholme Island (within the area of borehole Grid 5 and Grid 8) ranging from the mid-fifth millennium BC to the later third millennium BC. The earliest known sediment accumulation on Hatfield Moors is recorded here, although the refined basal model shows that this is topographically one of the highest parts of the Moors (see Chapter 4). The paludification of the basal sands began at HM21 (-0.11m OD) at 5470–5220 cal. BC (SUERC-8952) with sediment deposition in the form of a grey–brown, fine organic sediment which was deposited in open water within a shallow pool (see Chapter 2). Further samples from the base of this sediment unit, located on a rising altitudinal north–south transect, produced dates of 3770–3630 cal. BC (SUERC-8949) (HM18; 0.08m OD), 4350–4230 cal. BC (SUERC-8846) at HM2 (0.64m OD), 85m to the south-southeast. A date of 2920–2760 cal. BC (SUERC-8877) was available at HM24 (0.93m OD) some 35m to the south of HM2.

These data thus demonstrated that the pool expanded steadily across the landscape in the Kilham West area over a period of some 2000 years, until it was lapping onto the northern limits of Lindholme Island. It was at this location that the Neolithic trackway and platform was constructed (see Chapter 2 and Chapter 9). Beetle fauna from the basal 'pool' sediments at HM2 (see Chapter 7) were dominated by species indicative of deciduous woodland, dry *Callunetum* and bare, sandy areas with no evidence of raised mire environments at this time. Thus, these insect assemblages do not appear to reflect clearly the aquatic habitats that must have been present locally, but this appears to be a taphonomic issue. Macrofossil analyses of the corresponding deposits indicated that the sediment consisted of fine organic detritus with few identifiable remains other than *Sphagnum* and leaves of *Calluna* and *Erica tetralix*, suggesting acid heath habitats.

Certainly the coring transects across this area of the Moors (see Chapter 4) indicated that in places the 'pool' sediment, although widespread, was not continuous and in places is replaced by black, highly humified *Eriophorum* rich sediment, typical of that across much of the rest of Hatfield (see below). There was no direct evidence for ombrotrophic conditions during the earliest stages of peat development. The presence of stumps of *Pinus* and remains of *Betula* are common on the basal sands across this area and it seems likely that this pool was not a continuous sheet of water but perhaps a mosaic of shallow extents of open water and drier areas of heath and *Pinus–Betula* woodland.

It would appear that, by the mid-third millennium BC, peat had begun to accumulate in a range of contexts across the Moors, with a number of basal dates (see above) falling between 3300–3700 cal. BC. A basal date of 3350–2930 cal. BC (SRR-6119) was obtained from a location some 1.5km to the northwest of HM21 (HAT3, Kilham West; Whitehouse 2004). Although no altitudinal or detailed topographic data are available, the basal model indicated an approximate elevation of 0.5m OD on gently sloping ground. The basal organic deposit was described as highly humified peat, apparently indicating that sediment accumulation was under sub-aerial conditions rather than the sub-aquatic contexts discussed above. The insect record suggested that a *Calluna* dominated heathland was present during the earliest stages of peat development, whilst the remains of *Pinus* and *Betula* in the basal peats also reflected the presence of these trees. There was insect evidence for open water in the vicinity in the form of feeders of aquatic and semi-aquatic plants, including *Phragmites*, *Carex*, *Juncus*, *Typha latifolia* and *Iris pseduocarus*. This may be a reflection of the pool described above, but a more localised signal is perhaps likely, indicating a mosaic of habitats including *Betula–Pinus* woodland but with both dry heath and wet understorey contexts in the close vicinity.

West of Lindholme Island (western area of Hatfield Moors)

The LIND_A sampling site was at the western edge of the peatland, 500m west of LIND_B (Whitehouse 1998, 2004, see below), and the basal sediment at this location was described as a desiccated, humified peat with *Betula* remains, contrasting with the basal sediments at Kilham West and HAT3 described above. No radiocarbon dates were available from LIND_A, although Whitehouse (2004) tentatively suggested a date of *c.* 2000 BC for peat inception. The insect remains indicated the presence of a *Pinus* heath but with deciduous trees, as well as some open pools. The range of beetle species recorded indicated a subsequent progression to more open, dry heath, with the record of *Aphodius contaminatus* suggesting grazing animals in the near vicinity. Increasingly acid, but not necessarily very wet, habitats were apparent. This location is off the edge of the basal model (Chapter 4) and further topographic context is not available, but the impression is that this sampling site reflects a relatively dry environment at the wetland–dryland interface.

LIND_B was 500m to the east of the previous sampling site and a date of 2840–2300 cal. BC (Beta-91800) was obtained from 0.60m above the base of the organic deposits, providing a *terminus ante quem* for peat inception. The basal insect fauna indicated *Eriophorum–Calluna* heath with open, sandy areas but no trees or aquatic habitats. There was also evidence of ombrotrophic conditions present in the area by this time, including the endemic beetle *Bembidion humerale*. No detailed topographic data were available, although the refined basal model showed the site was located at around 0.6m OD on a gentle gradient rising to the west.

The HM5 (0.11m OD) sampling site was located within Grid 4 on the northwestern side of Lindholme Island (see Chapter 4), some 450m to the east of LIND_B and *c.* 150m from the edge of Lindholme Island. Peat accumulation commenced at a date of 3520–3340 cal. BC (SUERC-8850); therefore at a similar time to HAT3 but earlier than LIND_B. The refined basal model (Chapter 4) shows that the local topographic context of this sampling location seems to have been a depression between low north–south linear dunes around 1m high.

Four basal dates from the area to the southwest of Lindholme were arranged on an east–west transect falling slightly to the south. The sample at the east end, 40m from the western edge of Lindholme, was HM19 (0.59m OD) dated to 3630–3360 cal. BC (SUERC-8950). The next three samples were collected at 200m intervals: HM22 (0.36m OD) 3500–3340 cal. BC (SUERC-8953), and HM20 (0.31m OD) 3770–3630 cal. BC (SUERC-8951). Four dates on the macrofossil, humin and humic fractions from the basal sample at HM23 (0.37m OD) at the western end of this transect were statistically inconsistent (see above) 3770–3630 cal. BC (Twigs/*Carex* nutlets: SUERC-8875), 3640–3370 cal. BC (*Eriophorum* fibres: SUERC-8876), 3270–2900 cal. BC (humic: SUERC-9647) and 3270–2910 cal. BC (humin: SUERC- 9648). Interpreting these data is problematic but overall peat accumulation began in the fourth millennium BC.

The dates from the area to the west of Lindholme Island indicate that peat accumulation also commenced in this area during the mid-third millennium, perhaps within a fairly narrow range of 3300–3600 cal. BC (between 0.3 and 0.6m OD). The basal sediments at these locations were minerogenic, highly humified and *Eriophorum* rich, contrasting again with those at the north of the road or at the trackway site. Peat accumulation appears to have begun slightly later on the southwestern edge of Lindholme Island at a date of *2910–2580 cal. BC* (CAR-168) at the HAT1 sampling site (see above; Smith 2002). A basal height of 0.6m OD was originally recorded for this location, but the refined model indicates that the base of the sampling site may have been somewhat higher than this at around 2.2m (see Chapter 4). The precise location of the sampling site is unclear with respect to the refined model, but was described as situated in the drain around Lindholme and hence on the sloping western edge of the Island (Smith 2002).

Macrofossils from the basal deposits at HAT1 indicate that *Eriophorum* and *Calluna* were growing locally, with some *Sphagnum* (*S. cuspidatum* and Section Acutifolia). *Calluna* (10–15%) and *Sphagnum* were also relatively well represented (*c*. 20% TLP+spores) in the basal zone in the pollen diagram from this location (Phase A; Smith 2002, 57–58). Tree and shrub pollen also attained relatively high percentages (80%), probably reflecting the growth of woodland on the margins of Lindholme as well as the presence of trees on the mire surface itself. The initial character of the on-site vegetation at this location was therefore a relatively dry heath, with some evidence for wetter, acidic conditions.

Packards Southwest (southwestern area of Hatfield Moors)

The far southwestern corner of the Moors was the location of a detailed Coleoptera record (HAT4; Whitehouse 1998, 2004), on the very edge of the refined basal model, which indicates a basal height of 0.07m OD. The basal date for this sequence was 1520–1390 cal. BC (SRR-6127) and is thus the youngest available for the base of the peat deposits obtained for Hatfield Moors to date, although the problems with the overall chronology from this location (discussed above) means that the reliability of the determination is uncertain. The basal deposits at this location were markedly different to elsewhere on the Moors, with *Phragmites* rich peat underlying black, well humified sediment with wood remains and minerogenic matter. This is the only location on the Moors where *Phragmites* has been identified in the basal sediments.

The associated Coleoptera indicated fen conditions with evidence for drier heath as well as sizeable pools (>0.30m deep) with reeds and aquatic plants including *Lemna*, *Carex* and *Scirpus lacustris*. Mixed woodland including *Alnus*, *Pinus*, *Quercus* and *Betula* is represented by both macrofossil and beetle data. This woodland probably reflects the vegetation at the wetland–dryland interface. No evidence for ombrotrophic conditions is present during the early stages of peat accumulation and the general impression is of a minerogenic lagg fen community on the edge of the expanding mire.

Packards South (southern area of Hatfield Moors)

The Packards south borehole grid (Grid 1) was positioned over an obvious dune system on the southern edge of the Moors. The basal dates consisted of one determination from the peat overlying the top of a dune at HM11 (0.60m OD), 1690–1500 cal. BC (SUERC-8858) and another from towards the base of this dune 85m to the north at HM7 (-0.08m OD), which produced two statistically inconsistent dates of 3350–3020 cal. BC (humic: SUERC-9690) and 3490–3100 cal. BC (humin: SUERC-9691). At HM15 (-0.3m OD) a further 20m to the north, three replicate measurements from the basal deposits were also statistically inconsistent: 4990–4780 cal. BC (SUERC-8947), 4330–4040 cal BC (humic: SUERC-9640) and 4840–4780 cal. BC (humin: SUERC-9641). This area thus produced two contrasting basal dates in the form of the second oldest date (HM15) so far obtained for Hatfield Moors, whilst excluding that from HAT4 (discussed above), and the youngest basal date (HM11) obtained for Hatfield Moors, which is also the second highest dated elevation (0.7m OD).

Middle Moor (southeast of Lindholme Island)

Two basal samples were taken from the Middle Moor borehole grid (Grid 7), some 500m to the southeast of Lindholme Island: HM9 and HM13. HM9 corresponded to the highest point on the grid (0.43m OD) and produced a date of 1890–1680 cal. BC (Ericaceae stems: SUERC-8896). HM13, whilst sampled from the lowest point on the grid (0.07m OD), yielded samples of two macrofossils (charred twigs and charred bark material) which both produced dates of 3320–2910 cal. BC (SUERC-8860 and 8861). The basal deposits were highly humified with ericaceous remains and *Eriophorum*, again implying a relatively dry, heath like environment during the early stages of wetland development. The refined basal model shows a gently undulating pre-peat land-surface.

Southeastern area (north of Porters Drain)

One basal date was obtained from the borehole grid positioned to the north of Porters Drain (Grid 2) in the southeast of Hatfield and on the very northern edge of current peatland, 1000m north of Porters Drain. This sample location HM4 (-0.4m OD) was located on a local high point within an area of vertical variation of 0.90m. Two replicate dates on macrofossils of 3360–3020 cal. BC (charred grass stems: SUERC-8848) and 3640–3370 cal. BC (*Eriophorum*/Ericaceae stems: SUERC-8859) were statistically inconsistent, as were two determinations 3330–2920 cal. BC (humic: SUERC-9636) and 2910–2690 cal. BC (humin: SUERC-9637).

The Porters Drain area (southeastern area of Hatfield Moors)

The Porters Drain area is topographically one of the lowest areas of the Moors on the refined basal model (-1.2m OD). At HAT2 (Smith 2002) peat began accumulating at a date of *3340–3210 (9%) or 3180–3150 (1%) or 3130–2860 (83%) or 2800–2760 (2%) cal. BC* (CAR-254) (Phase A, Smith 2002). The basal height was recorded as 0.90m OD (Smith 2002) but the refined basal model suggests a height of -0.03m OD on ground sloping to the northeast. The basal sediment at this location was highly humified with few identifiable remains other than charred fragments of *Calluna vulgaris*. The corresponding pollen spectra are dominated by *Betula* (50% TLP) with *Alnus* (20% TLP), suggesting a fen community on or close to the sampling site with an understorey of heather and grasses. Fen persisted at this site until *2950–2580 cal. BC* (CAR-255) after which *Eriophorum* and *Calluna* dominate the macrofossil record with abundant charcoal fragments, suggesting the development of a dry acid heath.

A date of 2460–2133 cal. BC (NZA-18212) was available from the base of the Porters Drain sequence (Gearey 2005) 1200m to the east-southeast of HAT2. The chronology from this sampling site is not robust (see above) and hence the reliability of this date is uncertain. If the basal date is regarded as accurate, it demonstrates that peat accumulation began slightly later than at the HAT1 or HAT2 sites. The pollen and macrofossil data from this site (also see Chapter 2) provide some relatively detailed information on changes in hydro-edaphic conditions during the initial stages of peat accumulation. The basal peat at this location was black, highly humified and minerogenic and the pollen data suggested that during the earliest stages of peat accumulation (basal 0.05m of the sequence) woodland including *Corylus* (65%TLP), *Alnus* (20%) and perhaps initially also *Tilia* (10%), was growing locally, contrasting somewhat with HAT2 (described above). Again, the sampling site was possibly located close to the 'dryland' edge of the peatland at this time; hence the sequence may reflect vegetation at or close to the wetland–dryland interface, whilst vegetation at HAT2 is more typical of early communities in the more central area of the nascent peatland.

The corresponding plant macrofossil data at Porters Drain suggested that monocotyledons were the main peat forming taxa initially, although percentages of Poaceae (wild grasses) and Cyperaceae (sedges) remained relatively low in the pollen record. *Sphagnum* macrofossils were few and poorly preserved and no wood remains recorded. Testate amoebae were absent or recorded in only very low concentrations, a further indication that conditions were reasonably dry and not yet ombrotrophic. Initial increases in *Alnus* and *Quercus* suggested that fen woodland was expanding locally; hence edaphic conditions cannot at this stage have deteriorated too drastically. Subsequent increases in acidity seem to be reflected by rises in *Calluna* and *Pinus*. The expansion of *Pinus* was short-lived, with the demise of this tree probably related to increasing local wetness, although there was no stratigraphic evidence of this. A *Calluna–Eriophorum* heath community persisted on and around the sampling site prior to an increased representation of *Sphagnum* in the macrofossil record and the appearance of testates, albeit in low concentrations. This indicates enhanced wetness and acidity prior to a transition to ombrotrophy c. 1130 cal. BC, although the reliability of this date is uncertain (see above).

5.6 Discussion: patterns of early wetland development on Hatfield Moors

Whitehouse (2004) suggested that the eastern side of Hatfield Moors, around the Porters Drain area, might yield early dates for peat formation, relative to those dated contexts from elsewhere, given that this is one of the topographically lowest parts of the landscape (see above, refined basal model). Contrary to this, it is clear from the above discussion that the earliest dates available are from the area to the north of Lindholme Island. Sediment accumulation began in a shallow pool in this area during the mid-fifth millennium BC, indicating that for this part of the Moors at least ombrotrophic mire development followed the infilling of a basin (*cf*. Whitehouse, Buckland, Wagner *et al*. 2001). This pool spread across the Kilham west area and was encroaching on the northern edge of Lindholme by the early third millennium BC. The processes which could have resulted in peat inception commencing on the highest rather than the lowest part of Hatfield Moors will be discussed further below (Chapter 7).

The available palaeoenvironmental data clearly demonstrate that *Calluna* and *Eriophorum* heath were important habitats across the Moors during the initial phases of peat formation between 3300–3600 cal. BC. However, some variation in the spatial structure of the early environment is apparent. The pollen data (HAT1 and HAT2) in particular demonstrate that *Pinus* woodland persisted during the early stages of peat formation in some parts of the Moors. There is evidence from Porters Drain that, despite the beginnings of local peat accumulation, woodland was able to exist and perhaps expand in places. The pollen diagram from HAT2 also indicates that local fen woodland in which *Betula* and perhaps *Alnus* were significant was present close to, if not actually on, this site until its replacement by dry heath around *2950–2580 cal. BC (CAR-255)*.

By the mid to late third millennium BC, the pool deposit that had begun forming at Kilham west was encroaching on the northern limits of Lindholme and hence any woodland growing in this area must have been dying back, although trees no doubt persisted on the slightly higher and drier parts of the landscape (see Chapter 5). The timing of the expansion of wetland to the north, east and west is unknown. Hence, during this period (mid–late third millennium BC), a distinction may be hypothesised between a relatively wet northern section of Hatfield Moors with areas of open water, heath and scattered woodland, and a relatively dry central and southeastern area with heath and in places perhaps denser woodland where peat growth had not yet begun. Such drier habitats may likewise be evident at the western edge of the Moors at LIND_A, although the lack of a secure chronology hinders the understanding of peat growth and its relationship with other areas of the Moors.

Other sampling sites in the north of the Moors appear to reflect comparatively wetter conditions from the earlier stages of peat inception. Although different to that described for the pool, the initial character of sediment accumulation in the area on the northwestern side of Lindholme Island (Grid 4, HM5) at 3520–3340 cal. BC (SUERC-8850) was in the form of wetter heath with *Sphagnum*, apparently in a depression between the dunes, whilst there is also evidence in the insect record for aquatic and semi-aquatic habitats at HAT3. These sites were perhaps reflecting the mosaic of wet and dry habitats which resulted from slight topographic variation during early stages of wetland spread.

Slightly later at the LIND_B sampling site at a date of 2840–2300 cal BC (Beta-91800), the insect fauna indicated *Eriophorum–Calluna* heath. There was also evidence for the presence of wetter, ombrotrophic conditions nearby. Beetle samples from the Neolithic trackway site (Chapter 7) also suggest ombrotrophic peat growth in this area prior to *c*. 2900 cal. BC. These data would appear to imply that not only was the earliest wetland development in the northern part of the Moors, but the earliest centre of raised mire development may have been here also. This pattern has not been identified by previous study and is discussed further below.

The Porters Drain pollen diagram (discussed above) also indicates that *Tilia* might have been a component of the local woodland, although this pollen signal may also be reflecting trees beyond the edge of the mire. It has been suggested elsewhere that this tree might have had a preference for sandy soils during the mid-Holocene and *Tilia* appears to have been a significant component of the woodland in the Humberhead Levels (see below). Previously there has been little evidence for its presence as part of the local vegetation on Hatfield; Smith (2002) interpreted his pollen diagrams as indicating a prevalence of *Tilia* on Lindholme Island itself. There is no evidence in the beetle record from the peatland for the local presence of this tree. These issues extend beyond the strictly palaeoecological and have resonance for the landscape context of human activity on the Moors during early wetland development (see Chapters 8 and 9).

5.7 Ombrotrophic mire development on Hatfield Moors

The development of ombrotrophic mire marks the final stage in the succession of the Humber peatland systems (see Chapter 2). Understanding the timing, pattern and processes of the shift from fen to mire is an area of significant palaeoecological interest (Hughes *et al.* 2000, 2003) but is also important in terms of human activity (as outlined in Chapter 3) and this aspect of landscape change will be discussed further below. The transition to ombrotrophy marks the point at which the mire systems become directly 'coupled' to the atmosphere via the precipitation:evaporation ratio, meaning that the peat sequences may be regarded as containing a proxy record of the palaeoclimate (Chapter 2). The modelling of the Humberhead Levels 'recurrence' surfaces will be discussed further below. This section will briefly consider the evidence for the spatial and temporal patterning of ombrotrophic mire development at Hatfield Moors.

With the exception of Grid 4 (Chapter 4), the sites sampled as part of this current study do not record any extensive *Sphagnum* peat deposits on Hatfield Moors (Chapter 4). Therefore, the only detailed records of ombrotrophic mire development available are HAT4, with additional information from LIND_B and HAT3, HAT1 and HAT2 (Smith 2002) and Porters Drain (Gearey 2005). The transition to ombrotrophic raised mire is clearly resolved at HAT1 and 2 and Porters Drain in the form of stratigraphic changes and the associated pronounced increases in *Sphagnum* in the macrofossil and pollen records and the increases in testates indicative of acidic raised mire. The earliest date of *1665–1380 cal. BC (CAR-257)* is recorded at HAT2, with a later date of *1115–840 cal. BC (CAR-170)* at HAT1.

At HAT4 on the southwestern corner of the Moors, the stratigraphic transition reflects a 'classic' succession from eutrophic to mesotrophic to ombrotrophic *Sphagnum* dominated mire at a date of *c*. 1380–1030 cal. BC. However, the problems with the chronology for HAT4 (see above) makes it unclear how much reliance may be placed on this date. The estimated age for ombrotrophy at Porters Drain was *c*. 1130 cal. BC although again it is unclear how much confidence may be placed on this estimate (see above). At LIND_B, ombrotrophic conditions, as reflected by a decline in species diversity in the insect record, the appearance of aquatic taxa and the species *Bembidion humerale*, appear to have been present close to the sampling site from peat inception, which occurred prior to 2840–2300 cal. BC (Beta-91800).

It is unclear to what extent the relatively long period between peat inception and ombrotrophic mire development was typical of the development of the peatland in general or, as seems possible, reflects the better studied sites which tend to be on the higher, marginal areas of the current peatland. The problem in establishing any clear spatial pattern is that the available and most detailed data are heavily biased towards the southern end of Hatfield Moors. The refined basal model suggests that some of these sites are likely to have been distal to the locations of the initial peat growth and perhaps closer to the original 'edge' of the mire complex.

It could be argued that not only are the relatively long phases of pre-ombrotrophic peat development at the better studied sites representative of a particular landscape context which may not necessarily be 'typical', but that the southern end of Hatfield Moors was in any case peripheral to the earliest location(s) of ombrotrophic mire development. There is evidence that this took place to the north of Lindholme Island, around Kilham West, in the form of insects indicative of raised mire (e.g. *Plateumaris discolour* and *Octhephilum fracticorne*) in samples dating to shortly after c. 3350–3030 cal. BC.

The lowland peat 'specialist' *Curimopsis nigrita* (Whitehouse 1997) was also present in samples from the trackway site (see below, Chapter 7), shortly after peat inception dated to 4350–4230 cal. BC and predating c. 2900–2500 cal. BC. This implies that ombrotrophic conditions were present in the near vicinity of this site to the north of Lindholme from perhaps as early as the later fourth millennium BC. Therefore, not only might raised mire have initially developed in the north of the Moors, but these conditions might have been already established in the earliest stages of peat accumulation under relatively dry, *Calluna* and *Pinus* heath environments elsewhere on the Moors and before peat inception had even begun in some areas (see Chapter 8).

5.8 Wetland inception and ombrotrophic mire development on Thorne Moors

No additional radiocarbon dates were obtained from Thorne Moors (see Chapter 1). The following discussion therefore concerns the chronological modelling of previous palaeoenvironmental work discussed in Chapter 4 in the context of the basal model for this area (see Chapter 2). A basal date of *3500–3430 (6%) or 3380–2920 (89%) cal. BC* (CAR-221) from Rawcliffe Moor in the northwestern part of the Moors (Smith 2002, see Appendix 1) is the earliest so far obtained for peat formation. This date compares closely with that of 4458±40 BP (NZA-18334) from the base of the Middle Moor sequence, although the modelling indicates the chronology from this sequence is not robust. However, the data indicate that although *Betula–Alnus* fen communities were present at both sites, the shift towards ombrotrophic conditions was considerably quicker at Rawcliffe, with the growth of *Sphagnum* peat consisting mainly of *Sphagnum* Section *Acutifolia* and *Calluna vulgaris* occurring at a date of *3030–2630 cal. BC* (CAR-222).

A detrital mud began accumulating in a large basin at Crowle Moor at the edge of the current extent of the Moors around *3020–2580 cal. BC* (CAR-309), probably within a *Betula–Alnus* fenwood community (CLM-2), a similar environment to that implied at Middle Moor. The full environmental context of this date is unclear, however, since it is not associated with a complete pollen or macrofossil sequence. At Goole Moor (GLM1), an *Alnus* fen community was growing at *3500–3470 (2%) or 3380–3010 (90%) or 2980–2930 (3%) cal. BC* (CAR-232) and was replaced soon afterwards by a *Betula* fen.

The development of an acid *Calluna–Eriophorum* heath is evident by *1920–1680 cal. BC* (CAR-234) followed by the development of wetter conditions with *Scheuchzeria palustris* (Rannoch Rush), *Sphagnum cuspidatum* and *Carex* sp. with a shift to full ombrotrophic *Sphagnum* mire at *730–650 (7%) or 550–350 (88%) cal. BC* (CAR-238). At Crowle Moor (CLM1), pollen data reflects the growth of *Betula* fen from *770–390 cal. BC* (CAR-212), with the replacement of this community by ombrotrophic peat at *330–50 cal. BC* (CAR-215) via a vegetation community consisting of *Scheuchzeria* and *Carex*.

At the Thorne Moors site (TMT1), *Alnus–Quercus* fen was present in the earliest stages of sediment accumulation at *1500–1130 cal. BC* (CAR-180), with *Betula–Alnus* fen rapidly developing by *1250–910 cal. BC* (CAR-182) and ombrotrophic peat being evident shortly after this date. A later date of *cal. AD 1210–1310* (CAR-300) is available for ombrotrophic peat development from an acid heath community at Goole Moor 3. It has been suggested that mesotrophic fen expansion occurred in a series of wet–dry 'pulses', with at least three phases of *Pinus* growth, representing drier conditions, evident between 2916–1489 BC (Boswijk *et al*. 2001).

5.9 Modelling chronologies: assessing the evidence for climate change and human activity in previous palaeoenvironmental studies

One of the most significant conclusions in terms of the broader pattern of Holocene environmental change drawn from Smith's (2002) work on both mires was the identification of a series of 'recurrence surfaces', or apparently synchronous periods of increased mire surface wetness which were regarded as 'phase shifts' (*sensu* Barber 1981), representing a change from relatively dry to wet mire surface conditions and possibly linked to allogenic changes in climate and sea level (Chapter 2; Smith 2002). The chronological models (above) which were constructed for the previous palaeoenvironmental sequences from the Moors can be used to investigate the synchronicity of these five Humberhead Levels (HHL) 'recurrence surfaces'. The following section discusses the *Posterior Density Estimates* for each of these 'recurrence surfaces' from the sequences from Hatfield and Thorne Moors (Figure 5.8, Table 5.4). This discussion includes only those sequences which produced robust chronological models and hence excludes HAT4 (Whitehouse 2002), Porters Drain and Middle Moor (Gearey 2005).

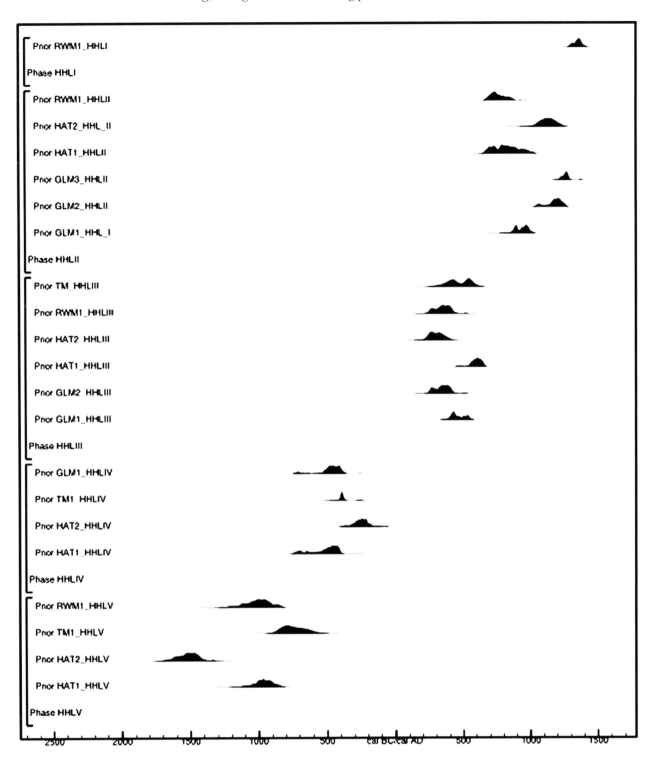

Figure 5.8 Probability distributions of radiocarbon dates associated with the Humberhead Levels (HHL) 'recurrence surfaces' on Hatfield and Thorne Moors

The Humberhead Levels 'recurrence surfaces'

RY. HHL/I: AD 1300–1400 (650–550 BP)
The youngest 'recurrence surface' identified (HHL I) was represented only at RWM1 (Smith 2002), with peat cutting having removed more recent peat. The estimated date for the growth of poorly humified *Sphagnum* peat identified as a 'recurrence surface' at this site is *cal. AD 1290–1410 (95% probability)* and probably *cal. AD 1330–1390 (68% probability)*, which can be compared to Smith's (2002) estimated date of between cal. AD 1275–1395. Peat cutting

Table 5.4 Posterior density estimates for 'recurrence surfaces' RY.HL/I, RY.HL/II, RY.HL/III, RY.HL/IV and RY.HL/V

Site	Posterior Density Estimate (68% probability)	Posterior Density Estimate (95% probability)
Posterior Density Estimates for recurrence surface RY.HL/I (650–550 BP)		
RWM1	cal. AD 1330–1390	cal. AD 1290–1410
Posterior Density Estimates for recurrence surface RY.HL/II (1300–1250 BP)		
GLM1	cal. AD 880–910 (22%) or 930–1000 (46%)	cal. AD 840–1030
GLM2	cal. AD 1160–1250	cal. AD 1040–1100 (10%) or 1120–1270 (85%)
GLM3	cal. AD 1230–1290	cal. AD 1200–1300
HAT1	cal. AD 690–750 (21%) or 770–910 (47%)	cal. AD 670–990
HAT2	cal. AD 1070–1190	cal. AD 1030–1250
RWM1	cal. AD 690–810	cal. AD 670–870
Posterior Density Estimates for recurrence surface RY.HL/III (1650–1550 BP)		
GLM1	cal. AD 400–470 (47%) or 500–550 (21%)	cal. AD 370–570
GLM2	cal. AD 260–280 (8%) or 320–410 (60%)	cal. AD 240–430
HAT1	cal. AD 570–650	cal. AD 460–480 (1%) or 520–670 (94%)
HAT2	cal. AD 240–350	cal. AD 210–410
RWM1	cal. AD 260–280 (6%) or 310–420 (62%)	cal. AD 230–450
TM1	cal. AD 370–470 (35%) or 500–580 (33%)	cal. AD 300–610
Posterior Density Estimates for recurrence surface RY.HL/IV (2500–2300 BP)		
GLM1	510–400 cal. BC	740–660 cal. BC (6%) or 560–360 cal. BC (89%)
HAT1	540–400 cal. BC	740–400 cal. BC
HAT2	310–200 cal. BC	400–110 cal. BC
TM1	410–370 cal. BC	480–350 cal. BC (92%) or 280–250 cal. BC (3%)
Posterior Density Estimates for recurrence surface RY.HL/V (2600–2800 BP)		
HAT1	1030–890 cal. BC	1120–840 cal. BC
HAT2	1580–1430 cal. BC	1690–1390 cal. BC
RWM1	850–670 cal. BC	900–570 cal. BC
TM1	1090–910 cal. BC	1210–830 cal. BC

has removed this section across much of the rest of the two study areas and hence it was not possible to determine the spatial extent of this transition.

RY. HHL/II: 630–740 AD (1300–1250 BP)
The second 'recurrence surface' (HHL II) was identified at all sites except TM1, with the relevant deposits at this site having been removed by peat cutting. The modelling indicates that HHL/II was not a synchronous event, with *posterior density estimates* ranging from *cal. AD 670–820* at Rawcliffe Moors to *cal. AD 1040–1100 (10%) or 1120–1270 (85%)* at Goole Moors. There may appear to be a pattern in the Thorne Moors sequences for the 'recurrence surfaces' becoming younger towards the margin of the peatland (RWM1, GLM1, GLM2 and GLM3) (*cf.* Smith 2002); this will be discussed further below.

RY. HHL/III: 300–400 AD (1650–1550 BP)
The third 'recurrence surface' (HLL III) was identified at all sites except GLM3 due to the persistence of mesotrophic, rather than ombrotrophic, conditions at this location. The six estimates are not consistent, with HHL III being of the same age (A_{comb} = 5.1% (An = 28.9%; n = 6); HAT1 in particular is clearly younger than the other estimates. If this site is excluded the other five estimates are in agreement (A_{comb} = 59.7% (An = 31.6%; n = 5) indicating that HHL III might mark a synchronous shift in BSW during the third to fifth centuries AD. Turner's (1962) dates on three 'recurrence surfaces' (see Appendix 1) yielded dates of between cal. AD 420–890, the broad range due to radiocarbon determinations with large standard deviations.

RY. HHL/IV: 350–250 BC (2500–2300 BP)
HHL IV, the fourth 'recurrence surface', was identified in both Hatfield Moors sequences (HAT1 and HAT2) and in two of the sequences from Thorne Moors (TM1 and GLM1). Again, it was not recorded at GLM2 or GLM3 due to the persistence of mesotrophic conditions, whilst no clear evidence of a shift in BSW was identified in the stratigraphy or biostratigraphy at the RWM1 site on Thorne (Smith 2002). The four estimates for this horizon are not consistent with being of the same actual age (A_{comb} = 19.7% (An = 35.4%; n = 4); HAT2 is clearly younger than the other estimates. If this site is excluded the other three estimates are in agreement (A_{comb} = 66.8% (An = 40.8%; n = 3) implying that the recurrence surface might be synchronous across three sequences (HAT1, TM1 and GLM1) at least.

RY.HHL/V: 650–850 BC (2600–2800 BP)
The final 'recurrence surface' (HHL V) was again identified in both Hatfield Moors (HAT1 and HAT2) and in two of the sequences from Thorne Moors (TM1 and GLM1 – Table 5.4). The four estimates are not consistent with

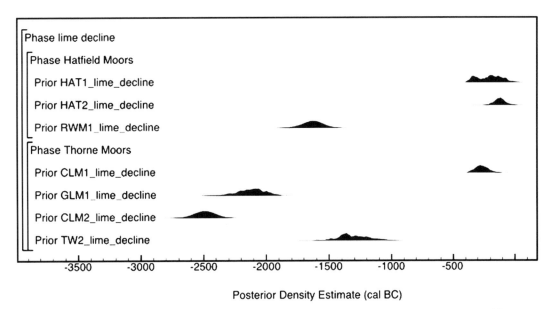

Figure 5.9 Probability distributions of radiocarbon dates associated with Tilia *declines. Sampling site abbreviations: TW = Thorne Waste, RWM1 = Rawcliffe Moor, HAT1 and HAT2 = Hatfield (Smith 2002), CLM1 and 2 = Crowle Moor 1 and 2, GLM1 = Goole Moor 1*

being of the same actual age (A_{comb} = 1.1% (An = 35.4%; n = 4). The estimated date from HAT2 (*1690–1390 cal. BC*) in particular is older than the other estimates. Even if this site is excluded, the other three estimates are still not in agreement A_{comb} = 40.0% (An = 40.8%; n = 3) indicating that it is unlikely that this 'recurrence surface' is synchronous.

The chronological modelling therefore indicated that of the five hypothesised Humberhead Levels 'recurrence surfaces', only two (HHL III and IV) may represent synchronous changes in BSW from relatively dry to wet conditions and in each case this necessitates the exclusion of *posterior density estimates* from HAT1 and HAT2 respectively. This suggests that BSW shifts in the mid-first millennium AD and the fourth–sixth centuries BC may be related to an allogenic forcing mechanism, possibly climatic deterioration (see below). Investigating the relationship between changes in BSW, possible periods of climatic change and human activity in the wider landscape is problematic, especially in the absence of well dated archaeological sequences in the wetland or dryland. However, one way of approaching this is to explore the evidence in the palynological record for episodes of vegetation change which can be related to human activities such as woodland clearance, settlement and agriculture on the dryland areas around the peatlands. The next section will assess this evidence from the palynological record.

Modelling the temporal range of human impact identified in the palynological records from Hatfield and Thorne Moors

The chronological models produced above can be used to refine the timing of the evidence for the vegetation change identified in the pollen records from the various sequences. These palynological data have been described in detail elsewhere (see Smith 2002) and it is not the purpose of this section to re-present these interpretations. Rather, the modelling process permits a comparison of the chronological patterning of events associated with anthropogenic activity on the dryland areas around both Hatfield and Thorne Moors and also allows an exploration of the temporal relationship between other 'events', such as the changes in BSW discussed in the previous section.

The pollen records from the Humber peatlands (see Chapter 2) reflect several periods of pronounced disturbance to the dryland woodland beyond the wetland edge, in the form of increases in pollen of *Plantago lanceolata* (ribwort plantain) with rising values for other taxa which are regarded as 'anthropogenic indicators' (*sensu* Behre 1981) namely *Rumex* (dock), *Urtica* (nettle) and *Pteridium aquilinum* (bracken), with concomitant falls in these taxa during periods of woodland recovery (Smith 2002). It is likely that *Tilia* was a major, if palynologically underrepresented, component of the mid-Holocene regional woodland (e.g. Schofield 2001, Schofield and Bunting 2005, Tweddle 2001). Phases of declining values for *Tilia* are therefore evidence for the clearance of tree cover on the dryland. The woodland included other tree and shrub taxa such as *Quercus*, *Ulmus* (elm) and *Corylus* (hazel) which also often fall during these periods of disturbance. *Posterior density estimates* for the *Tilia* declines and *P. lanceolata* rises and falls have therefore been calculated for each of the sequences (Figures 5.9, 5.10 and 5.11 and Tables 5.5, 5.6 and 5.7).

Discussion of spatial and temporal patterns of vegetation change

The pollen record indicates that the beginning of the

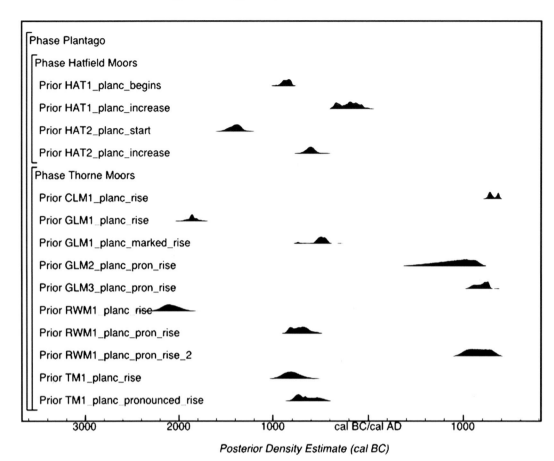

Figure 5.10 Probability distributions of radiocarbon dates associated with increased values of Plantago lanceolata. *Note: some sequences record more than one marked increase or a especially pronounced rise (coding shown thus e.g. planc_pron_rise_2)*

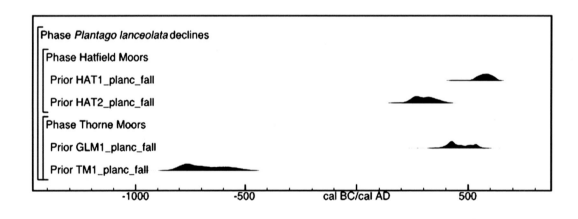

Figure 5.11 Probability distributions of radiocarbon dates associated with reductions in Plantago lanceolata *values*

first millennium BC saw a marked human impact on the vegetation around both Thorne and Hatfield Moors. The pollen records also reflect, in part, spatial differences in the extent of human activity relative to the pollen catchment areas for each of the sampling locations. It is possible to reconstruct something of this spatial variation in a fairly crude manner, although formal methods are now available for modelling pollen data (e.g. Fyfe 2006, Gaillard *et al.* 2008).

Certain of the pollen sequences (HAT1 and HAT2; RWM; GLM1; TM1) demonstrate two phases of increasing *P. lanceolata* values, an initial stage with low and sporadic percentages, often with little clear associated reductions in arboreal pollen, prior to a sustained rise accompanied by other anthropogenic indicators and pronounced falls in tree and shrub pollen percentages. In the case of HAT1 and CLM1, this event is synchronous with a decline in *Tilia*. Smith (2002) regarded the *Tilia* decline as a feature of the opening of zone HHL/C (see Chapter 2) and a reflection of a marked phase of human activity during the later Iron

Table 5.5 Probability distributions of dates associated with falls in Tilia values

Site	Posterior Density Estimate (68% probability)	Posterior Density Estimate (95% probability)
HAT1	360–290 cal. BC (20%) or 240–90 cal. BC (48%)	390–40 cal. BC
HAT2	170–80 cal. BC	210–20 cal. BC
RWM1	1710–1540 cal. BC	1800–1470 cal. BC
CLM1	320–210 cal. BC	370–170 cal. BC
CLM2	2580–2400 cal. BC	2660–2330 cal. BC
GLM1	2220–1980 cal. BC	2330–1920 cal. BC
TM2	1430–1190 cal. BC	1510–1070 cal. BC

Table 5.6 Probability distributions of dates associated with rises in Plantago lanceolata values. For certain sequences, multiple rises in P. lanceolata *are recorded and in others the increases are especially pronounced (see table codes in left hand column)*

Site	Posterior Density Estimate (68% probability)	Posterior Density Estimate (95% probability)
CLM1	cal. AD 1250–1310 (45%) or cal. AD 1350–1380 (23%)	cal. AD 1240–1390
GLM1	1900–1800 cal. BC	1980–1720 cal. BC
GLM1_marked rise_Plantago	540–430 cal. BC	770–700 cal. BC (6%) or 600–390 cal. BC (89%)
GLM2	cal. AD 760–1160	cal. AD 500–1210
GLM3	cal. AD 1110–1130 (4%) or 1150–1280 (64%)	cal. AD 1060–1290
HAT1_Plantago_Begins	900–810 cal. BC	980–790 cal. BC
HAT1	360–290 (20%) or 240–90 (48%) cal. BC	390–40 cal. BC
HAT2_Plantago_start	1460–1330 cal. BC	1540–1280 cal. BC
HAT2	660–550 cal. BC	730–480 cal. BC
RWM1	2190–1990 cal. BC	2290–1880 cal. BC
RWM1_pronounced_Plantago	840–640 cal. BC	870–540 cal. BC
RWM1_pronounced_Plantago_2	cal. AD 1040–1290	cal. AD 960–1360
TM1	900–720 cal. BC	970–620 cal. BC
TM1_pronounced Plantago	790–620 (54%) or 600–590 (1%) or 560–500 (13%)	810–440 cal. BC

Table 5.7 Probability distributions of dates associated with reductions in Plantago lanceolata *values*

Site	Posterior Density Estimate (68% probability)	Posterior Density Estimate (95% probability)
HAT1	cal. AD 540–620	cal. AD 500–650
HAT2	cal. AD 240–350	cal. AD 210–410
GLM1	cal. AD 400–470 (47%) or 500–550 (21%)	cal. AD 370–570
TM1	810–660 cal. BC (54%) or 630–570 cal. BC (14%)	840–500 cal. BC

Age. It is clear that the period between the eighth and fourth centuries BC was one of intensified woodland clearance and farming activity which resulted in the destruction of *Tilia* populations around both Hatfield Moors (HAT1 and HAT2) and Thorne Moors (CLM1 and TM1).

However, earlier declines in *Tilia* are dated at Goole Moor to *2330–1920 cal. BC* and at Rawcliffe Moor to *1800–1470 cal. BC* (both on Thorne Moors). Both these events are also associated with low, but persistent, values of *P. lanceolata* at or shortly after the fall in *Tilia*. There is also a slight fall in *Tilia* and trace values for *P. lanceolata* recorded at CLM2 at an estimated date of *2660–2330 cal. BC*, although the evidence for this reflecting the effects of human activity are somewhat more equivocal. The evidence for woodland clearance in the sampling sites from the northwestern segment of Thorne may imply that early human activity was taking place on the dryland close to this part of the growing peatland during the Bronze Age.

The high values in many of the pollen sequences for *P. lanceolata* and other herbs, including *Urtica* and *Rumex*, during the later Bronze Age and Iron Ages demonstrate that these periods saw the clearance of significant tracts of woodland beyond the mire edges. The predominance of 'pastoral' indicator pollen types is likely to be in part a reflection of an emphasis on pastoral farming during these periods (*cf.* Smith 2002), but it is also notable that the best represented herb taxa tend to be those plants which produce abundant quantities of wind-dispersed pollen. Very similar suites of herb pollen are recorded in sequences from raised mire deposits elsewhere during periods of prehistoric woodland clearance and it is likely that there is a taphonomic bias away from cereal pollen in such records (Caseldine *et al.* 2001).

The relatively limited areas of arable cultivation that might be anticipated during the earlier Bronze Age would probably not be clearly resolved by sampling sites from large wetland systems that were probably some distance from suitable locations for cultivation. There is some evidence from the western fringes of the Humberhead levels at Sutton Common that small scale cultivation within a largely wooded landscape was taking place during the early Bronze Age (Gearey *et al.* 2009). The possible role of human communities on the mires themselves, for which there is archaeological evidence for both the Neolithic and Bronze Ages (Chapter 2), will be discussed separately (see Chapter 8).

However, the role of human communities in modifying the landscape mosaic is somewhat unclear for the earlier prehistoric periods. It is apparent that by the Iron Age, much of the original woodland cover of the Humberhead levels area had been cleared. The dryland areas around both Hatfield and Thorne must have been largely open grassland, with reductions in trees such as *Alnus* indicating that marginal soils such as the carr fringing the mires were also being cleared during the later prehistoric period.

The cutting of peat means that the later Holocene is poorly represented across both sites. However, evidence of post-prehistoric human activity is recorded. Reductions in *P. lanceolata* at the dates of *cal. AD 500–650* and *cal. AD 210–410* are apparent at HAT1 and HAT2 respectively. This may suggest that human activity proximal to HAT1, perhaps on Lindholme Island itself, continued into the early Medieval period after it had ended during the later Romano-British period at HAT2 (see below). An intensification of human activity during the Medieval period is recorded at Thorne Moors with evidence for further woodland clearance and for agricultural activity. This included the cultivation of crops such as *Secale* (rye) and *Cannabis*-type (hemp, hops) which is recorded from an estimated date of *cal. AD 500–1210* in the central part of Thorne Moors at Goole Moor (GLM2) through to the latest date of cal. AD 1249–1390 at Crowle Moor (CLM1) in the southeastern corner of Thorne.

From local to regional patterns of change: interpreting BSW records and assessing implications for human activity

As discussed in Chapters 1 and 2, palaeo-records such as pollen, as a proxy for palaeovegetation, and BSW records, as a proxy for palaeoclimate, present a series of problems in terms of the interpretation of spatial patterns of change in the past. Recent research into modern peatland hydrology and physical climate parameters has indicated that summer moisture deficit may be the main 'driver' of the hydrological changes observed in BSW records from ombrotrophic peatlands, with temperature implicated to a lesser extent (Charman *et al.* 2004). BSW records by definition represent evidence of hydrological changes within a particular sampling location in a particular peatland system at a point in time although, as far as the latter is concerned, the chronological precision is generally defined in terms of centuries at best.

BSW records effectively acquire a spatial dimension as a result of the correlation of multiple records over spatial and temporal scales, as the local is linked to the regional and to the supra-regional, implying so called 'teleconnections' which may then be interpreted in terms of a single 'driver' of change in the form of climate change (see Chapter 2). Hence, the possible synchroneity across the peatlands demonstrated between HHL III and IV might indicate an underlying climate 'driver' of BSW across both peatlands. This, in turn, may be linked on a broader spatial scale to evidence from other peatlands across northwest Europe for 'key' climatic deteriorations *c.* 810 BC and around the middle of the first millennium AD (see Charman 2010). However, interpreting such potential climatic perturbations in terms of implications for human populations is problematic on a number of methodological as well as theoretical grounds.

However, the other HHL 'recurrence surfaces' do not imply synchronous changes across both peatlands, but this does not necessarily indicate that the HHL BSW records are of no interest or significance, nor for that matter that these observed changes are unrelated to climatic forcing. The evidence from HHL II in this case, may be interpreted as indicating something of the complexity of spatial and temporal patterns of BSW change, as the data imply increased wetness spreading from the oldest core of the peatland at *cal. AD 670–820* at Rawcliffe Moors to *cal. AD 1040–1100 (10%) or 1120–1270 (85%)* at Goole Moors close to the current edge of the system.

Establishing the potential significance of such patterns is not helped by the fact that there have been very few multi-sequence studies of individual peatlands, especially those of the size of Hatfield and Thorne Moors; although a recent study of multiple palaeoenvironmental sequences from a raised mire in the Netherlands has identified significant variations between proxy records concluding that "Raised bogs should not be seen as mere passive receptors of climate conditions but instead as relatively complex living systems ..." (Blauuw and Mauquoy 2012, 11). Understanding such spatial and temporal variations in BSW records from multiple sequences from the same peatland system is thus problematic.

This is further complicated by the fact that the original identification of the Humberhead Levels 'recurrence surfaces' depended upon a combination of plant macrofossil, stratigraphic and a limited testate amoebae dataset. The latter were derived from the same samples used for palynological analyses, and it has been demonstrated that the pollen preparation process leads to differential destruction of testates (Hendon and Charman 1997). Interpretation of hydrological changes in these sequences was thus approached in an effectively qualitative manner. Whilst apparently pronounced changes in the composition of the mire flora are recorded, no distinction between the magnitudes of the various hydrological shifts is possible, as the scale is effectively 'non-linear'.

An example of the problems of using stratigraphic and raw macrofossil data to infer climate change can be observed through a comparison of the HAT2 sequence with the testate amoebae derived watertables from Porters Drain (see Chapter 2, Gearey 2004). There is evidence for a significant 'wet shift' at Porters Drain After cal. AD 180, with rising watertables in *S. imbricatum* dominated peat. In the absence of these testate amoebae data it would be possible to suggest drier conditions at Porters Drain; as *Calluna* percentages are initially high in the pollen diagram. Indeed, the wet shift in the watertable at this point appears to be short lived. In other words, it may not be the case that the data from Smiths (1985, 2002) sequences are in direct disagreement, but that the testate derived watertable reconstructions are able to detect both more subtle hydrological events and those of a shorter duration than the macrofossil data.

5.10 Summary and conclusions

This chapter has described the chronological modelling of the sequences of radiocarbon dates from previous study on Hatfield and Thorne Moors. The results of the radiocarbon dating of samples from Hatfield Moors were also outlined. The current understanding of the timing and nature of wetland development on Hatfield Moors, based on previous palaeoenvironmental study and the associated radiocarbon dating, and the landscape context provided by the model of the pre-peat landscape, can be summarised as follows:

- The earliest peat accumulation on Hatfield was to the north of Lindholme Island at Kilham West, one of the topographically higher areas of the Moors. Paludification of the basal sands began c. 5200–5500 cal. BC as a pool or series of pools spread across the landscape. Mixed *Quercus–Tilia–Corylus* deciduous woodland was growing on the drier soils and on Lindholme Island.
- The local vegetation was *Calluna* heath, and probably mixed *Pinus* woodland. As the pool or pools expanded across the area to the north of Lindholme, the woodland must have started to die back.
- The palaeoenvironmental evidence indicates that much of the rest of the study area, including areas significantly lower than the northern part of the Moors, remained 'dryland' at this time. Again, the status of the vegetation at this time is unknown, but it is hypothesised that woodland in which *Pinus* and *Betula* were common would have been growing. The dendrochronological record from Tyrham Hall quarry indicates that *Pinus* was growing here at 2921–2445 cal. BC and probably died back as paludification affected the western edge of the peatland. Mixed deciduous woodland was present on the drier soils and also on Lindholme Island.
- Peat began accumulating in *Calluna* and *Eriophorum* heath environments in a range of topographic contexts from c. 3300–3700 cal. BC. The frequency of burnt material in this sediment is a further indication of a relatively low watertable. In the south of the Moors, woodland appears to have been present during the earliest stages of wetland development. *Betula* and *Alnus* fen in the central areas of the mire with mixed *Corylus–Quercus–Tilia* woodland persisting closer to the edges of the wetland and extending onto the drier soils.
- The transition to ombrotrophy occured between 1700 cal. BC and 800 cal. BC, with the earliest available date recorded in the southeast at HAT2 and later dates at sites HAT2, Porters Drain and HAT4 reflecting their proximity to the edge of the wetland. The transition appears to have been relatively rapid, with wet *Sphagnum cuspidatum* communities expanding at HAT2 but with evidence for slightly drier conditions elsewhere. Palynology records evidence for human impact on the woodland from the Bronze Age with marked phases of woodland clearance during the Iron Age. Woodland regeneration and reductions in human activity are attested for the Romano-British/early Medieval periods.
- There is indirect evidence that ombrotrophic conditions in fact developed earliest in the northern part of the Moors, in the vicinity of the Neolithic trackway site (see Chapter 7), before peat accumulation had begun elsewhere.

The model of the pre-peat landscape of Thorne Moors indicates much greater topographic variability in comparison to Hatfield Moors. The current understanding of the pattern and processes of mire development on Thorne Moors in the context of this basal model can be summarised as:

- The pre-peat landscape was probably one of mixed deciduous woodland on the heavy clay soils, with trees including *Tilia* and *Quercus* being important components of the vegetation.
- The earliest recorded peat accumulation was from the mid-fourth millennium BC at Rawcliffe Moor in the northwest of the current peatland, in a *Betula–Alnus* fen environment in a discrete basin which was one of the topographically lowest points of the pre-peat landscape.
- The pattern during the earliest phases of wetland development seems to have been one of peat formation within *Alnus* and *Betula* dominated fen growth in other topographic low points after c. 3150 BC, whilst mixed *Quercus* woodland, which was probably prevalent over much of the area prior to peat growth, persisted on the higher areas (e.g. Boswijk et al. 2001). There is potential evidence that early peat accumulation began in a palaeochannel in the Middle Moor area which may have extended across the central part of Thorne.
- The transition from fen to ombrotrophic mire seems to have been significantly time transgressive with early dates of c. 3000–2600 cal. BC at Rawcliffe through to the latest date of c. cal. AD 1200–1300 at Goole Moor 3. The palynological record implies spatial variation in environment, with possible low level disturbance close to Rawcliffe and Goole Moors during the Bronze

Age, with subsequent clearance intensifying in the Iron Age. As for Hatfield Moors, subsequent regeneration of woodland is apparent during the Romano-British/early Medieval periods, with evidence for later Medieval agriculture and cultivation.

The Bayesian modelling of the palaeoenvironmental chronologies has indicated that whilst HAT1 and HAT2 provided robust chronologies for the associated palaeoenvironmental records, those of HAT4 and Porters Drain were less reliable. The sequences from Thorne Moors were all chronologically robust with the exception of that from Middle Moor. The *posterior density estimates* derived from the models for palaeoenvironmental 'events' has indicated something of the spatial and temporal variation in human impact on the dryland areas around the peatlands and will be considered further in Chapter 8. The modelling of the Humber Head Level 'recurrence surfaces' indicates that only two of these (HHL III and IV), as originally identified at least (Chapter 2), may have been synchronous.

Note
1. A *posterior density estimate* is an interpreted estimate of date derived from the Bayesian mathematical modelling of the radiocarbon chronology (Bayliss *et al.* 2007)

6. Patterns of change: modelling mires in four dimensions

6.1 Introduction

The previous chapter demonstrated the complexity of patterns of environmental change across Hatfield and Thorne Moors during the mid-Holocene, with peat inception taking place in different wetland contexts at different times and palynological evidence for spatial and temporal variations in human activity on the dryland areas around the peatlands. This shifting environmental mosaic would have had implications for past populations living and moving through these landscapes, in the form of changes in resources, movement and access, all of which have relevance for archaeological understanding (see Chapter 7). The interpretation and modelling of peat inception necessitated data relating to the nature of the dryland landscape prior to wetland formation, and on the comprehensive dating of the basal organic deposits which represent the earliest peat formation.

The construction of models of the pre-peat landscapes of Hatfield and Thorne Moors (see Chapter 4) provided the first of these requirements. The chronological framework for the onset of wetland formation was provided through a combination of data from previous studies and the radiocarbon dating of basal deposits (see Chapter 5). The expansive nature of these landscapes means that a considerable number of radiocarbon dates would normally be required, but one intention was to generate robust and testable models of wetland inception and peat spread using a more limited chronological dataset.

In order to explore this pattern of peat growth and spread, it was necessary to establish the relationship between spatial and chronological aspects of change. The pattern and process of peat growth across both peatlands has been much debated, and these have been formulated into a series of testable hypotheses (Buckland and Smith 2003; see also Chapter 2; Figure 2.20). This chapter builds on the approaches outlined in chapter 3, detailing the methods and results used in the combination of spatial and temporal datasets for the testing of these hypotheses for wetland inception, and the generation of models of the timing and patterns of peat spread across them.

As outlined in Chapter 2, the current landscapes of Hatfield and Thorne Moors are the products of drainage, reclamation and peat cutting that began in the later and post-Medieval periods and continued throughout the twentieth century. Whilst an understanding of peatland formation processes is critical to the interpretation of Hatfield and Thorne Moors as archaeological landscapes, knowledge of the depth, extent and archaeological potential

Figure 6.1 The cut-over peatland landscape of Hatfield Moors. A sub-fossil stump of Pinus *is visible to the left*

of the surviving peat resource is required for the future management of the Moors. For example, the cutting of peat down to deposits dating to the Romano-British period (Figure 6.1) will mean that the entire archaeo-environmental resource which post-dates this period will have been destroyed, with the possible exception of archaeological material which may have been pushed or sunk into deeper, and thus earlier, levels of the mire.

Hence, the date of the base and top of the peat at any given location brackets the chronological 'envelope' for the preservation of the archaeo-environmental resource. Establishing this chronological span permits an understanding of the prospect for the preservation and survival of *in situ* archaeological sites and therefore assists a more informed future management of these landscapes from a heritage perspective. Hence, this chapter also describes an approach to the GIS deposit modelling of peat depth and chronological span aimed at quantification of the surviving peatland resource on a landscape scale.

6.2 Modelling Hatfield Moors in four dimensions

Wetland inception on Hatfield Moors

The DEM of the pre-peat landscape of Hatfield Moors (see Chapter 4) provided the foundation for testing the hypotheses relating to the processes of wetland inception and mire formation. In addition to this three-dimensional spatial dataset, chronological data were available from the previous studies and from the radiocarbon samples collected during fieldwork (see Chapter 5). The process of investigation began with the modelling of the different hypotheses relating to wetland inception and peat growth developed by Buckland and Smith (2003, see Chapter 2; Figure 2.20) through an analysis of the pre-peat landscape DEM in conjunction with the basal radiocarbon dataset.

In the first model (Figure 2.20; hypothesis A), wetland inception is seen as driven by paludification as a result of raised watertables. In this scenario, the topographically lowest areas of the pre-peat landscape would be expected to be the earliest *foci* of peatland growth. For this hypothesis to be supported, areas with the lowest gross elevation should correspond to the earliest dates for basal peat formation. The second model (Figure 2.20; hypotheses B and C) sees wetland development spreading 'inwards' from the river floodplains of the River Idle to the east and the River Torne to the south of the peatlands. In this instance, the earliest dates for peat formation should therefore be proximal to the river channels to the north, east and south of the current peatland with progressively younger dates towards the centre of the current peatland. The third model of wetland inception proposed peat spread following the infilling of one or more basins beneath the peatland, through processes of hydroseral succession (see Chapter 2) (Figure 2.20; hypotheses D and E). In this model, the earliest basal radiocarbon dates should correspond with either a stratigraphic sequence reflecting the infilling of a single lake basin (D) or alternatively with more discrete areas of impeded surface runoff in the form of small ponds or pools, the so called 'multi-focal' model (E). The following section presents details of the approach for testing these three models and an analysis of the results.

Modelling the timing of wetland inception in relation to elevation on Hatfield Moors

As described in Chapter 5, the sampling strategy for radiocarbon dating was designed to obtain age estimates from a range of basal elevations and contexts across Hatfield Moors. From the relationship between the radiocarbon dates and elevation as determined by the GIS basal deposit model (see Chapter 5), it is apparent that the earliest dates are from the north of Lindholme Island (Grid 1, HM17, 5470–5220 cal. BC at -0.11m OD) and Packards South (Grid 1, HM15, 4840–4690 cal. BC at -0.26m OD), whilst the lowest areas of the Moors lie in the area of Porters Drain in the southeastern region where the basal topography drops to -1.5m OD. From the area south of Porters Drain (Grid 3) a younger date of 2460–2135 cal. BC is available from a basal sample at -1.2m OD (Gearey 2005). Statistical analyses, comparing the calibrated radiocarbon dates from these samples and elevation, consisting of linear and second-order polynomial regression of the output produces R^2-values of 0.15 and 0.20 respectively, imply a weak relationship between age and altitude (Figure 6.2).

However, there appears to be a stronger relationship between age and elevation on a local scale, for example to the north of Lindholme Island (Table 6.1). This apparent local consistency in the age of basal samples in relation to elevation contrasts with the overall pattern. Thus, it appears that elevation might have had some significance locally, but in terms of the mire as a complete system there were evidently other underlying factors driving the process. Hence it was possible to discount hypothesis A (Figure 2.20) as the primary process underlying wetland inception on Hatfield Moors.

Table 6.1 Basal dates from the area north of Lindholme Island in relation to elevation

Site	Elevation (mOD)	Radiocarbon date (calibrated range BC at 95% confidence)
HM24	0.93	2920–2760 (SUERC–8877: macro)
HM2	0.62	2620–2460 (humin); 2470–2200 (SUERC–9689: humic)
HM18	0.08	3770–3630 (SUERC–8949: macro)
HM17	-0.07	4800–4610 (SUERC–9646: humin); 4600–4360 (GU–6365: humin)
HM21	-0.11	5470–5220 (SUERC–8952: macrofossils)

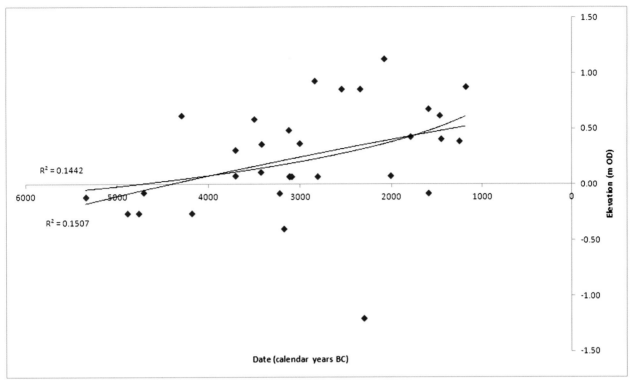

Figure 6.2 Age-altitude model for Hatfield Moors (humic fractions), comprising linear and second-order polynomial regression though the calibrated dates of samples from basal peats. For these analyses, radiocarbon dates have been calibrated using the maximum intercept method (Stuiver and Reimer 1993); the mid-point of these ranges are then used and the calibrated date is assumed to be normally distributed around this point

Modelling the timing of wetland inception in relation to proximity to watercourses on Hatfield Moors

The second model of wetland inception (Figure 2.20; hypotheses B and C) postulates wetland spread from the floodplains of the rivers adjacent to Hatfield and Thorne Moors. In order to test this model, the basal radiocarbon dates were analysed in relation to their proximity to these river courses or former courses, of the Rivers Torne and Idle. In order to support this hypothesis, locations closer to the rivers should demonstrate earlier dates for peat formation than those areas further away from the drainage network. A model of linear distance from the rivers (or former river channels) of the Torne and Idle was generated within the GIS in relation to the basal radiocarbon date sampling sites (Figure 6.3, Table 6.2).

For Hatfield Moors, the closest radiocarbon sample to the watercourses was HM11, 466m from the River Torne on the southern edge of the study area and provided a date of 1690–1500 cal. BC (SUERC-8858). Conversely, the sample furthest from the either of the rivers was HM5, which was over 2.6km away, towards the centre of Hatfield Moors. This sample provided a date of 3520–3340 cal. BC (SUERC-8850) hence considerably earlier than HM1. Furthermore, the earliest dated sample, HM21, was at a distance of over 1.9km from either river. Even accounting for local variations, such as HM7, HM11 and HM15 in the Packards South area where samples were taken from varying elevations across the dune system, the overall pattern indicates that there is no clear relationship between peat inception and proximity to watercourses. The outputs from this produced R^2-values of 0.28 and 0.30 respectively for dated macrofossils and humic fractions (Figure 6.4), and values of 0.88 and 0.16 respectively for dated macrofossils and humin fractions (Figure 6.5). This indicates a poor relationship between dates for wetland inception and distance from watercourses, and hypotheses B and C (Figure 2.20) are therefore not supported by these data.

Modelling the timing of wetland inception in relation to surface run-off on Hatfield Moors

The final two hypotheses relate to the infilling of one or more localised basins through the processes of hydroseral succession. Thus, the areas of earliest wetland inception should relate to either a stratigraphic sequence reflecting the infilling of a basin (Figure 2.20; hypothesis D), or to a series of discrete areas where impeded surface runoff would have formed small areas of wetland providing numerous *foci* of wetland inception from which peat would spread (Figure 2.20; hypothesis E). In other words, paludification would initiate in areas prone to accumulate rather than shed water.

In order to identify potential such locations of restricted drainage, the DEM of the pre-peat landscape (Chapter 4) was used to generate a *flow-accumulation* model in the GIS at 5m surface resolution. This function calculates the 'flow' as the accumulated weight of cells flowing into each 'downslope' cell in the output raster. Hence, the

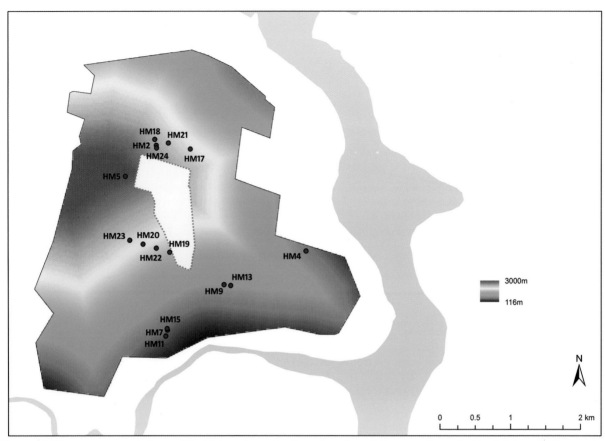

Figure 6.3 Watercourse proximity model with locations of radiocarbon dates (blue area shows the floodplains of adjacent rivers)

Table 6.2 Basal radiocarbon dates in relation to distance to watercourses

Site	Distance to watercourse (m) (GIS Output)	Radiocarbon date (calibrated range BC at 95% confidence)
HM11	466	1690–1500 (SUERC–8858: macros)
HM7	509	3490–3100 (SUERC–9691: humin); 3350–3020 (SUERC–9690: humic)
HM15	537	4840–4690 (SUERC–9641: humin); 1530–1410 (SUERC–9649: humic)
HM4	624	2910–2690 (SUERC–9637: humin); 3330–2920 (SUERC–9636: humic)
HM9	855	1890–1680 (SUERC–8856: macros)
HM13	868	3270–2910 (SUERC–8860: macros); 3270–2910 (SUERC–8861: macros)
HM19	1423	3630–3360 (SUERC–8950: macros)
HM22	1565	3500–3340 (SUERC–8953: macros)
HM20	1682	3770–3630 (SUERC–8951: macros)
HM17	1805	4800–4610 (SUERC–9646: humin); 4600–4360 (GU–6365: humic)
HM23	1836	3770–3630 (SUERC–8875: macros); 3640–3370 (SUERC–8876: macros); 3270–2910 (SUERC–9647: humic); 3270–2910 (SUERC–9648: humin)
HM21	1967	5470–5220 (SUERC–8952: macros)
HM18	2080	3770–3630 (SUERC–8949: macros)
HM2	2115	2620–2460 (SUERC–9689: humin); 2470–2200 (SUERC–9688: humic)
HM24	2141	2920–2760 (SUERC–8877: macros)
HM5	2656	3520–3340 (SUERC–8850: macros)

flow direction between adjacent cells is calculated by the relative elevation values within the input raster (in this case the refined model of basal topography) such that a cell of a higher elevation will flow into one of a lower elevation (Figure 6.6). The resulting *flow-accumulation* raster contains cells with values representing the number of other cells that flow into it, such that higher values represent greater accumulation. For example, cells that only flow outwards will have a value of zero; if four cells with a zero value flow into one cell, it will have a value of four, and so on. Areas of high cell values identify locations in the model of the pre-peat landscape which would have accumulated flow, whereas areas of low cell values may be interpreted as water shedding locations.

The resulting model of flow-accumulation for Hatfield Moors is presented in Figure 6.7. The process of flow-accumulation generates quantitative values for cells which may then be compared to the radiocarbon dates from each location. Hence, this process allows areas or 'nodes' of water accumulation and potentially poor drainage to be identified in the GIS basal topography DEM. It may be hypothesised that if initial processes of peat growth were related to the status of a particular area as being water shedding or water accumulating, then locations of greater flow-accumulation should be correlated with earlier dates for peat development. Conversely, later dates for peat growth should be correlated with lower values for flow-accumulation. In order to test this, values relating

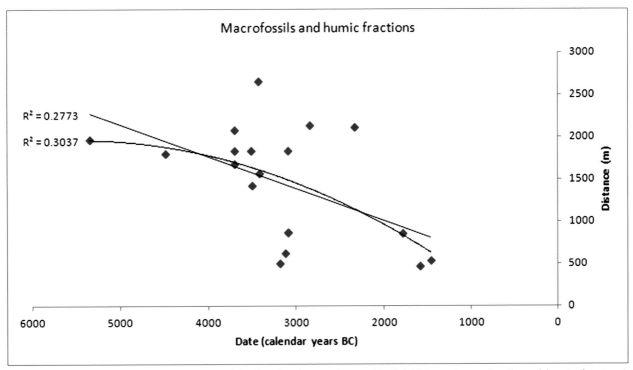

Figure 6.4 Age-distance to watercourse model of wetland inception on Hatfield Moors (macrofossils and humic fractions), comprising linear and second-order polynomial regression though the calibrated dates of samples from basal peats

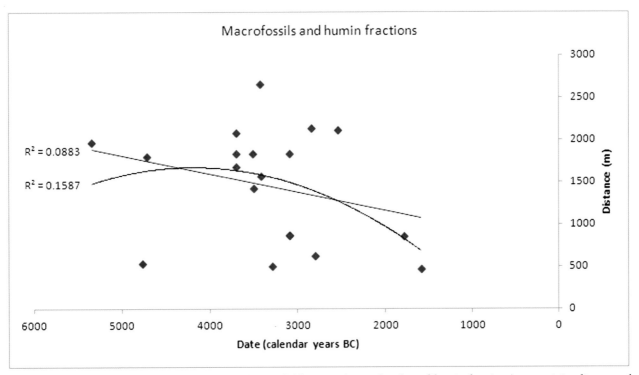

Figure 6.5 Age-distance to watercourse model for Hatfield Moors (macrofossils and humin fractions), comprising linear and second-order polynomial regression though the calibrated dates of samples from basal peats

to flow-accumulation were extracted from the GIS model and compared with the results of the radiocarbon dating (see Table 6.3).

The lowest recorded flow-accumulation value (0) corresponds with the youngest radiocarbon date for peat inception of 1690–1500 cal. BC (SUERC-8858; HM11), whilst the highest flow-accumulation value (a figure of 1083) relates to the oldest date of 5470–5220 cal. BC (SUERC-8952; HM21). This apparent relationship can be further considered on a local scale in the area

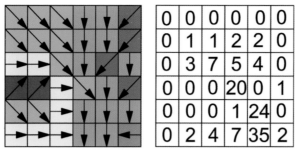

Figure 6.6 How flow-accumulation is calculated in a GIS. Slope dictates the direction of flow across each cell (left). Values on the right indicate the number of cells that 'flow' into each

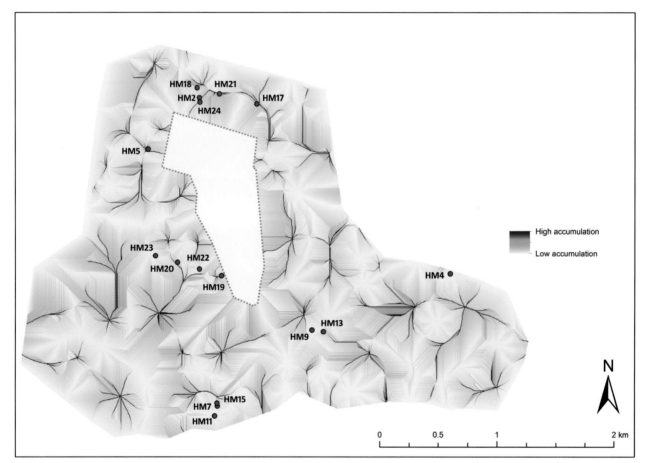

Figure 6.7 Flow-accumulation model of Hatfield Moors based on the pre-peat land-surface DEM and showing the locations of basal radiocarbon samples. Darker shades reflect areas of higher modelled flow-accumulation and hence those areas prone to waterlogging

to the north of Lindholme (Figure 6.8), where areas of modelled higher flow-accumulation correlate with the earliest radiocarbon dates. In other words, the dates for the earliest peat formation seem to correspond to poorly drained locations in the pre-peat landscape as defined by the flow-accumulation model. Statistical analyses consisting of linear regression of the output from the flow-accumulation model against the humic and humin fraction ages produced R^2-values of 0.39 and 0.41 respectively, whilst second-order polynomial regression produced R^2-values of 0.91 and 0.72 respectively (Figures 6.9 and 6.10), suggesting a good correlation between the model and humic age fractions in particular.

The data suggest that there is a relationship between the date of peat inception and locations of topographic variation in the pre-peat landscape prone to accumulating water. This implies that there was no single focus for the earliest wetland inception and hence peat growth on Hatfield Moors. The areas of earliest peat growth include the region to the north of Lindholme Island (HM21 and HM17), the very southern edge of Hatfield Moors around Packards South (HM15), and to the northwest and west of

Table 6.3 Flow-accumulation values in relation to basal radiocarbon dates

Site	Flow-accumulation Value (GIS Output)	Radiocarbon date (calibrated range BC at 95% confidence)
HM11	0	1690–1500 (SUERC–8858: macros)
HM4	1	2910–2690 (SUERC–9637: humin); 3330–2920 (SUERC–9636: humic)
HM23	1	3770–36 30 (SUERC–8875: macros); 3640–3370 (SUERC–8876: macros); SUERC–9647: 3270–2910 (SUERC–9647: humic); 3270–2910 (SUERC–9648: humin)
HM22	1	3500–3340 (SUERC–8953: macros)
HM9	4	1890–1680 (SUERC–8856: macros)
HM7	12	3490–3100 (SUERC–9691: humin); 3350–3020 (SUERC–9690: humic)
HM13	15	3270–2910 (SUERC–8860: macros); 3270–2910 (SUERC–8861: macros)
HM19	21	3630–3360 (SUERC–8950: macros)
HM24	30	2920–2760 (SUERC–8877: macros)
HM2	30	2620–2460 (SUERC–9689: humin); 2470–2200 (SUERC–9689: humic)
HM5	58	3520–3340 (SUERC–8850: macros)
HM20	65	3770–3630 (SUERC–8951: macros)
HM18	78	3770–3630 (SUERC–8949: macros)
HM15	94	4840–4690 (SUERC–9641: humin); 1530–1410 (SUERC–9649: humic)
HM17	205	4800–4610 (SUERC–9646: humin); 4600–4360 (GU–6365: humic)
HM21	1083	5470–5220 (SUERC–8952: macros)

Lindholme Island (HM18 and HM20). Radiocarbon dates from between these areas demonstrate that these initial areas of wetland development were not linked spatially, and so it seems that the process occurred at several separate *foci*. Hence hypothesis D (Figure 2.20) is not supported by these data. It seems most likely that wetland inception was related to paludification caused by impeded run-off and possibly earliest in the north of Hatfield Moors (see Chapter 5) linked to hydroseral succession. Hypothesis E is therefore supported by the integrated spatial and chronological analyses (Figure 2.20).

Summary: patterns of wetland inception on Hatfield Moors

Previous study has identified the significance of local topographic variation in the pre-peat surfaces of blanket mires, in terms of whether specific locations will tend to be water shedding or water accumulating contexts (e.g. Moore 1993, Charman 2002). The analyses presented above imply this was the case for Hatfield Moors, as the flow-accumulation model is the most robust in explaining the observed spatial and temporal pattern of peat inception. This supports the hypothesis that initial wetland inception and hence peat growth began in areas that were prone to accumulate rather than to shed water, irrespective of gross elevation within the landscape. Following initial growth in certain areas, peat would have spread gradually, vertically and laterally, with the different areas of growth subsequently

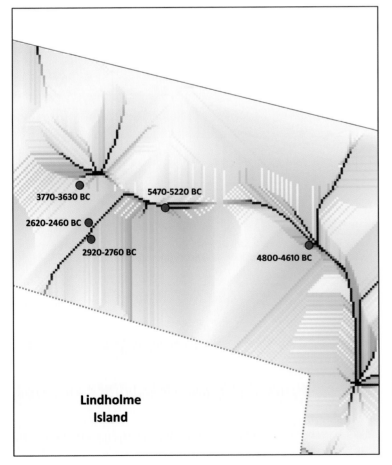

Figure 6.8 Flow-accumulation in relation to calibrated radiocarbon dates in the area north of Lindholme Island

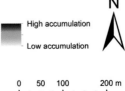

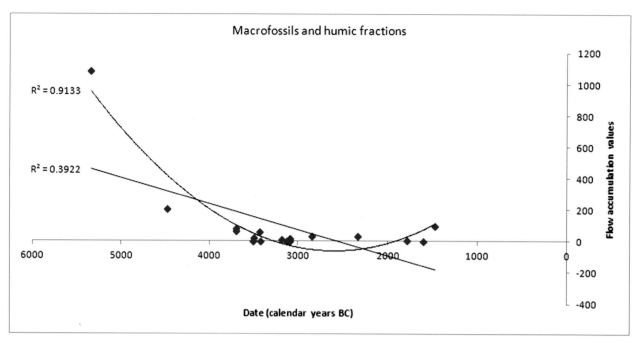

Figure 6.9 Flow-accumulation model for Hatfield Moors (macrofossils and humic fractions), comprising linear and second-order polynomial regression though the calibrated dates of samples from basal deposits (Stuiver and Reimer 1993)

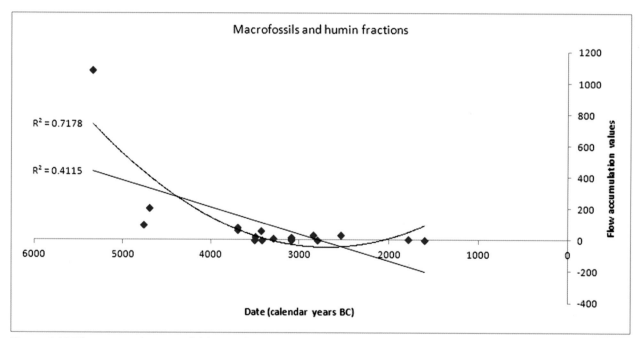

Figure 6.10 Flow-accumulation model for Hatfield Moors (macrofossils and humin fraction), comprising linear and second-order polynomial regression through the calibrated dates of samples from basal deposits (Stuiver and Reimer 1993)

coalescing. In this situation, multiple *foci* of peat growth rather than at a single cupolum are indicated (*cf.* Figure 2.20; hypothesis E).

Modelling peatland spread across Hatfield Moors

The flow-accumulation model can be employed as a foundation for exploring the spatial and temporal spread of peat across the pre-peat landscapes. The statistical correlation between cell-values in the flow-accumulation model and the basal radiocarbon dates provides a tool for estimating the date of peat inception across the entire pre-peat landscape. This model was generated through the reclassification of cell-values of the flow-accumulation model to represent dates, as derived from the statistical relationship between the flow-accumulation cell-values and the radiocarbon dates. This was calculated by applying the

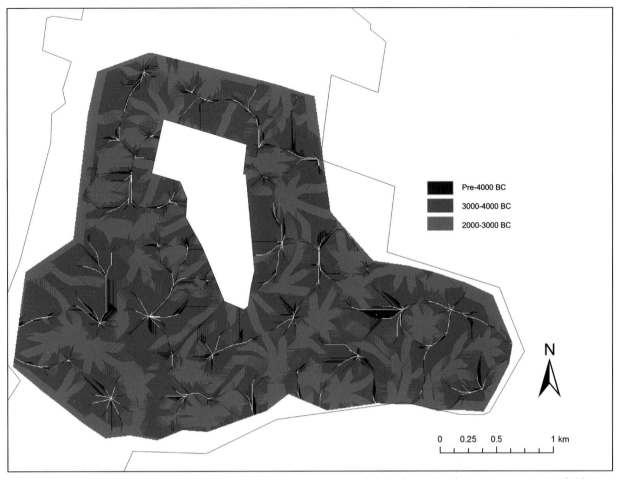

Figure 6.11 Flow-accumulation model with values converted to predicted dates for wetland inception across Hatfield Moors. This was derived from statistical regression comparing flow-accumulation values and radiocarbon dates. Darker blue indicates earlier dates

formula generated by the best fit (macrofossils and humic fractions) second-order polynomial regression to the cell values of the flow-accumulation model. The output of this process was a new GIS surface representing the modelled age of peat inception for each cell, effectively producing a 'chronozone' for the entire pre-peat landsurface (Figure 6.11).

A limited test of the output was undertaken using the legacy datasets which had not been incorporated into the model due to uncertainties in the elevation of the palaeoenvironmental sampling sites and their associated radiocarbon dated levels (see Chapter 5). This included five basal radiocarbon dates from Hatfield Moors (Chapter 5) (Smith 2002, Whitehouse 2004). A plot of the location of these basal dates is presented in Figure 5.1, with a comparison of the actual and modelled dates in Table 6.4.

Out of these five sampling locations and associated dates, three correlate closely in terms of actual and modelled age (LIND_B, HAT2 and HAT3), one more broadly (HAT1) and one shows poor agreement (HAT4). The lack of correlation in the case of HAT4 may be explained by the proximity of the location to the edge of the GIS model with subsequent potential for errors through the 'edge effect' (Wheatley and Gillings 2002), in addition to possible issues of the overall

Table 6.4 Comparing modelled dates with calibrated radiocarbon dates from legacy datasets

Sampling site (see chapter 5)	Modelled date BC	Radiocarbon date (calibrated range BC at 95% confidence)
LIND_B (Whitehouse 2004)	2790	2900–2350
HAT1 (Smith 2002)	3110	2910–2580
HAT2 (Smith 2002)	3210	3350–2700; 3130–2860
HAT3 (Whitehouse 2004)	3350	3350–2930
HAT4 (Whitehouse 2004)	3050	1520–1390

chronology of this sequence (see Chapter 5). Although this is clearly a restricted test based on relatively few samples, it supports the overall robustness of the model.

Given that the GIS model produces a relatively precise estimate of age for each cell, the output from the model was further subdivided into broad 'time-slices' relating to cultural periods for purposes of interpretation and presentation (Figure 6.12). Hence, whilst this model output provides a more useful visualisation of the results of the modelling process, data relating to specific dates from specific locations can be extracted from the original model.

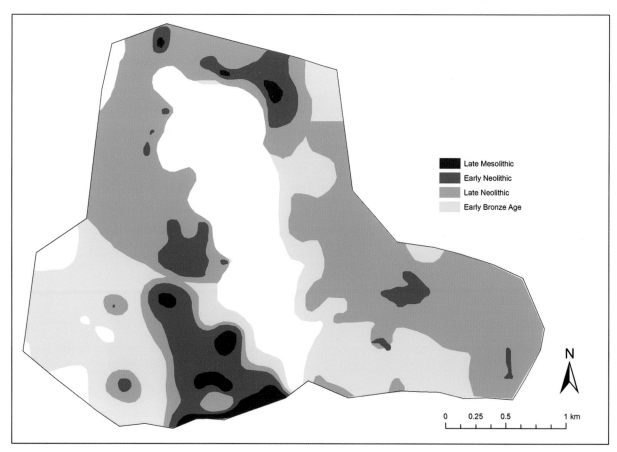

Figure 6.12 Resulting GIS model of wetland inception by cultural period on Hatfield Moors (based on the results shown in Figure 6.11)

Modelling the surviving peatland resource on Hatfield Moors

Peat cutting ceased on Hatfield Moors in 2004 and the borehole survey (Chapter 4) demonstrates that, an average of less than 1m of peat survives. Chapter 4 outlined the generation of a DEM representing the base of the peat and this chapter presented the modelling of this surface to represent dates for peat inception across the study area. The DEM provides a baseline against which the depth of peat can be calculated through comparison with the surface of the peatland. The model representing the dates of peat inception provides a *terminus post quem* for the potential for organic preservation of archaeological material and the palaeoenvironmental record. This section uses both models to define the depth of the surviving peat and the chronological span of these deposits across the study area.

Modelling the depth of surviving peat on Hatfield Moors

The depth of peat can be calculated by subtracting the elevation of the pre-peat landscape from that of the current peatland surface. Given the generation of a DEM of the pre-peat land-surface (Chapter 4), a DEM of the current land-surface was required in order to model peat depths. The Environment Agency LIDAR data (Chapter 2), collected in 2006 following the cessation of industrial peat extraction, provided an accurate record of the surface of Hatfield Moors. Hence, peat depths for all locations across Hatfield Moors were calculated by subtracting the pre-peat landscape DEM from this LIDAR DEM, and a new GIS model with cell-values representing depth of peat was generated.

The resulting model (Figure 6.13) indicates that organic deposits survive on Hatfield Moors to a maximum depth of 3.5m. However, these maximum depths are restricted to a relatively small area on the central southern edge of Hatfield Moors. Across the study area defined by the basal topographic model (see Chapter 4), the mean depth of organic deposits is 1.02m whilst peat depths across the majority of Hatfield Moors are less than 1m. In places, the pre-peat land-surface is entirely exposed. This raises significant challenges in terms of the potential for the preservation of archaeological sites *in situ*, as little protection is afforded even for potential archaeological sites towards the base of the peat (see Chapter 7).

Modelling the date of the cut-over peat surface of Hatfield Moors

Assessing the potential of the surviving peat in terms of the archaeo-environmental record requires an assessment of the chronology in the form of the earliest and latest date

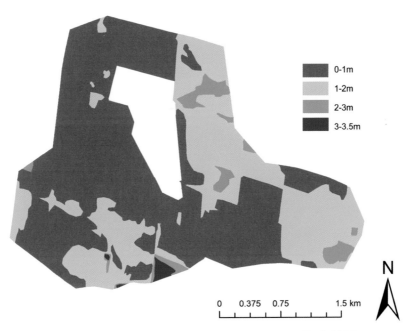

Figure 6.13 Modelled depths of surviving peat across Hatfield Moors

Table 6.5 Radiocarbon dates from surface deposits across Hatfield Moors

Location	Lab Code	Conventional age BP	Radiocarbon date (calibrated range BC at 95% confidence)
HM3 – Porters Drain (Grid 2)	SUERC-8847	3305±40	1690–1490
HM6 – northwest of Lindholme Island (Grid 4)	SUERC-8851	4410±35	3320–2910
HM8 – Packards South (Grid 1; surface 1)	SUERC-8855	2990±40	1390–1110
HM10 – Middle Moor (Grid 7; surface 1)	SUERC-8857	3200±40	1530–1400
HM12 – Packards South (Grid 1; surface 2)	SUERC-8859	2960±40	1320–1040
HM14 – Middle Moor (Grid 7; surface 2 - humic fraction)	SUERC-9692	3205±35	1530–1410
HM7 – Middle Moor (Grid 7; surface 2 – humin fraction)	SUERC-9696	3135±35	1500–1310
HM16 – Packards South (Grid 1; surface 3)	SUERC-8948	3630±35	2130–1890
HM24 – Trackway site (Grid 5)	SUERC-8878	3690±35	2200–1960

for the peat at any given location. However, no radiocarbon dating of upper peats has been carried out since the work of Smith (2002, chapter 2) and cutting has continued since the sampling for this study was carried out in the 1980s. Eight samples were taken from near to the surface for radiocarbon dating (Table 6.5) (see also chapter 5). Whilst this represents a very small number of dates with respect to the size of the study area, the aim was to explore methods for generating a testable model of the resource.

The results from the radiocarbon dating indicate that the area with the most extensive removal of peat is to the north of Lindholme Island where the surface of the peat dates to the Neolithic period (SUERC-8878; HM24). This is most graphically illustrated by the Neolithic Hatfield platform and trackway (see Chapter 7), which was discovered on the cut-over surface of the peatland with little *in situ* peat covering the archaeological remains.

The area with the youngest surface peat is in the southeast part of the Moors around Packards South, although the surface here still dates to the middle/late Bronze Age (SUERC-8855, 8859, 8948; HM8, HM12,

HM16) indicating that over three thousand years of the record has been removed. Whilst the radiocarbon dates represent only a very low concentration of dates relative to the total area of the peatland, none of the samples indicate the survival of peat post-dating the Bronze Age.

The radiocarbon dates from these peat surface locations were used in conjunction with estimated dates for the surface of peat from the work of the Humber Wetlands Project (Dinnin 1997) to generate a scatter plot of the relationship between elevation and age. Although the modelling used a limited number of just nine radiocarbon dates, linear regression applied to the data generated an R^2 value of 0.99, indicating a close correlation between surface elevation and radiocarbon date.

Whilst such a close correlation should be treated with caution due to the small number of radiocarbon dates, the process provided a formula that could be applied to the cell values of the LIDAR data to effectively convert elevation values to chronological values across space, with Lindholme Island excluded from the model (Figure 6.14).

Grouping the outputs from the surface modelling of dates

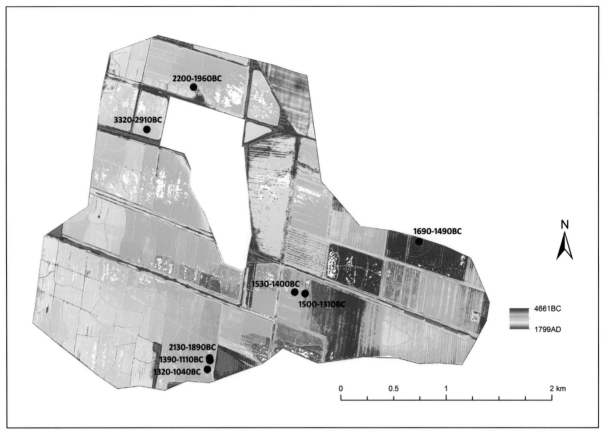

Figure 6.14 Resulting model of the dates of the surface peat across Hatfield Moors in relation to calibrated radiocarbon dates from surface peat samples

from across Hatfield Moors into time slices representing cultural periods, generates a model (Figure 6.15) which indicates that the majority of surface peat on Hatfield dates to the Bronze Age, with some areas having had peat removed to Neolithic levels. Surface peat dating to periods after the Bronze Age amounts to only 4.3 percent of the whole study area, with less than 1.5 percent of the deposits dating to the Medieval period or later. Thus for the vast majority of the Moors, any organic cultural remains post-dating the Bronze Age are unlikely to survive.

Summary: the peatland resource on Hatfield Moors

The remaining peat deposits on Hatfield Moors range up to 3.5m deep although, in the majority of areas, less than 1m of peat survives and in places none at all. In terms of cultural periods represented by this, and thus the potential for preservation of the archaeo-environmental record, the peat resource dates from the later Mesolithic through to the Medieval period. In the southern area of Hatfield Moors there is the potential for peat sequences representing this full time-span. However, in over 90 percent of the study area the peat has been cut down to Bronze Age and, in some places, to the Neolithic deposits, as reflected by the discovery of the Neolithic trackway and platform on and very close to the current surface of the peat (see Chapter 7). Overall, the peatland resource on Hatfield Moors has been largely removed with only a relatively shallow depth of organic deposits dating, for the most part, to the Neolithic and earlier Bronze Age.

6.3 Modelling Thorne Moors in four dimensions

Wetland inception on Thorne Moors

For Thorne Moors, a lower number of datasets were available, and the advanced rate of re-wetting of the landscape restricted the accessibility for gathering new data in the field. Hence, the model of the pre-peat landscape DEM incorporated fewer datasets than that for Hatfield Moors. However, the validation of the pre-peat landscape model demonstrated that it was generally robust in relation to the four areas of gridded borehole excavation and two transects, although localised topographic detail was shown to be poorly resolved (Chapter 4). Furthermore, fewer basal radiocarbon dates were available for Thorne Moors and no additional radiocarbon dates were obtained for this landscape. Hence the models for Thorne are regarded as less robust than those for Hatfield, although no statistical analyses of the output from the GIS models have been carried out. Within this context, the resulting DEM formed the basis of the analysis of the hypotheses of peat inception and spread across Thorne Moors.

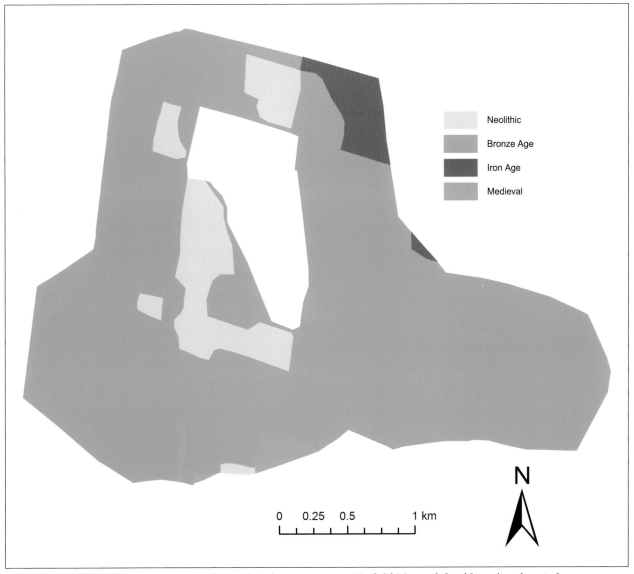

Figure 6.15 The modelled dates of surface peat across Hatfield Moors defined by cultural period

Modelling the timing of wetland inception in relation to elevation on Thorne Moors

The first hypothesis for wetland inception relates to the peat forming as a result of paludification following rising watertables (Figure 2.20; hypothesis A). Whilst there may be local variations in the shape of the groundwater (*cf.* Chapman and Cheetham 2002), at the landscape scale, this hypothesis would suggest that the earliest peat deposits should be located in areas of lowest elevations. Hence, the radiocarbon dates previously obtained from basal deposits across Thorne Moors were examined in relation to elevation (Table 6.6).

Whilst the level of vertical precision for the majority of the sites is based on modelled rather than surveyed data, it is clear that there is no strong relationship between elevation and date. The highest area, around Goole Moors in the northern part of the Moors, provided the earliest date, in the latter half of the fourth millennium BC. At a broad level, this date is not inconsistent with the date from the

Table 6.6 Basal dates from Thorne Moors in relation to elevation (*indicates elevations calculated from the basal model; all other elevations are surveyed; **whilst elevation data are not available for this sample, the lowest elevation encountered in this area during the gridded borehole survey was -0.10m OD; ***site with problematic chronological control following analysis presented in Chapter 5)

Site	Elevation (m OD)	Radiocarbon date BC (posterior density estimate)
Goole Moor Site 1	1.00 *	3500–3470 (2%) or 3380–3010 (90%) or 2980–2930 (3%) (CAR–223)
Crowle Moor Site 1	0.28 *	1970–1690 (CAR–208)
Trackway site	0.00 *	1740–1310 (BIRM–335)
Rawcliffe Moor	–**	3500–3430 (6%) or 3380–2920 (89%) (CAR–221)
Crowle Moor Site 2	-0.55*	3020–2580 (CAR–309)
Middle Moor***	-2.78	3345–2929 (NZA–18334)

lowest sample from Middle Moor, towards the centre of Thorne Moors. Hence, there is a vertical range of nearly three metres for the earliest dates for wetland inception. Similarly, there is no pattern in the relationship between the more recent dates and elevation. Hence, it appears that elevation was not the principal factor in the location of peat inception on Thorne Moors.

Modelling the timing of wetland inception in relation to proximity to watercourses on Thorne Moors

As outlined in Chapter 2, the second model (Figure 2.20; hypotheses B and C) links wetland inception and peat spread to proximity to watercourses. For these hypotheses to be supported, basal radiocarbon dates from closer to the rivers should provide earlier dates compared with those further away. As for Hatfield Moors, a model of linear distance from rivers and former river channels was generated for the River Don, the Old River Don and the River Went within the GIS in relation to basal radiocarbon date sampling sites (Figure 6.16, Table 6.7).

Crowle Moor Site 1 was 928m away from the Old River Don where the basal sample provided a date of *1970–1690 cal. BC*. The site furthest from watercourses was the Trackway site towards the centre of Thorne Moors at 3719m from any watercourses. This sample provided a slightly later date of *1740–1310 cal. BC* which fits with the hypothesis. However, the four sites within the intermediate areas (between 1291m and 3699m from watercourses) are all considerably earlier, with the earliest date, Rawcliffe Moor, being the second furthest from watercourses. Hence, it appears that the hypotheses relating to proximity to rivers do not hold up for Thorne Moors. However, the proximity of the earlier samples from Middle Moor and Crowle Moor Site 2 to the channel feature identified in Grids 1 and 2 of the gridded borehole survey (see Chapter 4) raised the question of whether other channel features (see C-hapter 2) might have provided the focus for early wetland inception.

Modelling the timing of wetland inception in relation to surface run-off on Thorne Moors

The final two hypotheses (Figure 2.20; hypotheses D and E) suggest that wetland inception was related to localised paludification caused by the poor drainage of surface water, in conjunction with hydroseral succession. This was again tested through the analysis of surface flow patterns based on the DEM of the pre-peat landscape (see Chapter

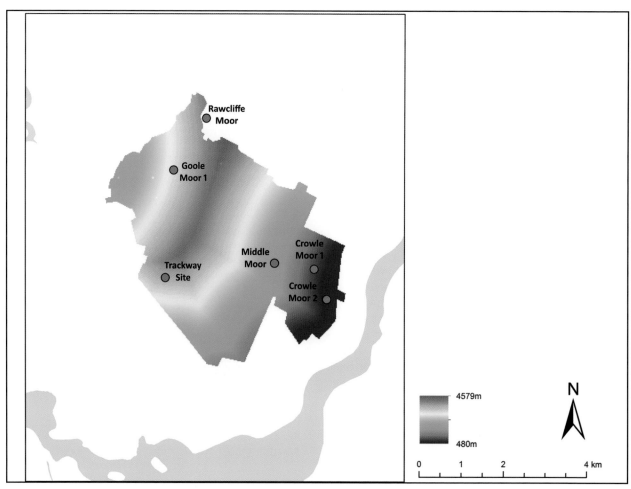

Figure 6.16 Watercourse proximity model for Thorne Moors with the locations of radiocarbon dates (blue areas show the floodplains of adjacent rivers)

6. Patterns of change: modelling mires in four dimensions

Table 6.7 Basal radiocarbon dates in relation to distance to watercourses

Site	Distance to watercourse (m) (GIS output)	Radiocarbon date (calibrated range BC at 95% confidence and posterior density estimates)
Crowle Moor Site 1	928	*1970–1690* (CAR–208)
Crowle Moor Site 2	1291	*3020–2580* (CAR–309)
Middle Moor	2271	*3345–2929* (NZA–18334)
Goole Moor Site 1	3357	*3500–3470 (2%) or 3380–3010 (90%) or 2980–2930 (3%)* (CAR–223)
Rawcliffe Moor	3699	*3500–3430 (6%) or 3380–2920 (89%)* (CAR–221)
Trackway site	3719	*1740–1310* (BIRM–335)

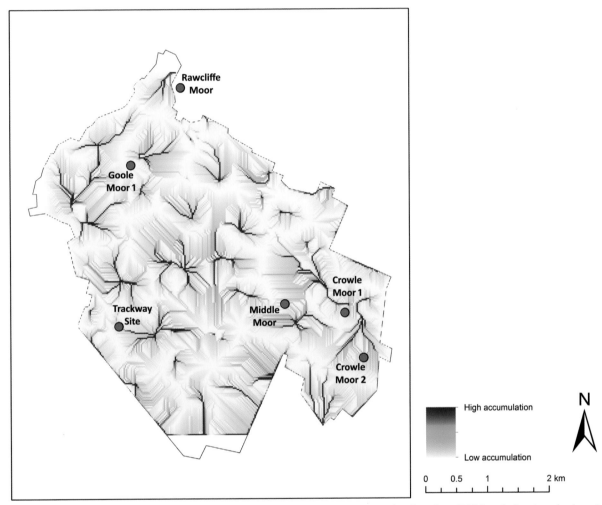

Figure 6.17 Flow-accumulation model of Thorne Moors based on the pre-peat land-surface DEM and showing the locations of basal radiocarbon samples. Darker shades reflect areas of higher modelled flow-accumulation and hence those areas prone to waterlogging

4) with a flow-accumulation model generated for Thorne Moors, (Figure 6.17). Values from the resulting model were sampled on the basis of the locations of the radiocarbon samples. Table 6.8 provides a comparison between these flow-accumulation values and the calibrated radiocarbon dates for each of the sample sites of basal locations on Thorne Moors.

The earliest date for peat formation correlates with the highest level of flow-accumulation, in the area of Middle Moor towards the centre of Thorne Moors. However, the general trend is less convincing than that for Hatfield Moors, with early dates also equating to areas of very low accumulation, as at Goole Moor. In part, this lack of any distinct trend might be due to the relatively small number of radiocarbon dates available which makes discerning any pervasive pattern problematic. In addition, the basal topography input layer is of a lower resolution compared to that of Hatfield Moors, and the relatively poor positional accuracy of the sample locations from previous research further hampers this process. Whilst

Table 6.8 Flow-accumulation values in relation to basal radiocarbon dates (*site with problematic chronological control following analysis presented in chapter 5)

Site	Flow-accumulation value (GIS output)	Radiocarbon date (calibrated range BC at 95% confidence and posterior density estimages)
Goole Moor Site 1	1	3500–3470 (2%) or 3380–3010 (90%) or 2980–2930 (3%) (CAR–223)
Crowle Moor Site 2	4	3020–2580 (CAR–208)
Trackway site	8	1740–1310 (BIRM–335)
Crowle Moor Site 1	14	1970–1690 (CAR–208)
Middle Moor*	18	3345–2929 (NZA–18334)
Rawcliffe (Smith)	–	3500–3430 (6%) or 3380–2920 (89%) (CAR–221)

the modelling of the available data indicates that it is unlikely that wetland inception on Thorne Moors was primarily driven by patterns in local surface drainage, there are reasons to regard the results of this analysis as inconclusive. However, the flow-accumulation model does provide a model that can be tested through future fieldwork and dating.

Summary: patterns of wetland inception and peat spread on Thorne Moors

The results from the testing of hypotheses relating to the primary drivers of wetland inception were less conclusive than they were for Hatfield Moors. In part, this is likely to be due to the relatively low resolution of the DEM of the pre-peat landscape, and the low number of basal radiocarbon dates with accurate georeferencing. In comparison with the study of the Hatfield Moors landscape, this has significant implications regarding the quantity and quality of data required for undertaking such landscape-scale analyses of 'hidden' environments (see Chapter 8).

Analyses of the available data do not support the hypothesis that elevation was the principal driving force in wetland inception and it is inconclusive whether, as for Hatfield Moors, localised surface run-off played a significant role in early peat formation. Whilst proximity to watercourses did not provide a correlation with the dated samples from across the Moors, it is possible that early wetland inception was related to palaeochannels, such as that identified in the gridded borehole survey of Grids 1 and 2. The process of generating a model of wetland development across Thorne Moors was consequently much more subjective than for Hatfield Moors and relied upon a combination of input data sources including localised basal topography and previous radiocarbon dates, but was rendered in relation to generic period categories as at Hatfield (Figure 6.18).

The model indicates that the earliest wetland development within the northern part of the study area was on Rawcliffe Moor in a pronounced topographic depression. Following this initial spread of wetlands during the early Neolithic period, localised areas of peat formation are indicated. It is likely that early accumulation took place during the late Neolithic within a channel that ran broadly east–west across the central area of Thorne Moors. By the early Bronze Age it appears that the areas of wetland had coalesced across the majority of Thorne Moors except for the 'islands' in the south, which may have remained dryland until the later Bronze Age.

Modelling the surviving peatland resource on Thorne Moors

Determining the quantity of surviving peat on Thorne Moors was more problematic than for Hatfield Moors, partly due to the reduced quantity of data available for this landscape (see Chapters 2 and 3), but it is also due to the greater complexity in the pre-peat land-surface. The basal model indicates a range of topographic variation with pronounced 'low points', which are in contrast to that of the generally more gently undulating pre-peat landscape of Hatfield Moors.

Modelling the depth of surviving peat on Thorne Moors

Unlike Hatfield Moors, data were not available for the generation of peat depth modelling. However, the same dataset obtained from Natural England for the basal topography also provided low-resolution contours relating to the elevations of the land-surface. Whilst this dates from the mid-1990s, it provided an opportunity to explore the potential depths of peat across the Moors which could then be compared with data from boreholes (Chapter 4). It may be assumed that there would be less depth of peat identified in the boreholes due to the effects of subsequent peat cutting during the intervening period between each of the datasets being gathered.

Initially the land-surface contours were modelled in the same way as the basal topography contours to generate a DEM (see Chapter 3). The basal topography DEM was then subtracted from this new land-surface DEM to create a third model. This third model relates to depths of peat, being the difference between the top of peat and the base of the peat, providing a range in depth of between 0m and 3.83m, with a mean depth of 1.50m. The depth model may be assessed in comparison with the depth values generated from the borehole grids and transects across Thorne Moors. Table 6.9 provides a summary of the statistics from these data sources.

The comparison of depth values between the Natural England (NE) contour derived depths and the peat depths

6. Patterns of change: modelling mires in four dimensions

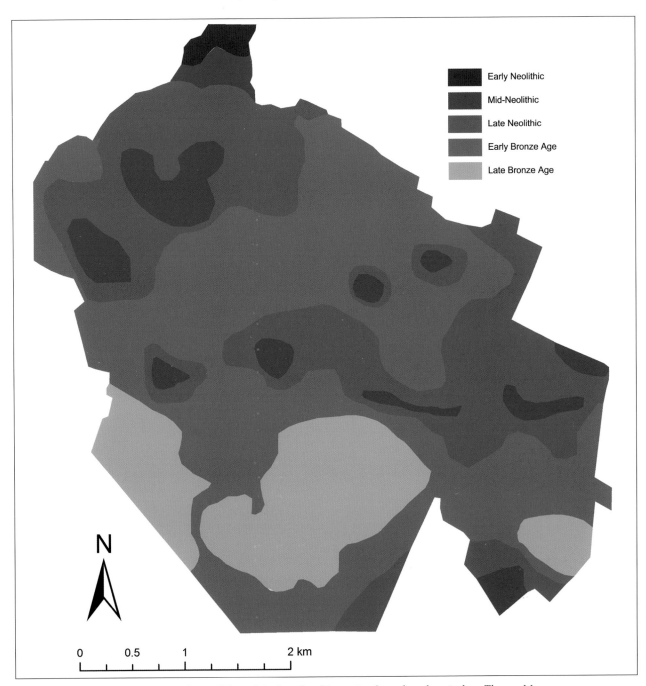

Figure 6.18 Resulting GIS model of wetland inception by cultural period on Thorne Moors

Table 6.9 Comparison of depth statistics from the boreholes and Natural England (NE) data

Site	Data Source	Min depth (m)	Max depth (m)	Mean depth (m)
Thorne Moors	NE Data	0.00	3.83	1.50
Grid 1 (west)	Borehole grid	0.15	3.62	1.48
	NE Data	1.31	2.34	1.98
Grid 2 (east)	Borehole grid	0.02	4.13	1.98
	NE Data	1.92	2.33	2.11
Grid 3 (north)	Borehole grid	0.18	1.90	0.96
	NE Data	1.49	2.05	1.86
Grid 4 (Rawcliffe)	Borehole grid	0.46	2.84	1.34
	NE Data	0.70	2.07	1.26
Cottage Dyke	Boreholes	1.63	2.97	2.16
	NE Data	1.54	2.49	2.20
Birtwistle	Boreholes	2.22	2.77	2.42
	NE Data	1.99	2.35	2.17

from the boreholes on Thorne Moors indicates some correlation given the difference in resolution of the two datasets. Values from the NE dataset for Grids 1, 2, 4 and Birtwistle are contained within the range of depths provided by the borehole data. The results from Area 3 are also comparable although the NE data indicate a greater depth of peat surviving, which was to be anticipated given the timing of the data collection in the mid-1990s. Conversely, at Cottage Dyke, the minimum depth of peat determined from the Natural England model is less than that observed from the borehole survey, (Grids 1, 2 and 3 proved to be locally variable topographically). In terms of mean values, there is a maximum difference of 0.9m for Grid 3. However, the average difference between the mean values is 0.21m, indicating some level of reliability in the model. This does suggest that the values of peat depth overall provided give a reasonable indication of the broad peat depths across Thorne Moors (Figure 6.19).

The relatively large range in peat depths, demonstrated by the borehole data in particular, indicates that peat depth is determined to some extent by the nature of the basal topography. The data indicate that there is between 0.02m and 2.22m of peat remaining on Thorne Moors, with up to 4.13m of deposit within topographic low points such

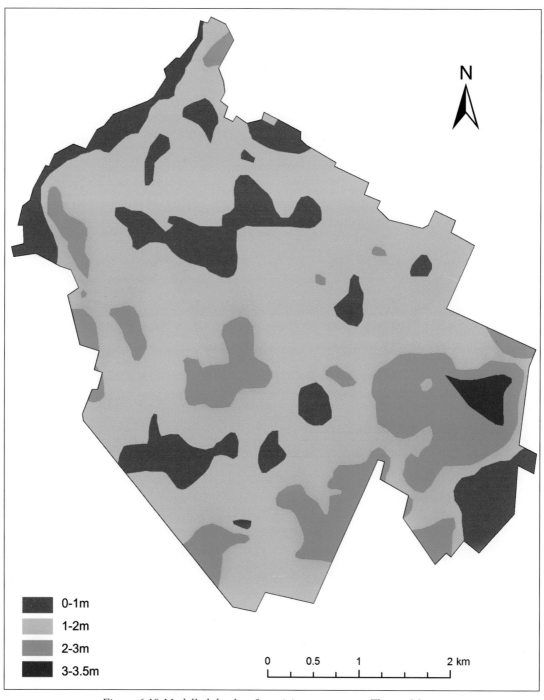

Figure 6.19 Modelled depths of surviving peat across Thorne Moors

as those at Rawcliffe Moor (Grid 4) and Middle Moor (Grids 1 and 2).

Modelling the date of the cut peat surface of Thorne Moors

Modelling the age of the upper peats for Thorne Moors is problematic due to the more limited range of datasets available for this peatland in comparison to Hatfield. Other than the relatively low resolution Ordnance Survey datasets and the Natural England contours, there are no accurate datasets relating to the surface topography of Thorne Moors. Furthermore, radiocarbon dating of the upper deposits is limited. There is only one recent study (post-dating the cessation of peat extraction) which has one radiocarbon dated near surface sample (Gearey 2005), from a depth of 0.24m (NZA-18329; 1614 ± 40BP, cal. AD 349–552) from the Middle Moor site in the central area of Thorne Moors (Area 2), indicating that the upper peat in this area dates to the later Romano-British–early Medieval period.

Summary: the peatland resource on Thorne Moors

Despite the relatively small quantity of data relating to the quantification of the peat resource on Thorne Moors, a number of conclusions can be drawn. Overall, there is a range of between 0.02m and 2.22m of peat remaining, with a maximum depth locally of 4.13m. In terms of the chronology this peat represents, the earliest organic deposits on Thorne Moors date to the early Neolithic period, at around 3640–3370 cal. BC. Away from the Rawcliffe depression, the earliest dates for peat accumulation centre in most areas around 3000 cal. BC, with wetland spreading across the southern half of the Moors during the middle Bronze Age. The surviving upper age of peat on Thorne Moors is very poorly known, with only a single recent date which indicates the potential survival of peat up to the early post-Roman period. It is possible that later peat deposits exist locally, but the effects of re-wetting and vegetation re-colonisation mean that access for future survey and sampling will be variable across the Moors.

6.4 Summary: modelling Hatfield and Thorne Moors in four dimensions

Wetland inception and peat spread

The GIS modelling process described in this chapter was intended to manipulate and analyse topographic and chronological data with the aim of testing previous hypotheses regarding the development and growth of the peatlands both spatially and temporally. Understanding wetland inception and peat spread across raised mire landscapes is only possible through an appreciation of the principal drivers behind the process. A series of five hypotheses generated from previous work (Buckland and Smith 2002; Figure 2.20) were tested through the combined spatial modelling with the GIS, related to chronological data available from previously existing and new radiocarbon dates from across both Hatfield and Thorne Moors. For both landscapes, the DEMs of the pre-peat land-surfaces (Chapter 4) provided the foundations for modelling and comparison between these datasets. In each case, the processes underlying peat formation were examined through the testing of these five hypotheses. Peat spread was then examined, in relation to both the drivers for inception and other factors, in order to generate models of the development of both peatlands.

For Hatfield Moors, the available datasets for both the refined pre-peat landscape and the chronology of wetland inception were relatively detailed. These allowed for a robust interrogation of the relationship between the topography of the pre-peat landscape and chronological information on peat inception provided by radiocarbon dating. The analyses demonstrated that it was unlikely that factors including elevation and proximity to rivers played a primary role in the onset of paludification and subsequent peat growth. The results indicated that factors relating to surface hydrology played a greater role in the formation of wetlands in multiple foci across the Moors. Initial peat inception was apparently related to the microtopography of the basal sands of the pre-peat landscape with wetland development commencing in areas liable to accumulate water, rather than related to the absolute height of different parts of the landscape or proximity to the rivers.

The model does not of course explain what might have caused paludification. Whilst this appears to have had the initial significant influence on peat formation, secondary factors will presumably also have controlled the subsequent patterns of growth. For example, wetland inception might also have been linked with other landscape-scale changes, such as climate change, rising base levels associated with rising sea level, and the impact of these factors on fluvial systems around the Moors, which might have effectively exacerbated the impact of flow-accumulation through poor drainage. The relationship between these landscape-scale changes remains unclear. Furthermore, the role of human agency might also have influenced environmental changes that could have exacerbated waterlogging and the spread of wetland.

The lower resolution of the data relating to the pre-peat landscape, coupled with the fewer dated basal samples, has restricted the power of this approach for Thorne Moors. The testing of the hypotheses for wetland inception was inconclusive for this study area and so the resulting modelling of the evolution of the landscape during its earlier phases is less robust. The difference in the quality of outputs for both of the landscapes indicates the required levels of data for such a study. The modelling of Hatfield Moors demonstrates the potential for generating testable models from relatively small quantities of input data relating to both the pre-peat landscape and the chronology of wetland inception and peat spread. However, the modelling of wetland inception indicates the influence of micro-topography in the early stages of peat accumulation. In contrast, modelling wetland inception and peat spread on Thorne Moors was more complex in part due to the apparently greater topographical variability of the pre-peat landscape.

Peatland resource modelling

The results of the modelling of wetland inception and peat spread across Hatfield and Thorne Moors provide a foundation for contextualising archaeology (see Chapter 7) and for understanding patterns of environmental change through time. These models also provide an indication of the earliest date of potential wetland preservation of organic material at any specific location across the study areas, which have implications for their future management. The archaeo-environmental resource of Hatfield and Thorne Moors was also modelled in terms of peat depth, since the peat holds the potential of both containing archaeological remains in addition to protecting them. This modelling demonstrated that, for Hatfield Moors, an average depth of 1.2m of peat remains, with a maximum depth of 3.5m over a small area on its southern edge, and some other areas where the pre-peat land-surface is entirely exposed. For most areas of Hatfield Moors, less than 1m of peat remains. Whilst the data were more limited for Thorne Moors, the modelling indicates an average depth of 1.50m, with a maximum depth of 3.83m, but with some areas having no peat at all. The gridded borehole surveys on Thorne Moors confirm this large range, with values from the sample areas of 0.02m of peat and up to 4.13m in specific points such as Rawcliffe Moors to the north and Middle Moor towards the centre of the Moors.

For Hatfield Moors, it was possible to model the likely dates represented by the cut-over peat surface across the landscape. This process revealed that, for the majority of the landscape, no peat dating to later than the Bronze Age survives and, in many central areas, the most recent peat dates to the Neolithic period. Towards the southern part of the Moors, Medieval and perhaps even more recent peat survives at ground surface. In terms of the date range represented, and hence the potential for the survival of wet-preserved archaeological remains, most areas equate to the later Neolithic through to the Bronze Age, but there are some areas which contain a wider chronological span. On the southern edge and to the north of Lindholme Island it is likely that organic deposits dating from the later Mesolithic through to the Medieval period survive, although such locations are very restricted. Much of the peat loss might have been relatively recent: for example, work undertaken in the early 2000s during peat cutting indicated that peat depths in the southeastern area of Porters Drain at that time extended up to 2.5m deep, and represented a chronological span of between *c.* 2300 cal BC and perhaps as late as AD 700 (Gearey 2005). In contrast, the subsequent analysis has shown that peat depths in the same area are between 1m and 2m, with surface peats dating to perhaps the Bronze Age.

Summary

The integration and modelling of spatial and temporal datasets, each collected at a range of resolutions, has provided a considerably improved understanding of the evolution of Hatfield and Thorne Moors, and of what the resource might represent in terms of the depths of peat and the associated chronological span. For Hatfield Moors it seems that the principal driver for wetland inception was impeded drainage and surface water accumulation within localised hollows, with peat growth commencing in numerous separate locations before spreading across the landscape. Within these localised hollows, wetland inception probably commenced towards the end of the Mesolithic period and by the early Bronze Age almost the entire landscape, other than Lindholme Island, would have been wetland.

For Thorne Moors the analysis was less conclusive. It is unlikely that wetland inception was driven by factors relating to elevation, nor to proximity to the main watercourses surrounding the Moors. However, there is some indication that inception might have been related to the palaeochannels beneath the Middle Moor and Rawcliffe areas, but these features are poorly investigated. It seems that for this landscape a much higher resolution dataset would be required compared with Hatfield Moors due to the subtle variations in the pre-peat landscape. It seems that wetland inception commenced a little later than for Hatfield Moors, with an initial focus in the northern part of the Moors during the early Neolithic period, although it is possible that earlier inception might have occurred in channels towards the centre of this landscape. Some areas in the southern half of Thorne Moors probably remained relatively dry until the later Bronze Age.

The analysis of the surviving resource on both Hatfield and Thorne Moors has shown that the extensive extraction of peat has reduced most of the area of Hatfield Moors to less than 1m of peat, with most areas representing an overall chronological span from the later Neolithic through to the Bronze Age. It seems that more peat survives on Thorne Moors, although the majority of the landscape consists of less than 2m of peat, with an overall mean value of 1.5m.

7. Archaeological Investigations of a late Neolithic site on Hatfield Moors

with Nicki Whitehouse, Pete Marshall, Maisie Taylor, Michael Bamforth and Ian Powlesland

7.1 Introduction

Whilst there have been reports of archaeological material from Hatfield Moors (Chapter 2), no sites or firmly provenanced artefacts have been identified on the peatland, in spite of recent ditch survey and fieldwalking programmes. Hence, the discovery of a cluster of worked wooden poles on the cut-over surface of the Hatfield Moors just to the north of Lindholme Island by Mick Oliver of the Thorne and Hatfield Moors Conservation Forum (Figure 7.1) in November 2004 was highly significant. The exposure of the site was a result of peat milling activity which had cut to within 0.50m of the base of the organic sequence, uncovering and in part removing the archaeological remains. On the basis of the peat milling records for the area, it is possible that the site had been exposed for up to ten years prior to its discovery.

The initial discovery (Figure 7.2) was located in an area directly north of a peat bund, although a walk-over survey identified a series of additional exposures of archaeological wood to the south of the peat bund over an area approximately 40m long and over 5m wide. The likelihood that the site was actively deteriorating resulted in three phases of investigation between 2004 and 2006, consisting of excavation, survey, radiocarbon dating and palaeoenvironmental analyses. This chapter outlines these investigations and discusses its form and possible function within the context of the modelling of wetland inception and peat growth described in Chapter 6.

7.2 Methods

Borehole survey

In order to establish the local context and to determine the depth and character of the peat deposits a grid of 121 boreholes (Grid 5; see Chapter 4) was excavated around the site, over an area measuring 100 × 100m, with boreholes positioned at 10m intervals and set out using GPS (*cf.* Chapman and Gearey 2003). This demonstrated that the maximum depth of peat was at most 0.7m and with basal sands between 1.72m OD and -0.39m OD locally. Very little ombrotrophic peat survived, indicating that peat milling had cut right down to the very earliest stages of peat accumulation (Figure 7.3).

Excavation, sampling and fieldwalking

The excavation of the site took place over three seasons in the winter of 2004 and the late summers of 2005 and 2006 when local groundwater levels were at their lowest. In total, an area of 230m^2 was investigated, covering the whole of the site except for a strip 4.5m wide that lay beneath the peat bund. Two trenches were also excavated at the northern end of the site and one at the southern end to investigate whether archaeological features extended in this direction. An additional two trenches were excavated in 2006 on the western and eastern sides of the area to investigate slightly raised areas of basal sands identified from the modelling of the borehole data (Figure 7.4).

All areas were excavated by hand to the level of the archaeological remains where present, or to the basal sandy topography where archaeological remains were absent. In four areas of the site, sondages were cut through the site to the basal sand to examine the possibility of earlier structures; to investigate the stratigraphic context of the site and to collect samples for palaeoenvironmental analyses.

Variable preservation across was mostly a result of damage caused by peat milling with a drain cutting through the southern end of the site. In some areas the structure was exposed at the ground surface and thus only required cleaning prior to planning, whilst in other places the archaeology had been completely destroyed by milling. Two areas were better preserved, concealed beneath thin layers of undisturbed peat; the southern end of the site was preserved beneath up to 0.20m of peat, and the area directly north of the bund was covered by over 0.10m of peat. Thicker deposits were present below the archaeological remains and a sequence of three bulk sediment samples (GBA/BS *sensu* Dobney *et al.* 1992)

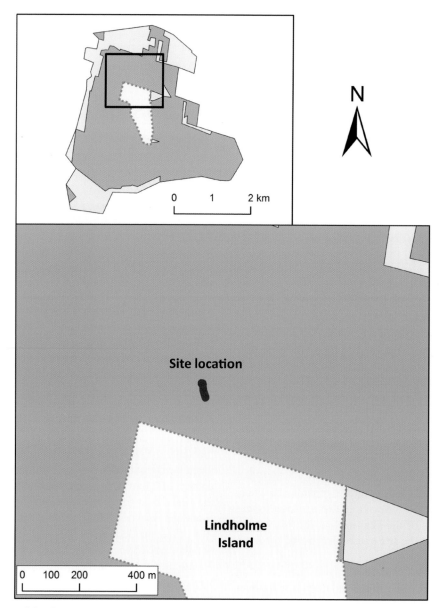

Figure 7.1 Location of the discovery of a cluster of worked wooden poles on the northern side of Lindholme Island on Hatfield Moors

were recovered from the thickest *in situ* organic deposits beneath the archaeological structure at the northern end of the site (a depth of 0.28m), with a fourth sample taken from a patch of *in situ* peat overlying its southern end. Additional samples of sediment were collected for radiocarbon dating.

Fieldwalking and drain face survey was undertaken to the north of Lindholme Island over an area of 120.6ha, with assistance from members of the Thorne and Hatfield Moors Conservation Forum and Natural England. This resulted in the discovery of a concentration of struck lithics on the exposed pre-peat sands at the northwestern edge of Lindholme Island, approximately 300m to the west of the site (see below; Figure 7.5).

7.3 Results

The trackway and platform (wood analyses by Maisie Taylor and Michael Bamforth)

The excavations revealed a timber trackway leading from 'landfall' on the northern tip of Lindholme Island and extending for 45m to a structure interpreted as a platform (Figure 7.6), although this part of the site had been damaged by peat milling and probably also by subsequent exposure (Figure 7.7). The trackway was of a corduroy design (*cf.* Raftery 1996), consisting of transverse poles overlying two 'rails'; poles aligned in the direction of travel. The width of the structure narrowed considerably between its southern, landfall end and its junction with the platform. Along this route, the distance between the underlying rails missing or had never been put in position. In each of these areas, the gaps were filled by a layer of birch (*Betula*) bark laid directly onto the surface of the peat. Two of these were close together and located at the transition or threshold between the trackway and the platform. The third lay approximately a third of the way along the trackway from its southern terminus (Figure 7.8a and b). In four cases, roughly cut pegs had been positioned vertically between the transverse poles of the trackway, presumably to add stability to the site by preventing structural movement. The trackway also appeared to incorporate the surviving stumps of trees that had been growing on the basal sands prior to plaudification (Figure 7.8a).

The upper-structure of the platform was of a similar construction to the trackway, although aligned obliquely to it, with the poles positioned at an angle of approximately 45° relative to those of the trackway to the south. The surviving remains of the platform covered a total area of approximately 10 × 5m, although it had been damaged by peat milling and probably also by exposure, with the eastern part of the structure showing significant deterioration. From the positions of the fragmentary remains of the sub-structure, and assuming some level of symmetry, the total width of the platform is likely to have been perhaps over 13m. The underlying structure of the platform was less regular than that of the trackway, with a mixture of brushwood and long poles following a near perpendicular alignment to the upper-structure.

Figure 7.2 The initial discovery of wooden poles on Hatfield Moors (peat bank in background, looking southwest)

The analysis of 35 samples of archaeological wood from the structure indicated that scots' pine (*Pinus sylvestris*) was used exclusively for rails, transverse pole and pegs. All of the wood was roundwood or slightly modified roundwood, and the size of the transverse timbers from the trackway ranged between 26/38mm and 128/119mm diameter, with the rails slightly smaller at between 34/39mm and 70/75mm. There was very little indication of surviving bark on the timber, and no clear evidence for modification except for the tips of the pegs which had been roughly sharpened to points (Figure 7.9a and b). Traces of shaping on two stakes were indicative of stone rather than metal tools with small and 'ragged' facets. The better preserved poles showed little evidence of decay in antiquity on their upper sides, suggesting that the site was submerged by the expanding peatland reasonably rapidly or had sunk into the waterlogged deposits shortly after construction.

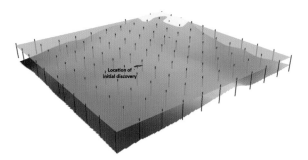

Figure 7.3 The gridded borehole survey in relation to the current land-surface, the pre-peat land-surface sands (in yellow–brown) and the locations of the initial discovery, looking northwest (100 × 100m)

Site chronology (radiocarbon and Bayesian analysis by Pete Marshall)

Five samples from timbers forming the trackway and platform were submitted for radiocarbon dating; three to Beta Analytic Inc. Florida, USA (Beta) and two to 14CHRONO Centre, Queen's University, Belfast (UB). Three of these samples were taken from the best preserved section of the site towards the southern end of the trackway and comprised one rail and two of the transverse timbers. The other two samples were taken from the platform poles at the northern end of the site. In addition, a further five

narrowed from 1.90m to approximately 0.90m, with the overall trackway structure narrowing from 3.10m to around 1.00m wide. This tapering of the structure was perhaps also reflected in the dimensions of the poles used for its construction. The poles at the southern end of the structure were around 0.13m in diameter, but less than 0.04m at its northern end. The overall plan of the trackway also reveals a slight but clear 'dog-leg' in its route, which appears to avoid a slightly higher area of basal sands identified during the gridded borehole survey.

In three areas of the site, narrow gaps were identified in the trackway superstructure where a single pole was

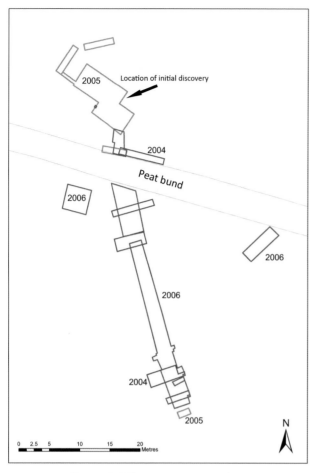

Figure 7.4 The excavation trenches

samples of peat were submitted to the Scottish Universities Environmental Research Centre, East Kilbride (SUERC). This included four from on-site contexts consisting of two from the base of the peat below the site and one from the peat directly below the structure. Two samples of basal peat recovered from locations to the north and east of the excavated area were also submitted for radiocarbon dating using the AMS method (see Chapter 3) and the results are given in Table 7.1.

All five of the measurements from the archaeological wood (Figure 7.10) were found to be statistically consistent (T'=3.9; v=4; T'(5%)=9.5; Ward and Wilson 1978) and could therefore be of the same actual age. This strongly suggests that the structure was built in a single phase of construction. The three measurements from the base of the peat below the northern end of the trackway (SUERC-9688-9689 and SUERC-8846) are not statistically consistent (T'=1265.5; v=2; T'(5%)=6.0); Ward and Wilson 1978) and thus the sample contains material of various ages.

The radiocarbon dates from the basal peat demonstrate that paludification of the sands began in the topographically lower areas of the landscape (-0.11m OD) some 180m to the north of the platform at a date of 6355±35 BP (SUERC-8952; 5470–5220 cal. BC). By 4905±35 BP (SUERC-8949; 3370–3630 cal. BC) this had spread to a location 70m north of the platform (0.08m OD). Paludification of the northern margins of Lindholme Island at the southern terminus of the trackway (0.94m OD) was dated to 4250±35 BP (SUERC-8877; 2920–2760 cal. BC).

It would therefore appear that increasingly wet conditions in the vicinity of the site began in the lower parts of the landscape around the mid-fifth millennium BC. By the early third millennium BC the wetland had spread southwards towards Lindholme and it was around this time that the site was constructed (see also Chapter 6). Bayesian modelling provides a *posterior density estimate* for the construction of the trackway of *2730–2450 cal. BC (last trackway constructed; 93% probability)* and probably *2620–2470 cal BC* (68% probability) and reinforces the indication that the structure was built in a single phase of construction in the late Neolithic period.

Finds (lithics analysis by Ian Powlesland)

The concentration of struck lithics discovered during fieldwalking on an exposure of the basal sands approximately 300m to the west of the site (Figure 7.11) consisted of a total of 13 pieces of flint (16.77g), mainly of debitage in the form of chips and chunks, broken flakes and one retouched piece. Two complete flakes were both identified as tertiary flakes, of a size indicative of the working of small nodules or prepared cores to fabricate flake tools. No obvious blades were present, and the majority of identifiable flakes were generally small and irregular in form with plain platforms, indicating the use of hard-hammer techniques. This form of flake fabrication is more typical of later prehistoric flint working, often of late Neolithic or Bronze Age date.

Over a third of the assemblage had some form of patination and many pieces displayed edge damage, indicating that the material has been disturbed from its original context. The only retouched piece was a fragment of probable gravel flint with fine invasive retouch across both faces to create a curving edge that had evidence for a small secondary flake removal along one edge. This is likely to have been a small, finely worked leaf-shaped arrowhead of earlier Neolithic date, possibly broken during an unsuccessful attempt to re-sharpen the edge (Figure 7.11; B).

The sources of flint used were fairly uniform, indicating either the repeated use of readily available local sources or roughly contemporaneous flint working. Two main sources seem to have been employed using broadly similar techniques and a third slightly different source. The first source (60% by number) comprised light grey flint with a slightly abraded chalk cortex, possibly brought to the area in the form of partially prepared nodules or cores, probably from the Yorkshire Wolds. This material was used for the fabrication of the larger flakes. The second source was a light grey/brown flint with a thin and much abraded white cortex derived from gravel flint, presumably from riverine or beach pebbles on the basis of abrasion. This flint used was in the form of small nodules, and the debitage from this source was smaller. A third source was represented by

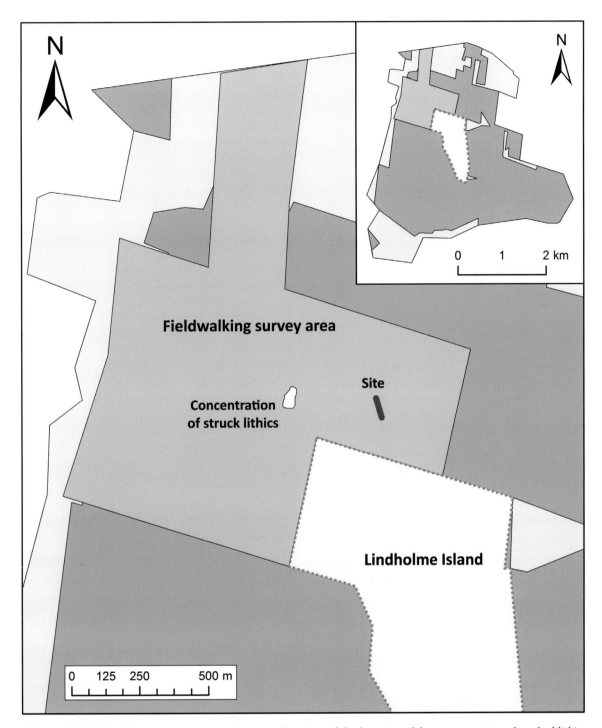

Figure 7.5 Fieldwalking survey area in relation to the site and the location of the concentration of worked lithics

a single piece of dark grey granular chert which might have been imported from a location outside the Yorkshire area.

The similarity of flint working techniques and flake forms indicate that the assemblage is broadly contemporaneous in date and function. The flint working strategies were mainly reliant on hard-hammer techniques using plain platforms, typical of later Neolithic and Bronze Age flint working. The general absence of retouched pieces and the irregular form of the few surviving flakes also indicate a later Neolithic date (possibly mid–late Neolithic), considering the tentative identification of the leaf-shaped arrowhead. The consistent use of specific flint sources and the importation of prepared cores from a chalk land source are also indicative of Neolithic rather than Bronze Age flint working when more *ad hoc* use of flint sources was undertaken. Although not definitive dating evidence, this is indicative of a more structured approach to flint procurement which is typical of Neolithic use of lithic resources.

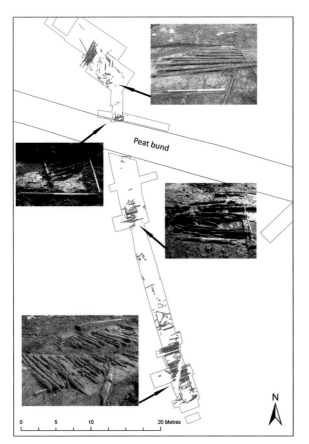

Figure 7.6 Overall site plan drawing together all phases of excavation

Coleoptera and macrofossils
(by Nicki Whitehouse)

Six bulk samples were analysed from undisturbed peat both beneath, and hence pre-dating the trackway and platform, and above the site (see Figure 7.7). Three were taken directly from beneath the platform (samples 1–3) and a further two from the edge of the landfall end trench of trackway, adjacent to the terminus of the site and on the fringes of the dune system to the north of Lindholme (samples 4 and 5). This material is probably contemporary with samples 1 and 2 beneath the platform, consisting of organic 'pool' mud. A sixth sample from an area of undisturbed peat which sealed the archaeological structure was also analysed (sample 6). These samples were processed following the techniques outlined by Kenward *et al.* (1980, 1986). The macrofossil analyses showed that few identifiable vegetative remains were present. Remains of *Sphagnum* mosses were common in the samples, with unidentifiable fragments of wood, bark and some charcoal. Leaves of *Calluna vulgaris* (heather) and *Erica tetralix* (cross-leaved heather) were also present. Hence, palaeoenvironmental interpretation relied predominantly on the coleopteran record (Figure 7.12). Beetle nomenclature follows Lucht (1987). The full taxonomic list is presented in Appendix 2.

Sample 1: base of peat (0.67–0.60m OD)
The sedimentology of the basal unit directly overlying

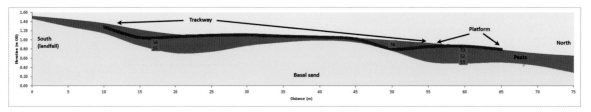

Figure 7.7 Profile through the site showing the exposed areas of the structure following peat cutting, with the locations of samples (S1–6) taken for Coleopteran and macrofossil analysis (see below)

Figure 7.8a and b One of the birch bark layers laid onto the surface of the peat between the horizontal poles of the superstructure (scale on 'a' is 1m, scale on 'b' is 0.2m)

7. Archaeological Investigations of a late Neolithic site on Hatfield Moors

Table 7.1 Radiocarbon dates from the site

Laboratory Code	Context	Material	Radiocarbon Age (BP)	$\delta^{13}C$ (‰)	Calibrated date BC (95% confidence)
Archaeological wood					
UB-6927	Platform pole	*Pinus sylvestris*	4145±35	–	2880–2570
UB-6928	Platform pole	*Pinus sylvestris*	4165±35	–	2890–2620
Beta-200619	Trackway: transverse pole	*Pinus sylvestris*	4160±70	–	2910–2490
Beta-200620	Trackway: rail	*Pinus sylvestris*	4150±70	–	2910–2490
Beta-200621	Trackway: transverse pole	*Pinus sylvestris*	4030±60	–	2860–2450
Samples from top of peat directly below the archaeological structure					
SUERC-8878	Top of peat directly below S (dryland) end trackway (1.14m OD)	Ericaceae stems/roots	3690±35	-26.7	2200–1960
Samples from the base of peat					
SUERC-9688	Base of peat below N end trackway (0.63m OD)	Peat, humic acid	3880±35	-27.6	2470–2200
SUERC-9689	Base of peat below N end trackway (0.63m OD)	Peat, humin	4010±35	-27.1	2620–2460
SUERC-8846	Base of peat below N end trackway (0.63m OD)	Charred twigs & stems, unident.	5430±35	-26.0	4350–4230
SUERC-8877	Base of peat below S (dryland) end trackway (0.94m OD)	Ericaceae stems/roots	4250±35	-26.8	2920–2760
SUERC-8949	Base of peat 70m N of platform (0.08m OD)	Charred Ericaceae twig & unident. Twig	4905±35	-26.9	3370–3630
SUERC-8952	Base of peat 180m E of platform (-0.11m OD)	Herbaceous charcoal, small twigs	6355±35	-26.0	5470–5220

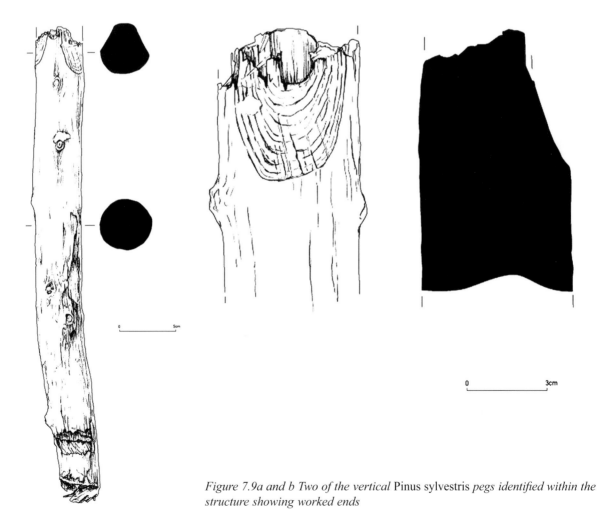

Figure 7.9a and b Two of the vertical Pinus sylvestris *pegs identified within the structure showing worked ends*

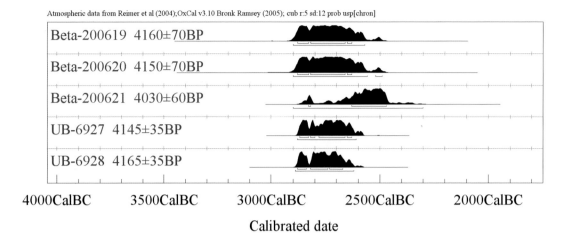

Figure 7.10 Calibration of the radiocarbon dates from the five samples taken from the trackway and platform structure

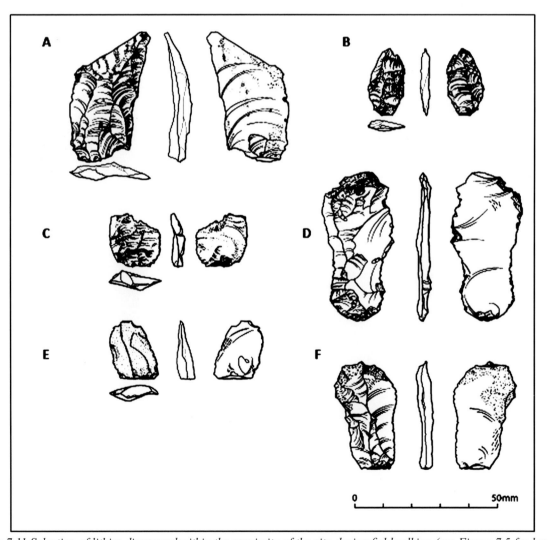

Figure 7.11 Selection of lithics discovered within the proximity of the site during fieldwalking (see Figure 7.5 for location)

Figure 7.12 Fossil beetles from the Hatfield trackway and platform site

the sands of the pre-peat land-surface indicates that initial sediment accumulation was in a shallow pool environment. These aquatic conditions were not clearly reflected in the Coleoptera record, which was dominated by species reflecting the landscape around the pool, which seems to have been a relatively dry, open heath with pine, oak and heather. The fauna from this sample is very similar to that recovered from other nearby basal deposits on Hatfield Moors (e.g. HAT3; Whitehouse 2004).

A number of beetles associated with trees include *Tomicus piniperda* which often attacks *Pinus sylvestris* (Koch 1992), a habitat which would also have suited the other wood borers identified (*Pityopthorus pubescens*, *Pityogenes bidentatus*). Of particular biogeographic interest is the non-British cossinine beetle, *Rhyncolus sculpturatus* which is an old forest or "wildwood" relict taxon, being associated largely with *Pinus*, but today restricted to some localities in central and southern Europe, being seldom found north of southern Fennoscandia. Whitehouse (2006) has argued that it is likely that climate change could at least partially explain its demise in Britain, although the disappearance of its undisturbed habitat and host is likely to have been the most influential part in its disappearance. *Sphindus dubius* indicates the presence of rotting fungoid wood (Palm 1959), whilst the leaf-miner *Rhynchaenus quercus* indicates the presence of *Quercus* trees. Heathland was also represented through species such as the heather weevil *Micrelus ericae* and *Olophrum piceum*. No strictly ombrotrophic taxa were identified in the sample, which implies eurytopic to mesotrophic conditions and a rather dry environment overall. However, a small number of aquatic taxa were identified including *Hydroporus obscurus* and *H. pubescens*, which are both found in bogs and standing water. The recovery of the dung beetle, *Aphodius sphacelatus*, which is found in all types of animal dung, indicates that animals periodically used the area for grazing.

Sample 2: (0.78–0.67m OD)
After this relatively dry heath phase, in which Pine and heather were important components, conditions abruptly became wetter and more ombrotrophic. There is just one beetle associated with deciduous trees (*Rhynchaenus fagi*, associated with *Fagus sylvatica*), probably originating from beyond the mire margins, although *Agonum obscurum*, found especially in deciduous and mixed forests, living in damp, shaded places among litter and moss, suggests stands of *Alnus* and *Fraxinus* habitats (*cf.* Lindroth 1986).

Several ombrotrophic taxa indicate the rise in the importance of this ombrotrophic conditions, including *Plateumaris discolor* and *Acidota crenata*. The former is associated with ombrotrophic conditions, living on *Eriophorum*, *Sphagnum* and on *Carex* species (Stainforth 1944, Bullock 1993), habitats also suitable for the latter taxon. *Pterostichus diligens* and *P. nigrita* are characteristic of wetland sites with peaty substrates. The *Calluna* feeding chrysomelid *Lochmaea suturalis* and heather weevil *Micrelus ericae* highlight the continuing importance of ericaceous habitats

Whilst heathland conditions are indicated, the sample contained a high number of aquatic and hygrophilous species including *Hydroporus nigrita*, *H. pubescens* and *Enochrus affinis*. Both members of the *Hydroporus* genus can be found in all types of stagnant water, often in shallow, temporary pools (Nilsson and Holmen 1995), whilst *Enochrus affinis* is more typical of bog pools, especially shallow acid waters (Friday 1988, Hyman 1992).

Sample 3: below archaeological structure (0.88–0.78m OD)
After the early phase of wetland development indicated by sample 2, there is evidence for a limited drying phase, with a number of heather indicators (e.g. *Micrelus ericae*). The number of water beetles remains small and is confined to just two taxa (*Hydroporus nigrita, H. angustatus*), corroborating the idea of a drier phase associated with this level. The presence of *Dryophthorus corticalis* would lend support to this, since it is an ancient woodland indicator usually associated with oak (Donisthorpe 1939, Hyman 1992), suggesting the presence of old trees growing on drier areas of the mire or in close proximity at this time. The state of preservation of the material also supports the suggestion of a drying phase, since some specimens showed differential preservation levels (e.g. pitting and fungal attack), typical of material that had been subjected to drying out in the past.

Samples 4 and 5: beneath the southern terminus of the trackway
This material was sampled from the landward edge of the trackway and reflects assemblages adjacent to the drier landscape of Lindholme Island. Taking the upper-most sample first (sample 4; 0–10cm depth), this includes a diverse range of ground beetles (*Dyschirius globosus, Pterostichus diligens, Agonum fuliginosum, A. obscurum*), many of which are typical of wetlands and peatlands. However, the features that distinguished this assemblage are elements associated with sandy heath conditions. This is represented by the presence of the heather weevil in large numbers, *Micrelus ericae,* but *Bradycellus harpalinus* and *Olisthopus rotundatus* are also present in small numbers. *O. rotundatus* is stenotopic of *Callunetum*, and is only found where there is a developed humus layer under *Calluna* (Lindroth 1945). *Bradycellus harpalinus* is an exclusively sandy ground species, also under *Calluna* (Lindroth 1986). These dry habitats would also have been suitable for *Bembidion lampros*. Nearby, habitats supporting dryland trees were present, as indicated by the weevil, *Rhynchaenus quercus*. A very diverse range of water beetles indicates that aquatic habitats remained important, however, perhaps in the form of vegetation-rich, standing-water pools. *Plateumaris discolor* indicates at least locally ombrotrophic conditions, living on *Eriophorum, Sphagnum* and *Carex* species (Stainforth 1944, Bullock 1993).

There is not a huge difference in the material across the two samples, except that the lower-most sample (sample 5; 10–20cm depth) includes a wider range of species associated with trees, including the pine-associate *Tomicus piniperda* and *Leperisinus varius*, which lives on *Fraxinus excelsior*. Other species typical of ancient woodland and rotting wood include *Dryophthortus corticalis, Cerylon histeroides* and two species of *Rhyncolus* spp. It is also noticeable that although dry heath habitats were important, the range of taxa that characterises the later material is in less evidence. Mesotrophic conditions are suggested by the presence of both *Plateumaris sericea* and *P. discolor*.

The material suggests that the lower samples were characterised by a more important woodland habitat (sample 5) whilst, later, this appears to have developed into more extensive *Calluna* heath (sample 4). Trophic conditions appear to have become more ombrotrophic by the time the upper sample was deposited. To some degree, both these samples bear a reasonable similarity to sample 1 (above), especially in terms of the tree-associated species that were clearly living in close proximity. However, the important sandy-heath community emphasises the localised conditions of the site close to sand-dunes, on the margins of Lindholme Island.

Sample 6: from above the archaeological structure
The sample from the peat associated with the surface of the archaeological structure is presumably contemporaneous with the site or the period shortly after. Surprisingly, there are relatively few species that appear to have arrived on the site transported in the wood used for construction, with no species indicative of rotting wood such as those that have been found in association with other prehistoric trackways (e.g. Thorne Moors; Buckland 1979, Sweet Track; Girling 1984). The exception is a specimen of the extirpated *Rhyncolus sculpturatus* that might have inhabited the pine wood of the trackway. Its presence is significant in that it indicates that the trackway was constructed of "wildwood" wood, as this ancient woodland relict is a poor flier. Heather and other dry sand indicators were present, as indicated by the heather feeder, *Micrelus ericae* and the ground beetles *Cymindis vaporariorum*. This often occurs in sandy moraine (Lindroth 1974), although Hyman (1992) lists its preferred habitat as peat and heather moorland, especially on sandy soils. Exposed sandy dunes would have created suitable habitat for this species.

It was also apparent that mesotrophic to ombrotrophic mire had begun to form by the time that the site was constructed. An increasing range of small aquatic taxa indicates an increasingly wet and ombrotrophic environment, including water beetles commonly found in *Sphagnum* pools (e.g. *Hydroporus scalesianus, H. obscurus*). Ombrotrophic environments are also reflected by the presence of the reed beetle *Plateumaris discolor*, which is associated with *Sphagnum* and *Eriophorum* (Stainforth 1944) and is typical of raised bogs (Koch 1992), although the presence of *Agonum fuliginosum*, which is typical of minerotrophic to mesotrophic bogs (Koch, 1989) suggests that conditions were not yet fully ombrotrophic at this time.

Summary
The fauna represented in the four samples from the site of the Hatfield trackway and platform bear striking similarities with those from Lindholme B (Whitehouse, Buckland, Wagner *et al.* 2001, Whitehouse 2004) (see also Chapter 5). This suggests that peat development in this area of Hatfield Moors was characterised by an initial dry- and then wet-heath phase (*cf.* Whitehouse 2004). The importance of heath habitats on Hatfield Moors has been emphasised by Whitehouse and suggests that the underlying sand dunes

remained significant exposed habitats during these early stages of peat development. As indicated by modelling (see Chapter 6), the evidence suggests that the trackway and platform were constructed at a time of increasing paludification of the local landscape and a transition from a predominantly dryland to a wetland environment; this will be discussed further in the next chapter (see Chapter 8).

7.4 Discussion

This is the first first archaeological site excavated on Hatfield Moors, and consisted of a corduroy trackway and platform built in a single phase towards the end of the Neolithic period, around *2730–2450 cal. BC*. The site comprised a 45m long trackway extending from the dryland of the northern edge of Lindholme Island across the edge of the peatland to a 13m wide platform of similar corduroy design. There is no surviving evidence that the site extended beyond the platform, although it is possible that peat cutting has removed any part of the structure which extended further to the north. It was constructed from *Pinus sylvestris* (Scots' pine) wood and *Betula* (birch) bark which was almost certainly acquired from the woodland which was growing locally.

The corduroy style of construction of the trackway and platform using transverse wood supported (or sometimes unsupported) by wooden rails is common on many sites dating from the Neolithic through to the Medieval period (Raftery 1996). The Hatfield Trackway and Platform is the earliest structure of this type in Britain, with only two earlier sites known elsewhere; the first in Lower Saxony in Germany (Hayen 1985, Metzler 1993) and the second in Nieuw Dordrecht in Holland (Casparie 1982, 1987, Casparie and Moloney 1992). These earlier examples, dating to the later fourth and early third millennium BC, were both more substantial than the Hatfield site, being between 3.5m and 4m in width respectively but without pegs, and were much longer (Raftery 1996). The surfaces of these sites were both relatively uneven and there is some evidence for the laying of turves over them to provide a more stable surface.

Other younger examples include the trackway from Cloonbony, Co. Longford, Ireland, dating to 2850–2470 cal. BC (GrN-16874; 4035±25 BP; Raftery 1996) and the Abbot's Way in Somerset, dating to around 2000 BC (Coles and Orme 1976, Coles 1980). The former was a 2.5m wide structure that consisted of pegged split timbers, whereas the Abbot's Way was just over 1m wide and over 2.5km long, constructed of split wooden planks and logs. The Bronze Age examples are generally narrower than the earlier structures, with a flatter, probably more stable surface, and it has been suggested that the changes in trackway technology on the Continent reflected an increased sophistication in the design of wheeled transport (Raftery 1996). The dating of the Hatfield structure places it later than the Continental examples, but earlier than these examples from the Somerset Levels and Ireland.

The unusual architecture and fragmentary preservation of the Hatfield trackway and platform presents something of a challenge for archaeological interpretation. The combination of archaeological, chronological and palaeo-topographic datasets, integrated together within the GIS, provided a foundation for its interpretation through the creation of a three-dimensional model of the local context of the site in the first half of the third millennium BC (Figure 7.13).

Comparison of the plan of the site and local topography revealed that the slight 'dog-leg' in the trackway corresponds with a small 'island' within the peat. Excavation over this area did not identify any associated structures or other features (such as tree stumps rooted in the underlying sands) which might indicate why a change of alignment might have been required. Instead it would appear that the orientation of the trackway was intended to avoid this 'island' rather than to take advantage of it. Given the proximity of the trackway to this slight 'island' in the wetland and the platform, it is curious that it was not used as the location for the platform, as arguably it would have provided an appropriate position for a functional structure, providing a stable footing for hunting activities such as fowling. The apparently deliberate avoidance of this feature indicates that the platform was sited with other considerations in the minds of its builders.

Secondly, the trackway runs almost parallel to a spur of what would have been dryland at the time of construction.

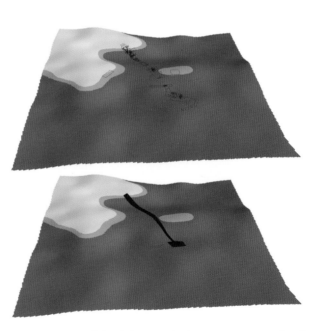

Figure 7.13 Model of the site in its contemporary local environmental context. The top image shows the excavated trenches and the remains of the trackway and platform, whilst the bottom image shows the interpreted shape of the monument (based on stratigraphic survey and palaeoenvironmental data; yellow indicates dryland, blue indicates pool deposits, and green indicates the wetland edge)

Significantly, this means that the shortest route between the platform (the destination of the monument) and nearest area of dryland was much less than the length of the trackway. The direct route from the edge of contemporary dryland along the trackway was 42m, whereas the closest distance between contemporary dryland (on the spur) to the platform was just 25m. If the sole purpose of the trackway was to provide access from dryland to the platform, then it could have been achieved with less effort and fewer raw materials. Instead, a longer route was chosen, running almost parallel to the spur of land to the east. This apparent lack of practical considerations echoes the interpretation of the dog-leg route of the structure above.

The lack of direct parallels, in addition to the range of unique structural and architectural details of the site, therefore means that the Hatfield Trackway and Platform is difficult to interpret from a purely 'functional' perspective. It was not a routeway linking areas of dryland (*cf.* Crockett *et al.* 2002), and the lack of appropriate finds makes it hard to interpret as a hunting platform (*cf.* Ellis *et al.* 2002). Elsewhere in a wetland context, the combination of a trackway leading to a platform has been interpreted as reflecting the need to first cross an area of fringing fen woodland to enable activities such as fowling and hunting from a platform on the edge of an open mire (e.g. Derryville, Co. Tipperary, Ireland; Gowen *et al.* 2005), However, no such fringing fen woodland was present at the site. The trackway terminates at the platform but this does not mark the edge of the pool within which it is situated, which extended some way to the north.

7.5 Conclusions

The Hatfield trackway and platform is the earliest corduroy structure outside of Germany and Holland, and is the first well-provenanced site to be recorded on Hatfield Moors. It was constructed at the end of the Neolithic period across and into a shallow pool of water. The unique architecture of the site cannot be explained in terms of traditional cost-effort considerations: the position of the platform avoids a natural 'island', the trackway is not straight, and it follows a considerably longer route than if it were merely to link the platform to dryland. The trackway seems to follow a route that is almost parallel to the drylands; close enough to maintain visual and audible links, but separated by the emerging wetlands – something with makes little sense in terms of hunting or similar activities. The trackway route was designed to define both how the platform was approached and also perhaps how it was seen from the dryland end. However, interpretation of the site must also draw on the wider context of rates and processes of environmental change and this will be discussed in the next chapter.

8. The 'hidden landscape archaeology' of Hatfield and Thorne Moors

8.1 Introduction

As outlined in Chapters 1 and 2, understanding peatlands such as Hatfield and Thorne Moors as archaeological landscapes is problematic since earlier phases are 'hidden' by the successive phases of peat accumulation. This is made all the more challenging by the fact that the conventional 'tool kit' of landscape archaeology cannot be easily applied to peatland landscapes. In this book an alternative methodological approach, combining various archaeological, palaeoenvironmental and chronological datasets and centred on GIS modelling, has been presented. The first two sections of this chapter present a discussion of the results of this approach, providing an overview of the evolution of the landscapes of Hatfield and Thorne Moors and exploring the implications of the rates and patterns of environmental change for past human activity and the archaeological record. The models of wetland inception, peatland development and the surviving resource provide a foundation for considering the future heritage management of these two landscapes. It concludes with a discussion of the results of these studies of Hatfield and Thorne Moors within the regional archaeological context for the periods represented by the archaeo-environmental record.

8.2 An integrated landscape archaeology of Hatfield Moors

The late Mesolithic period (pre-4000 BC)

Peat inception began much earlier on Hatfield Moors than previously identified (*cf.* Whitehouse 2004), with the accumulation of sediment within a shallow waterbody or pool from as early as 5470–5220 cal. BC (HM21; SUERC-8952) to the north of Lindholme Island (Figure 8.1). Predicting the lateral extent of these waterlain deposits is difficult, although the data (Chapter 5) indicate that the pool expanded steadily southwards such that by the Bronze Age it was lapping onto the northern edge of Lindholme Island. This shallow pool, or perhaps series of open pools, within a heathland vegetation mosaic is evidenced in the north of Hatfield Moors during the Mesolithic (see Chapter 2), but it is not known whether initial peat accumulation, which the peat inception model (Chapter 6) predicts for other locations across the Hatfield Moors, was also of this form. Rates of sediment accumulation during this period appear to have been slow, with perhaps just a few centimetres of vertical accretion over the period from the mid-sixth to mid-fifth millennium BC, although these deposits are likely to have undergone significant compression. Further peat formation had perhaps begun by the end of the Mesolithic in the area of the dunes system in the southern part of Hatfield Moors.

The character of the vegetation around the area or areas of open water during this period is provided by palaeoenvironmental evidence from the Neolithic trackway site (Chapter 7), with the beetle record indicating the presence of *Pinus* and *Quercus* woodland and *Calluna* heath with areas of exposed sandy substrate. It is unclear to what extent the perhaps relatively open, heath environment, which appears typical of the Moors during these earliest stages of wetland spread, was also that of the pre-peat landscape.

By the time wetland development had begun, the character of the vegetation across Hatfield Moors may have already been changing as a result of the rising watertables. The sub-fossil *Pinus* stumps rooted in the underlying sands and the woodland at Tyrham Quarry (Boswijk and Whitehouse 2002) demonstrate the presence of this tree as well as earlier populations of *Quercus* (see Chapter 2). A micromorphological study of pre-peat soil horizons from Hatfield described these as 'brown podzolic soils' and Dinnin (1997) suggested that the early Holocene vegetation across the area consisted of *Pinus* heath with some *Quercus*. A parallel may be drawn with the site of Routh, near Beverley, in the Hull valley, where pollen data indicate that the relatively open mixed woodland, in which *Pinus* and *Betula* seem to have been more prevalent in comparison to other sites in the region, persisted on sand and gravel-rich soils adjacent to the River Hull throughout much of the early to mid-Holocene (Gearey and Lillie 1999).

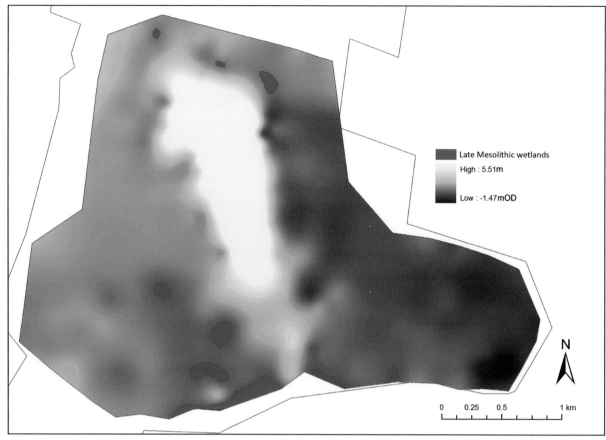

Figure 8.1 Wetland inception on Hatfield Moors towards the end of the Mesolithic period in the second half of the sixth millennium BC

The Mesolithic landscape of Hatfield was thus perhaps one of *Pinus* and *Calluna* on the relatively dry, sandy soils, but with the some areas of shallow acidic pools which appear to have gradually encroached from poorly drained parts of the landscape. The surrounding dryland areas were probably dominated by dense *Tilia–Quercus–Corylus* woodland, although data specific to this period and location are also lacking. Human activity may have been focussed on early locations of wetland development, as has been identified at sites elsewhere within the Humber wetlands where lithic scatters suggest a concentration in areas adjacent to water bodies, such as rivers and pools, which has been argued to reflect a 'specialist' use of wetland resources (Van de Noort and Ellis 1997, Dinnin and Van de Noort 1999, Van de Noort 2004). However, there is currently no archaeological evidence of human activity on Hatfield Moors during the Mesolithic period, although there is potential for the preservation of such sites, perhaps in the areas to the north and south of Lindholme Island.

The evidence for the local burning of the mire surface of Hatfield Moors during the earliest stages of peat accumulation has been much discussed (e.g. Dinnin 1997a and Whitehouse 2000) and consists of both direct (charred macrofossils) and indirect evidence (the presence of pyrophilous coleoptera). Direct evidence of burning was represented most dramatically by the areas of burnt *Pinus* and *Betula* trees, which have been estimated to cover an area of 5–6km^2, mainly in the Tyrham Hall quarry area, but also at Kilham West and Packards North. Charred trees have also been reported on the margins of Lindholme Island (see Whitehouse 2000). None of the dendrochronologically dated *Pinus* from Tyrham Quarry had fire scars (Boswijk 1998), suggesting none of these trees had survived earlier fires. However, it is clear that at least some of this material was burnt in antiquity, prior to submergence in the peat (either during the trees' lives or shortly after their deaths), rather than as a result of more recent peat fires. It would appear likely that the woodland was dead or dying by the time it was burnt. There is no direct dating of the main burning events which resulted in the charred trees reported across Hatfield Moors.

The clearest and most closely dated indirect evidence of burning is in the form of the macrofossil material preserved in the peat which has generally been either radiocarbon dated directly or is in a stratified, dated sequence. The macrofossil samples from the basal sediment units at HAT1, HAT2 and at Porters Drain were characterised by abundant charcoal fragments. Likewise, many of the radiocarbon samples for the current study (see Chapter 5; Appendix 1) were obtained from charred macrofossil material, including ericaceous material, twigs and herbaceous material at the base of the remaining peat archive.

The range of dates that directly relate to burning or burnt material at Hatfield Moors demonstrates burning

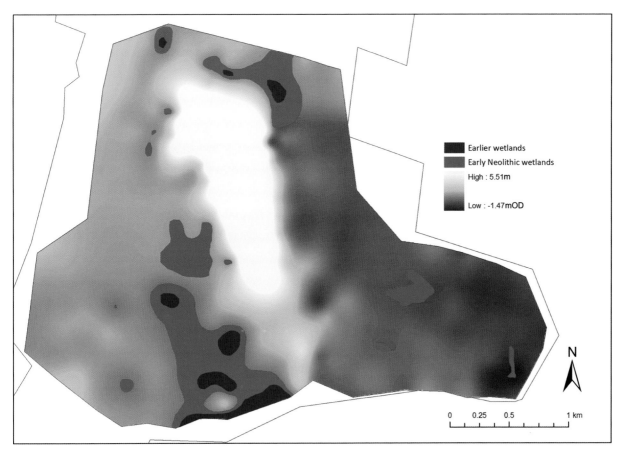

Figure 8.2 Wetland inception and peat spread on Hatfield Moors by the early Neolithic period in the mid- to late fourth millennium BC

from the earliest stages of wetland formation in the north of the Moors in the fifth millennium BC through to the main phases of peat growth at around the middle of the fourth millennium BC (*c.* 3700–3300 cal. BC). It is therefore clear that the burning of the heath and woody vegetation must have occurred on the mire surface during the Mesolithic period and the earliest stages of peat accumulation up until the widespread development of ombrotrophic conditions in the Bronze Age, although the relative frequency of such events is difficult to discern. Smith (2002) regarded the evidence for burning at both Hatfield and Thorne Moors as indicating the deliberate use of fire by human communities to clear the woodland, a contention which may be partially supported for later periods at least by a radiocarbon date of 2120–1691 cal. BC (CAR-313; 3545±70 BP) from a trunk of *Pinus sylvestris* from peat in the vicinity of Crowle Moor which is described as having been burnt and chopped. De la Pryme also reported the discovery of burnt and chopped trees associated "with great wooden wedges and stones in them, and broken Ax-heads" (de la Pryme 1701).

The earlier Neolithic period (*c.* 4000–3300 BC)

The model of peat inception indicates that peat spread during the earlier Neolithic period (Figure 8.2) was primarily across areas to the north and south of Lindholme Island, perhaps as the localised pools of the Mesolithic landscape coalesced. The palaeoenvironmental record implies peat accumulation under relatively dry *Calluna–Eriophorum* heath (see Chapter 5), with wetter areas in localised depressions such as between the sand dunes in the south. Higher and drier areas of sandy soils remained peat free at this time, probably accounting for the combination of both wet and dry habitats reflected in the palaeoentomological record (Whitehouse 2004, Whitehouse *et al.* 2008).

Trees such as *Pinus, Betula* and *Alnus* were apparently also growing in places on the pre-peat landscape with mixed *Tilia–Quercus–Corylus* woodland on the dryland beyond and probably also Lindholme Island itself (see Chapter 5). Woodland must have been under increasing stress from the rising watertables and, as peat spread, a steady process of die-back can be envisaged with trees increasingly restricted to the dryland fringes and any other areas of topography that were sufficiently elevated to permit continued tree recruitment. It is in this period that a distinction seems to emerge between relatively wet northern and southern sections of Hatfield Moors and drier western and southeastern areas, with limited peat spread and hence possibly greater tree cover (Chapter 6).

The mosaic of woodland, heath and wetland environments during this period would have provided a range of different affordances to human communities (*cf.* Ingold 1992). The contrast between the naturally more open vegetation on Hatfield Moors compared with the denser deciduous

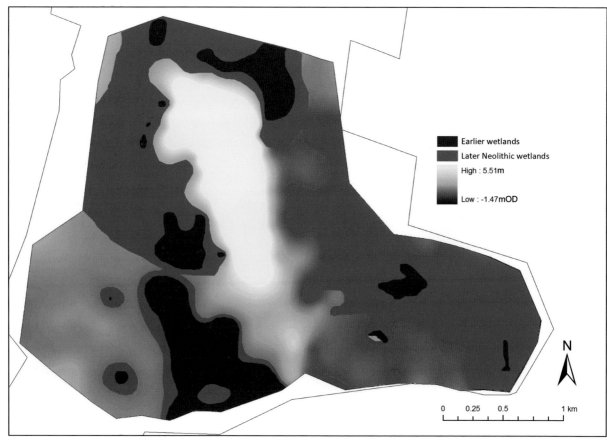

Figure 8.3 Wetland inception and peat spread on Hatfield Moors by the later Neolithic period in the first half of the third millennium BC

woodland of the wider region would have made this area readily distinctive and might also have proved attractive to local early Neolithic populations. Reports of "axes" and other finds recovered during peat cutting in the past (Chapter 2) provide further evidence of a human presence at this time, but otherwise the precise nature of human activity on and around Hatfield Moors at this time remains somewhat elusive.

The relatively dry nature of the peat accumulating environment is reinforced by the evidence for the burning of this surface which continued until at least 3700–3300 cal. BC (see above). It is equivocal whether this represents deliberate burning by human communities, but it is likely that the surface of the peatland was dry enough, seasonally at least, to permit human movement and access. Hence, from a purely functional perspective, it is possible that other than in areas such as to the north of Lindholme Island and the southern area around Packards South where open water was present, there was no need for structures such as trackways to access or exploit the resources of the expanding wetlands. As for earlier periods, the archaeological evidence from the wider region again indicates a focus on locations closely associated with watercourses, areas of open water and wetlands (Van de Noort *et al.* 1997, Van de Noort 2004).

The later Neolithic period (*c.* 3300–2500 BC)

The model implies that, by the later Neolithic, peat had spread across much of Hatfield Moors (Figure 8.3) with dryland probably limited to the southwestern and southeastern corners, and the eastern edge of Lindholme Island. The HAT2 sequence in Porters Drain in the southeast of the study area (Smith 2002; chapter 5) indicated peat inception in a *Betula–Alnus* fen environment from the later fourth millennium (CAR-254; *3340–3210 (9%) or 3180–3150 (1%) or 3130–2860 (83%) or 2800–2760 (2%) cal. BC*) with a transition to *Calluna–Eriophorum* heath around the early–mid third millennium BC (CAR-255; *2950–2580 cal. BC*, Chapter 5). Whilst there are few continuous sequences to reconstruct the spatial variation in vegetation, heath of this character appears to have been typical of the later Neolithic environment across much of the study area.

Whilst the southwestern corner of Hatfield Moors probably remained peat free towards the end of the Neolithic, peat was expanding towards and beyond the western edge of the study area. The *Pinus* woodland growing at Tyrham Hall quarry was in the process of dying back due to paludification from before 2921 BC, with an episode of stress on the woodland identified at 2857–2855 BC. Mire deposits appear to have spread rapidly in this area, resulting in the death of woodland by the end of this period, by *c.* 2445 BC, and its subsequent preservation in the basal peats (Boswijk and Whitehouse 2002, chapter 2). There is also evidence that ombrotrophic mire may have begun to develop somewhere to the north of Lindholme Island (Chapter 5) but the extent and precise location of such environments remain unclear.

The area directly to the south of Lindholme Island probably remained dry and hence any routeways extending southwards would probably have remained open, although increasingly restricted by the rising watertables. Rates of peatland spread to the east of Lindholme Island are less clear, although the data from adjacent areas imply that peat growth was underway in the region of Packards South and the southeastern corner of the Moors. It was during this period of landscape change that the trackway and platform was constructed just to the north of Lindholme Island, the only excavated archaeological site on Hatfield Moors (see Chapter 7).

To fully appreciate the significance of the form and location of this site it is necessary to consider the processes and patterns of environmental change across this period. Previous palaeoenvironmental study on the Moor had indicated that the initial peat growth at Hatfield Moors began around 3000 cal. BC, with the earliest date of 3350–2930 cal. BC from Kilham West (Whitehouse 2004, Whitehouse, Buckland, Boswijk *et al.* 2001; see Chapter 5). The radiocarbon date from the base of the peat to the east of the site of 5470–5220 cal. BC (SUERC-8952; 6355±35 BP) is therefore the earliest available from Hatfield Moors for wetland inception and indicates that this area in the northern part of Hatfield Moors was one of the earliest foci for peat formation (see Chapter 6). By the later Neolithic, this pool was encroaching on the northern extent of Lindholme Island (SUERC-8877, 4250±35 BP; 2920–2760 cal. BC). Dendrochronological analysis shows that *Quercus–Betula–Pinus* woodland was growing on the western edge of the moors at Tyrham Hall quarry between 3618–3418 BC (Boswijk and Whitehouse 2002). Elsewhere, *Calluna* and *Eriophorum* rich basal peats indicate areas of drier heath (Whitehouse 2004), which is reflected in the coleoptera from the basal deposits at the trackway site. From around the fifth millennium BC onwards, it is thus clear that a shifting mosaic of habitats developed across Hatfield Moors as the effects of rising watertables affected different areas of the landscape at different rates.

The site was constructed during a period in which the area of Hatfield Moors was undergoing a transformation. The evidence from the coleopteran analyses demonstrated that the site was constructed at a time of increasingly wet and acidic conditions, with ombrotrophic mire possibly also developing close to this location. Modelling of the wider landscape (Chapter 6) indicates that, by the later Neolithic, continued paludification in the vicinity of the trackway site had led to the spread of wetlands across greater areas of the Moors (Figure 8.4).

The modelling (Figure 8.5) also illustrates that the structure was built across and into the pool that marks the earliest known area of wetland development, at what was then the northernmost edge of Lindholme Island. The *Pinus* dominated woodland was in places dying back by this time, although some of the higher areas probably supported *Pinus* heath, probably with mixed deciduous woodland on Lindholme itself (Figure 8.6). It could be suggested that local Neolithic populations must have been aware of the changing landscape, both in a practical sense, in terms of changing availability of resources and movement through the landscape, and in terms of their perception of place. The emerging wetlands would have created a more open landscape which would have been distinctive regionally.

Consideration of the form and function of the site must therefore be made within the context of these processes of change. The trackway and platform were constructed of locally available wood, but the selection of this material probably had a deeper resonance to the builders. Birch bark is attested in a wide range of contexts elsewhere during the Neolithic, and was used to make containers such as bowls, whilst birch bark tar was used as a waterproofing agent, and as chewing gum (Pollard and Heron 1996). It also appears to have had particular symbolic and ritual associations; deposits of birch bark were also found in the ditches of the Etton causewayed enclosure (Taylor 1998) and there is evidence that it was possibly burnt during the Neolithic as a form of incense, and although its odour is unpleasant, this may have served to mask the odour of decomposition during funerary rituals (Lucquin *et al.* 2007). In addition, *Betula* is a seral species, and is often one of the first taxa to re-colonise previously cleared or disturbed areas of woodland; peaks in *Betula* are recorded in the pollen diagrams from the Humberhead Levels following prehistoric human disturbance to woodlands (e.g. Gearey *et al.* 2009).

After the site was built, the whiteness of the bark must have been quite visible against the greys and browns of the pool muds, the peat and the wood. Anthropologically, 'whiteness' has been identified as relating to notions including social cohesion and continuity, unconcealed openness and birth (Turner 1966). The materials used clearly reference the previous dryland woodland which had been dominated by *Pinus*, with the *Betula* bark perhaps representing the stands of birch occurring naturally within the woodland, especially on areas of early wetland spread (Chapter 5) but probably had additional resonance. It could be argued that the use of birch bark of this tree might have had strong symbolism connected to the "re-establishment" of the natural order following disruption, whether by human or natural agencies.

The trackway was also designed to narrow – both in width along its length and also in the diameters of the poles used in its construction. The narrowing of the trackway might also have provided a specific visual impression of the site from its dryland terminus, making it appear longer and "grander" (so called "forced perspective"). In addition, the narrowing would have controlled the movement of people along it. Effectively, whilst the landfall end of the trackway would have allowed several people to walk abreast, the width of the structure where it meets the platform would

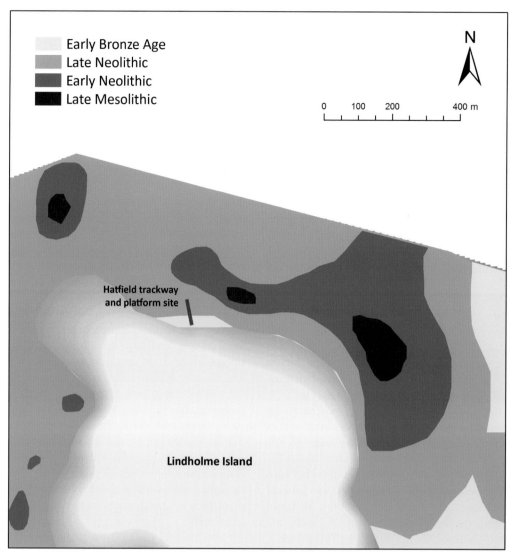

Figure 8.4 Location of the Hatfield trackway and platform in relation to wetland inception and peat spread within the local area

have enabled single file access to the platform. The choice of location for the platform was significant, avoiding the more "practical" island of drier land and instead located at the edge of a deeper area of the pool. The whole structure was situated almost parallel to a dryland ridge which would have allowed people on this area to view activities on the structure, whilst being physically separated from it by the wetland.

This interpretation therefore stresses the site as an arena for performances or ceremonies. The site was located and designed within the context of a community's response to the changing landscape within which it had lived for generations. The platform was reached from what at that time remained the edge of the dryland, which had once been the "natural" order but which was steadily being lost to the inexorable rise of the waters and the development of raised mire perhaps to the north of the site (Chapter 5).

This destination was reached along a trackway which was not just a mode of access but represented a journey through the local landscape, as it once was, into what it was becoming: a liminal place at the transition between wet and dryland (see Bond 2010). The order of arrival of the participants at the platform was strictly controlled by and marked by stepping over a "threshold" marked by birch bark which may in itself have represented renewal or re-birth. "Passage across this threshold may have constituted a movement between arenas of value" (Edmonds 1993). Those excluded from whatever ceremony or ritual was being enacted were physically separated by the waters around the platform, but could however observe from the overlooking dryland adjacent to the structure.

Whilst no structural parallels seem appropriate for the interpretation of the Hatfield trackway and platform, a number of thematic parallels may be drawn from interpretations of contemporary dryland monuments across Britain and northwest Europe. A common theme is that of public ritual and performance (*cf.* Sahlins 1985), and the creation of public arenas to both facilitate and validate such rituals (Bradley 1998). Similarly, monuments interpreted in this way commonly seem to provide some level of separation between the "performers" and the "audience", defined by the architecture of the site. It has been argued

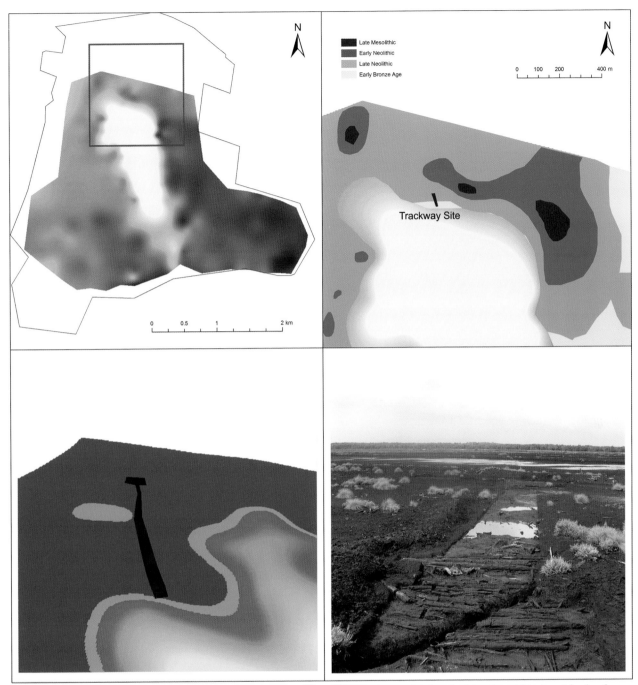

Figure 8.5 The Hatfield trackway and platform site. The upper-right image provides detail of the site within its wider landscape context. The lower left image shows the site in its local context, with areas of the wetland (dominated by pool deposits) and fringe areas

that the internal ditches on many henge sites provided just such a separation between the rituals that take place within the interior and the audience occupying the surrounding banks (e.g. Bradley 1998, 2003).

A further common theme in the monumental architecture of the period is the construction of avenues leading into such areas for performance, often with significant "thresholds" dividing the inside from the outside of the monument. Furthermore, these can have dog-legs and elements of visual impressiveness such as through forced perspective. For example, the West Kennet Avenue's southern approach to the Sanctuary at Avebury, Wiltshire, includes both a dog-leg in its route and a significant narrowing as it reaches its destination (Ucko *et al.* 1991).

Whilst the site is very different in form and scale to other structures, these themes are arguably reflected within the Hatfield trackway and platform. The structure provides an avenue leading to an area for performance, facilitated by a visually impressive and narrowing monumental approach with thresholds marked by birch bark. Audiences would have been facilitated by the adjacent dryland, running parallel to the trackway, but

Figure 8.6 Pinus *trunk preserved towards the base of peat on Hatfield Moors. Although this has not been dated, it is likely that this tree was part of the woodland that was subsumed during wetland inception and peat spread across the northern part of Hatfield Moors during the later Mesolithic or Neolithic periods*

physically separated from the performance by the water. The broader context is one of a landscape in longterm flux and change.

A parallel may be drawn with the interpretation of Picts Knowe henge, Dumfriesshire, Scotland (Tipping *et al.* 2004). In this case, it was argued that the henge was constructed in the early Bronze Age at the juncture of an expanding wetland, the "wildscape", with the "tamed" agricultural land. As at Hatfield, the role of water is implicated in this scenario, with rising groundwater, flooding, the spread of peat and the loss of land. Richards (1996) has argued that henges were built to include water within their ditches, both as a means of stressing "inside" and "outside" of the monument, as well acting as a linking metaphor with nature.

Bond (2010, 51) has also postulated (in the context of the Somerset Levels) that during this period access to a supernatural ancestor may have been "... granted at a watery margin, set between the two worlds: wet and dry." In this context, the apparent location of the site close to what may well have been the earliest development of raised mire during the later Neolithic, not only on Hatfield Moors but in the wider region might be highly relevant. This environment must therefore have been very distinctive and rather alien to local populations. The possible later perception of raised mire environments as liminal and dangerous and associated ritual deposition of metalwork in this area (Van de Noort 2004) may therefore be rooted in these formative stages of mire inception and spread which ultimately led to the submergence of significant areas of dryland.

The site helps to fill a gap in understanding of the later Neolithic in this region which currently polarises between the monuments of the Wolds (e.g. Manby 1988, Chapman 2003, 2005) and the ephemeral traces of communities in the lowland (e.g. Van de Noort and Ellis 1997). Furthermore, the site appears to represent a new type of monument, arguably echoing themes observed in contemporary megalithic sites elsewhere, but of a specific form in response to local conditions and context. The architecture of the site cannot be fully understood without reference to the contemporary Neolithic landscape and the patterns and processes of environmental change during this period. This also highlights the difficulties of fully understanding monuments on dryland areas for which such contextual environmental data are often lacking.

The earlier Bronze Age period (*c.* 2500–1500 BC)

Peat had spread across most of the pre-peat landscape of Hatfield Moors by the earlier Bronze Age, with few areas of dryland surviving above the wetlands, although a few isolated drier 'islands' may have been present along the western edge of Hatfield Moors (Figure 8.7). Lindholme Island had been severely encroached upon by the growing mire. By the end of the earlier Bronze Age, movement across the landscape of Hatfield Moors would have become problematic except for southward from Lindholme Island, where the modelling suggest that dryland had narrowed to a spit perhaps 150m across leading towards the River Torne, although it would appear that by this time that floodplain wetlands had spread across the floodplain of the river Torne, further hindering any movement south, towards the higher and presumably drier land currently occupied by the village of Wroot on the southern side of the River Torne.

Peat accumulation continued under relatively dry conditions in this period, at least in the southern section of Hatfield Moors for which palaeoenvironmental data are

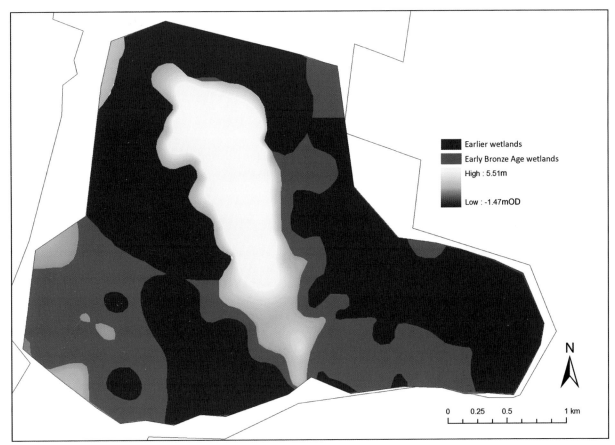

Figure 8.7 Wetland inception and peat spread on Hatfield Moors by the early Bronze Age period in the second half of the third millennium BC (derived from the data presented in Figures 6.11 and 6.12)

available, although wetter habitats are indicated for some locations. Access on foot for parts of these wetland areas might therefore have been possible, although increasingly problematic as wetter and deeper peat continued to accumulate. Once full ombrotrophic mire growth had commenced, from the middle of the second millennium BC with a relatively rapid transition from fen to bog, it can be hypothesised that the surface would have been too wet and unstable for unimpeded movement through the landscape.

The GIS model of the age of the contemporary peatland surface indicates that the majority of the surviving deposits do not extend much beyond the middle of the second millennium BC, although the data are of relatively low resolution and areas of later peat may survive locally. It is possible that any archaeological remains dating to the later part of the Bronze Age will be located at or near to the current surface of the peatland, although no archaeology was discovered during previous fieldwalking programmes across Hatfield Moors (Van de Noort and Ellis 1997, Gearey 2005).

The later Bronze Age period (*c.* 1500–700 BC)

The modelling of peat inception (Chapter 6) indicates that, by the later Bronze Age, the earlier heath and woodland habitats of Hatfield Moors had been entirely replaced by ombrotrophic mire, with woodland excluded to the dryland fringes and the higher parts of Lindholme Island. The palaeoenvironmental records and associated chronological modelling (Chapter 5) demonstrate something of the spatial and temporal variation in this process. The earliest date for raised mire growth of *1670–1380 cal. BC (CAR-257)* was recorded at HAT2 in Porters Drain, followed by that at HAT1 at *1120–840 cal. BC (CAR-170)* on the western edge of Lindholme Island. The longest extant record for pre-ombrotrophic vegetation was at HAT1 (Smith 2002) on the western edge of Lindholme Island where the *Calluna* and *Eriophorum* heath had persisted for *c.* 1800 years. At HAT2 (Smith 2002) *Betula* fen was replaced by a similar *Calluna* heath community within perhaps less than 200 years of the beginning of peat accumulation, with ombrotrophic peat development *c.* 1280 years later. The chronology is problematic at HAT4 (Whitehouse 2004, chapter 5) on the southwestern fringes of the Moors, but the transition from mesotrophic to ombrotrophic conditions appears to have been relatively rapid here as well.

These palaeoenvironmental sequences illustrate something of the spatial and temporal variation in vegetation during mire development but it is not possible to model the transition to ombrotrophic mire on a landscape scale in any meaningful way. It is clear that at the southern end of the Moors, the "switch" to ombrotrophic conditions took place within a reasonably narrow period of between 1000–1300 cal. BC and this may have been a relatively rapid transition.

The pollen, macrofossil and testate amoebae sequences at HAT1, HAT2 and Porters Drain (Smith 2002, Gearey 2005; see Chapter 2) show a change from macrofossil samples dominated by *Calluna, Eriophorum* remains to *Sphagnum* over two sample levels (0.05m interval at HAT1 and HAT2). The palaeoentomological record (Whitehouse 2004) also suggests an abrupt transition, although identifying the precise rate is more difficult given the relatively wide range of sampling intervals for Coleoptera analyses.

Other data indicate the transition to ombrotrophy indirectly; no radiocarbon dates are available from HAT3 (Whitehouse 2004), but this is recorded at c. 0.10m above the basal date for peat accumulation (3350–3030 cal. BC; see above), which may also suggest a relatively abrupt transition. Smith (2002) regarded ombrotrophic mire development at both Hatfield and Thorne in the mid-second millennium BC as evidence for climatic deterioration. The modelling of the palaeoenvironmental sequences (Chapter 5) indicated that the four horizons regarded as demonstrating this HHL V 'recurrence' surface were not synchronous, although this does not necessarily demonstrate that climatic factors played no role in this development.

The earliest *direct* evidence (in the form of *Sphagnum* deposits with associated dating and palaeoenvironmental analyses) for ombrotrophic mire development is therefore in the southeastern section of the Moors (HAT2) although, given the comparatively long period of heath vegetation recorded here, it would seem that earlier dates for peat inception do not necessarily equate to the most *rapid* transition to ombrotrophy. A comparison on the basis of altitudes or precise topographic contexts was not possible, but the earliest date is in the central part of the peatland at HAT2, with the slightly later dates at HAT1, Porters Drain and finally HAT4.

There are insufficient data to postulate any clear relationship between the age of peat inception and the subsequent transition to ombrotrophic raised mire. Therefore, it is difficult to assess the implications of these landscape changes for human populations. On the basis of the current understanding, it seems likely that, during the later Bronze Age, any routeways north towards or from Lindholme Island would have been lost as ombrotrophic mire began to grow. However, to the south of the island, a route towards the floodplain of the River Torne may have remained open, although this area was certainly dominated by floodplain wetlands by this time (*cf.* Mansell 2011a, 2011b). Hence, Lindholme Island would have remained accessible towards the end of the Bronze Age but perhaps became finally 'cut off' entirely during this period. Elsewhere, the wider landscape of Hatfield Moors probably became less and less accessible as ombrotrophic peat continued to accumulate (see Chapter 6). Human activity is thus most likely to have been confined to the contemporaneous margins of the peatland but any archaeological sites representing past human activities at this wetland-dryland interface will have probably been lost to peat cutting or are now concealed beneath alluvial deposits associated with later flooding or agricultural improvement ('warping', see Chapter 2).

The pollen data (Chapter 5) indicate that human clearance activity began to have a detectable impact on the woodland around Hatfield during the later Bronze Age with increases in *Plantago lanceolata* recorded from an estimated date of *1540–1280 cal. BC* (95%; see Chapter 5) in the HAT2 pollen sequence (Smith 2002), although there is no evidence for large scale clearance of woodland until later. Any settlement and farming activity in the close vicinity was thus taking place on the edges of a large raised mire system that would have been largely inaccessible without the construction of trackways. Elsewhere, such human activity on the margins of raised mires can be seen in the spatial distribution of trackway structures within landscapes such as Derryville, Co. Tipperary, Ireland (Gowen *et al.* 2005). The apparently rapid shift to ombrotrophic mire at Hatfield Moors may be highlighted here as this transition may have been sufficiently rapid to have been perceptible to human communities, over the course of few generations if not an individual's lifespan (see Van de Noort 2004).

In this context, finds of possible Bronze Age artefacts from Lindholme (Eversham *et al.* 1995; see Chapter 2) may raise other questions. If there was human activity on Lindholme Island at this time, then it is perhaps most likely that the area was accessed from the south, assuming that the floodplain of the River Torne was traversable. Alternatively, given the lack of data regarding the spatial extent of ombrotrophic deposits at this time, it is possible that there were open routeways across the mire, possibly with the assistance of structures such as trackways, although it must be stressed that there is no evidence of this.

The Iron Age and later

For the majority of Hatfield Moors, peat cutting has removed post-Bronze Age layers and hence the potential for the preservation of cultural remains dating to corresponding periods must be regarded as extremely low (see Chapter 6). The modelling indicates that some Iron Age peat might survive within the northeastern corner of the Moors and that deposits extending into at least the earlier Medieval period survive within the southern part of the Moors. The focus in terms of a landscape archaeology, therefore shifts to the evidence for landscape changes derived from previous palaeoenvironmental study (Chapters 2 and 5). The Bayesian modelling (Chapter 5) of the chronologies of these previous studies has provided information regarding the development of the mire system and the wider landscape during this period.

In terms of Hatfield Moors, all of the existing palaeoenvironmental data imply that raised mire would have been well established across much, if not all, of the study area by the end of the Bronze Age, although as the evidence from Thorne Moors (see below) demonstrates, ombrotrophic mire continued to expand at the peripheral areas of the peatlands into the later prehistoric period. It is therefore difficult to assess the potential implications of mire development for human activity on Hatfield Moors during this period. Elsewhere finds such as bog bodies have been made in raised mires, many of which date to the Iron Age and later (see Chapter 2). Given the very low occurrence of these discoveries generally and the limited amount of Iron Age peat surviving, it seems unlikely that any such finds will be made on Hatfield Moors in the future.

Previous finds of bog bodies during the eighteenth and nineteenth centuries, possibly dating to later prehistory and subsequent periods, are poorly provenanced (see Chapter 2) but are clearly associated with the cutting of the shallower peats. Future discoveries of archaeological remains on Hatfield Moors dating to the post-Iron Age period are unlikely for two main reasons. Firstly, the landscape was extensive ombrotrophic mire by this time, as indicated by the isolation of Lindholme Island, apparent in references to the inaccessibility of this area during the Post-Medieval period (*cf.* Whitehouse *et al.* 2001). Secondly, aside from limited deposits on the southern edge of the Moors, very little post-Iron Age peat survives. Whilst there is some potential for the preservation of archaeology in these areas, extensive wetlands existed at the interface between the raised mire and the floodplain of the River Torne (Mansell 2011a, 2011b). It seems likely that by this date these southern and eastern fringes of the wetlands remained effectively impassable.

8.3 Thorne Moors: complex patterns of mire development

The programme of work on Thorne Moors was more limited than on Hatfield Moors. A model of the pre-peat landscape was generated using peat depth data, but no new radiocarbon dates were obtained to assist in modelling the spatial and temporal pattern of peat inception (see Chapter 6). The following section therefore presents a general summary of previous palaeoenvironmental studies in the light of the contextual information provided by the model of the pre-peat landscape (Chapter 4) and the results of the chronological modelling (Chapter 5). The results are considered in relation to evidence for human activity in both the palaeoenvironmental and archaeological records. Figure 8.8 presents a summary of previous palaeoenvironmental evidence for mire development.

The Mesolithic period (pre-4000 BC)

Prior to the onset of peat accumulation, the landscape would probably have been one of mixed woodland in which *Quercus* was perhaps the predominant tree on the clay rich soils, although *Betula* and *Alnus* fen woodland is also recorded for this early period with a range of other trees and shrubs indicated (Smith 1985, 2002). However, the precise character and spatial variation of the vegetation at this time is unknown. Archaeological finds from this period include a tranchet axe and lithic flakes from beneath peat deposits at Nun Moor to the south of Thorne Moors, on the edge of the River Don, outside of the study area (see Chapter 2). As for the Mesolithic finds from the wider region, these appear to be associated with watercourses and areas of either open water or wetland (e.g. Van de Noort and Ellis 1997, Van de Noort 2004). The probable palaeochannel at Middle Moor/Crowle Moor may have been taking flow during this period and, as such, could have provided one such focus, although any potential archaeological evidence would be deeply buried beneath the peat.

The Neolithic and Bronze Age periods (*c.* 4000–700 BC)

Peat accumulation began within a *Betula–Alnus* fen woodland community during the earlier Neolithic in the topographically low area of Rawcliffe Moor in the northern part of the study area (Figure 8.9). By the end of the Neolithic period, much of the northern and central areas of Thorne Moors were becoming wetland (Figure 8.10). The basal date of *3500–3430 (6%) or 3380–2920 (89%) cal. BC* (*CAR-221*; Smith 2002) is the earliest available for peat formation on Thorne Moors and compares closely with that from the base of the Middle Moor sequence (Gearey 2005; Figure 8.8), although the chronological modelling of the latter sequence suggests that the reliability of this estimation is uncertain (Chapter 5). At Middle Moor, the accumulation of sediment within an *Alnus* fen system continued, apparently uninterrupted, for perhaps some 1200 years, demonstrating that *Alnus* fen carr was a longer lived stage in the succession to raised mire at Thorne Moors than previously identified (Smith 2002). The sequences from Crowle and Middle Moor are stratigraphically and biostratigraphically similar, both indicating initial sediment accumulation within shallow, standing or sluggish water flow.

The pre-peat landscape model indicates that the peat accumulation at Middle Moor might have begun within a palaeochannel, possibly linked to a similar feature at Crowle Moor (Smith 2002). These may represent palaeochannels, perhaps of the River Don, which cuts through the Magnesian Limestone ridge west of Doncaster and hence would have introduced calcareous waters into the southern and central areas of Thorne Moors, delaying the onset of acidification. Previous researchers have drawn attention to the influence of calcareous waters in the early stages of Thorne Moors: Rogers and Bellamy (cited in Smith 2002) reported the presence of a basal fen peat in the "Canals" region of the Moors with abundant remains of both *Phragmites* and the calcium demanding *Chara*. However, the context of these deposits is unclear and neither this project nor other recent work (Smith 2002) on the Moors has encountered any similar sediments.

The evidence from Middle Moor demonstrates that the transition to fen then raised mire vegetation followed the infilling of the depression. At present there is no direct evidence from the basal model to indicate that these features were linked to one another or to either the adjacent rivers Don or Went. They may represent two of

Date (Cal. BC)	Crowle Moor (CLM1)*	Thorne Moors (TM1)*	Middle Moor**	Goole Moor (GLM1)*	Rawcliffe Moor (RWM1)*	Dendro-evidence***
0	Ombrotrophic mire 340-65 cal. BC					
500	Betula fen	Ombrotrophic mire	Ombrotrophic mire			
1000		1250-910 cal. BC	(1070BC)	Ombrotrophic mire		
1500		Betula-Alnus-Quercus Fen 1425-1275 cal. BC	Betula-Pinus fen (1600BC)			Pinus (1690-1489 BC)
2000	Quercus-Alnus fen 1670-1245 cal. BC			1920-1680cal.BC	Ombrotrophic mire	Pinus (GM01) 2227-1810BC
2500					2790-2415 cal. BC	Pinus (PISY) 2921-2445BC
3000			Alnus-Quercus fen (3140BC)			
3500				Betula-Alnus fen 3500-3430 cal. BC (6%) or 3380-2920 cal. BC (89%)	Betula-Alnus fen 3380-2920 cal. BC	Quercus woodland 3777-3017BC
4000						

Figure 8.8 Summary of selected previous palaeoenvironmental evidence for mire development on Thorne Moors. ---------------- = transition to ombrotrophic mire, * = Smith (2002), ** = Gearey (2005) *** = Dendrochronological data (Boswijk 1998). Dates in italics are posterior density estimates and dates in brackets indicate uncertain accuracy (see Chapter 5)

a number of localized depressions in the surface of the pre-peat landscape, which paludified as a result of rising local base levels. The later accumulation of sediment within *Betula–Alnus* fen is recorded at Thorne Moor (TM1) at *1425–1275 cal. BC*, as paludification affected the higher parts of the landscape.

The dendrochronological study of Thorne Moors (Boswijk 1998, Boswijk *et al.* 2001) illustrated the growth of mature, mixed oak woodland between 3777–3017 cal. BC (TM01) which apparently overlapped with the spread of *Alnus–Betula* fen across the northern part of the Moors. The pattern during the earliest phases of wetland development thus seems to have been one of *Alnus* and *Betula* dominated fen growth in topographic low points after *c.* 3150 BC, whilst mixed *Quercus* and perhaps *Tilia* woodland, which was probably prevalent over much of the area prior to peat growth, persisted on the higher areas (e.g. Boswijk *et al.* 2001). It has been suggested that mesotrophic fen expansion occurred in a series of wet–dry "pulses", with at least three phases of *Pinus* growth, representing drier conditions, evident between 2916–1489 cal. BC (Boswijk *et al.* 2001).

Typically, initial *Alnus–Betula* fen was replaced by mesotrophic fen consisting of *Scheuchzeria* and *Sphagnum cuspidatum*, and then often apparently drier heath communities with *Calluna, Eriophorum* and occasional

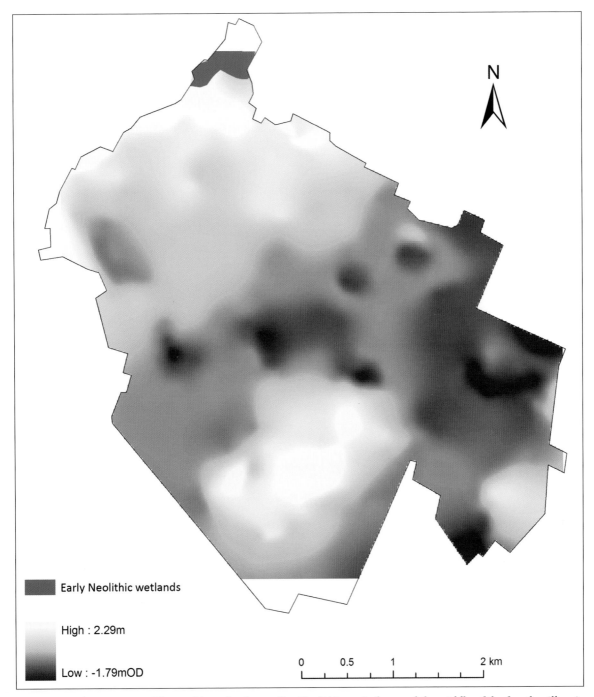

Figure 8.9 Wetland inception on Thorne Moors by the earlier Neolithic period around the middle of the fourth millennium BC

remains of *Sphagnum* (Smith 2002). However, in common with the nature of mire development across Thorne Moors, the transition was strongly diachronous. As observed above, peat accumulation began at similar dates at Middle Moor, Goole Moor and Rawcliffe Moor, but at Middle Moor ombrotrophic peat did not begin to grow until *c.* 1070 cal. BC, hence 1500 years after it is recorded at the other two sites to the north (Figure 8.8).

Initial peat accumulation began in both the central (Middle Moor) and northern (Rawcliffe Moor) parts of Thorne Moors around the same time in the late third–early fourth millennium BC. However, the transition to ombrotrophic mire was considerably later in the central part of the study area (see Figure 8.8), with the latest date for this transition recorded at Crowle Moor (Smith 2002) at the eastern edge of Thorne Moors. This pattern may be a reflection of the influence of calcareous ground water at these locations which would have delayed the onset of ombrotrophic mire growth.

A complex mosaic of environments must therefore have developed across Thorne Moors by the early Bronze Age (Figure 8.11), with the entire study area affected by wetland inception and peat growth by around 1500BC (Figure 8.12). In terms of implications for human activity during the Neolithic period and Bronze Age, a shifting range of different resources would have been available

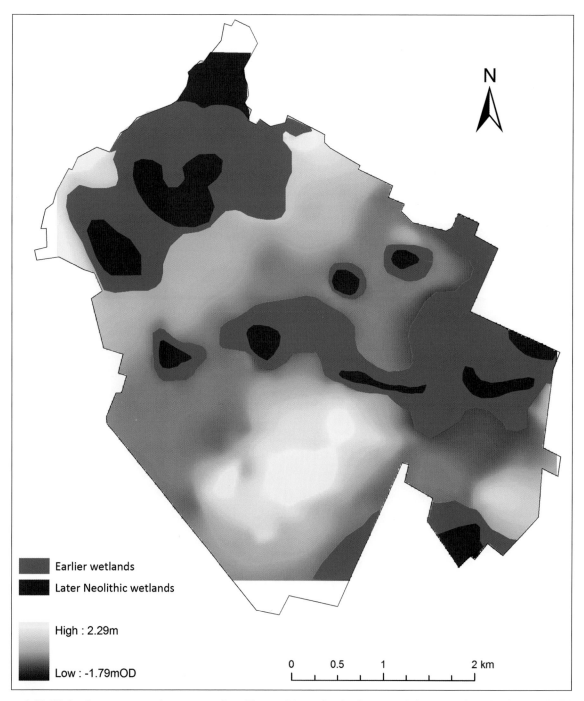

Figure 8.10 Wetland inception and peat spread on Thorne Moors by the later Neolithic period in the first half of the third millennium BC

as mire began to spread from early centres of growth at Rawcliffe Moor and Goole Moor throughout this broad period. This would also have had an impact on movement across and through the landscape as areas became wetter and woodland started to die back. Higher areas within the southern and southwestern part of Thorne Moors may have remained drier until the second half of the second millennium BC and it is possible that any human activity during the Bronze Age would have focussed on these peripheral dryland 'islands', but the chronology of peat spread over the highest parts of the study area remains unclear. This interpretation may be indirectly supported by the location of the Thorne trackway (see Chapter 2) which may have connected Pony Bridge Marsh and Pighill Moor (now under Thorne Colliery) (Buckland and Dinnin 1997).

The precise form and hence function of the Thorne trackway remains open to debate, but it may either represent one section of an originally longer structure designed to cross an extensive area of unstable mire, or a shorter structure across a particularly wet area. Either way, the preservation of this site does reflect both a later spread of peat across the more peripheral areas of the Moors and

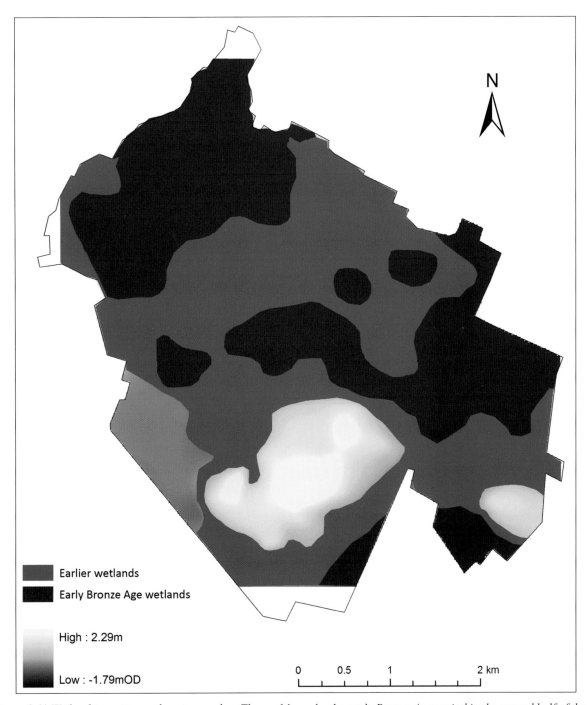

Figure 8.11 Wetland inception and peat spread on Thorne Moors by the early Bronze Age period in the second half of the third millennium BC

some form of human interest in the peatland during the Bronze Age. It had previously been speculated that the evidence for burning adjacent to the trackway reflected human clearance of woodland (Buckland and Kenward 1973) and whilst this must remain a possibility, there are currently no unambiguous supporting data.

In terms of the pollen records, an early decline in *Tilia* has been identified (Smith 2002) on the eastern side of Thorne Moors at CLM2 (Crowle Moor) at a date of *2660–2330 cal. BC* (95%; Chapter 5). This event is not clearly associated with increases in any anthropogenic indicators other than trace values for *P. lanceolata* but may reflect limited Neolithic disturbance to the local vegetation. Clearer evidence for human activity is recorded in the pollen sequences from RWM1 (Rawcliffe Moor) in the northwestern corner of Thorne Moors with *P. lanceolata* and other anthropogenic indicators including *Rumex* and *Urtica* increasing from *2290–1880 cal. BC* (see Chapter 5). Values of *Tilia* tend to be relatively low but consistent at this site, although there is a clear fall at an estimated date of *1800–1470 cal. BC*. Sustained rises in *P. lanceolata* and other indicators of open grassy areas are recorded at

Figure 8.12 Wetland inception and peat spread on Thorne Moors by the middle Bronze Age in the middle of the second millennium BC

870–540 cal. BC alongside falling percentages of tree and shrub taxa. Thus, human activity during the middle and later Bronze Age period appears to have intensified around Thorne Moors.

At GLM1 (Goole Moor; Smith 2002) in the northwestern segment of Thorne, a reduction in *Tilia* at an estimated date of *2330–1920 cal. BC* is followed by low and sporadic values for *P. lanceolata* and other indicators of open, grassy habitats from *1980–1720 cal. BC*. It is only after approximately *770–700 cal. BC (6%)* or *600–390 cal. BC (89%)* that high percentages of *P. lanceolata*, alongside ruderal herbs including *Rumex* and *Urtica*, are associated with falls in arboreal pollen. The latter two taxa disappear from the pollen record at approximately *cal. AD 370–570*, whilst the *P. lanceolata* curve shows a fall but remains steady after this time. At TM1 (Thorne Moors 1; Smith 2002) in the southwestern corner of Thorne Moors, *P. lanceolata* increases at *970–620 cal. BC* prior to a pronounced rise at *810–440 cal. BC*. At CLM1 (Crowle Moor site 1) on the eastern edge of Thorne Moors a decline in *Tilia* and the beginning of a steady *P. lanceolata* curve is recorded from *370–170 cal. BC*. In general, the first clear palynological evidence for human clearance of woodland therefore began in the Bronze Age and increased in the Iron Age.

After the Bronze Age

By the end of the Bronze Age, around 700 BC, it is likely that much of the study area of Thorne Moors would have been covered by ombrotrophic peat. Any human activity at this time is perhaps likely to have been on the peripheral drier and more easily accessible areas such as the former 'islands' in the south. Later evidence for landscape change is available in the pollen diagrams (Smith 2002). The record at GLM2 (Goole Moor site 2) towards the centre of Thorne Moors does not include the deeper sediments at this location but demonstrates reductions in trees and shrubs and clear rises in Poaceae, *Cannabis*-type (hops, hemp) and *Secale* (rye) from a date of *cal. AD 500–1210*. A later intensification of human activity in the form of increasing percentages of *Cannabis*-type, the appearance of a range of ruderal herbs, and clear reductions in tree and shrubs, are recorded at a date of *cal. AD 1240–1390*. At GLM3, just to the east of the centre of Thorne Moors, a very similar pollen assemblage zone to that at the previous site opens at an estimated date of *cal. AD 1060–1290*. In the Rawcliffe Moor diagram, increases in *P. lanceolata* and the appearance of a range of agricultural pollen taxa including *Secale* and *Cannabis*-type at consistently high percentages is estimated to date to *cal. AD 960–1360*.

8.4 Peat inception and mire development: the regional context

Whilst the GIS deposit modelling and radiocarbon dating programme carried out on Hatfield Moors in particular have shed significant light on the pattern of peat inception during the mid-Holocene, questions remain regarding the factors that led to these initial processes of paludification. This is a particular issue for Hatfield Moors since the substrate of this peatland is mainly free draining sands and gravels as opposed to the heavier, less well drained clays and silts of Thorne Moors. Whilst it was not the purpose of this project to investigate these "drivers" in peatland development directly, a brief summary of previous discussion in the light of the revised pattern of peat inception for Hatfield Moors in particular will be presented here.

The role of rising sea levels in the development of the peatlands has been much discussed with study in the Humberhead levels elucidating the marine and perimarine records and a Holocene relative sea level curve available for the Humber estuary (Metcalf *et al.* 2000). The broad outline of changes may be summarised thus. During the early Holocene, the lower reaches of many rivers were fluvial systems with channels incised down to depths as low as -15m OD due to low base levels. Peat inception due to paludification under the influence of relative sea level rise (RSL) led to the spread of *Alnus* dominated fen carr communities in many river valleys in the Humberhead levels from *c.* 6000 cal. BC. Dates as early as *c.* 7250 cal. BC are available from Bole Ings in the lower Trent (Brayshay and Dinnin 1999) and *c.* 5000 cal. BC at Brigg in the Ancholme valley (Neumann 1998).

Peat formation is apparent at many sites from *c.* 4050 cal. BC, as the accumulation of wood peats in backswamp floodplain environments kept pace with the rate of rising relative sea levels (Long *et al.* 1998, Lilllie 1997, 1998, Neumann 1998, Brayshay and Dinnin 1999, Lillie and Gearey 1999, 2000). An increased influence of estuarine expansion at sites in the lower Aire valley is apparent between *c.* 3050 cal. BC and 850 cal. BC in the form of raised waterlevels associated with rising tide levels (Kirby 2001). Detrital peat formation on the floodplain of the River Idle at Misterton had begun by *c.* 2550 cal. BC (Buckland and Dinnin 1997). The effects of rising RSL are generally time transgressive, with sites higher up the river reaches registering the effects later than those lower in the drainage network.

It has previously been hypothesised that peat inception was connected to rises in relative sea level which would have had the effect of the backing up of seaward draining freshwater courses, such as the rivers Torne and Idle which form the eastern and southern boundaries of Hatfield Moors, leading to rising local watertables (Whitehouse, Buckland, Boswijk *et al.* 2001, Kirby 2001). The results of the radiocarbon dating and models presented in Chapter 5 indicate that the interrelationship between these landscape scale events of sea level change and peat spread were fairly complex.

At Hatfield Moors, the character of the nutrient poor sand substrate underlying most of the area was probably an important factor in the character of initial peat development, with the lack of fen development, except for on the southern fringes of the site, a feature of this (e.g. Whitehouse 2004, Whitehouse, Buckland, Boswijk *et al.* 2001, Whitehouse, Buckland, Wagner *et al.* 2001; chapter 5). Whitehouse, Buckland, Boswijk *et al.* (2001) have observed that the *Calluna* and *Eriophorum* deposits, typical of the early peat forming environment at Hatfield (see Chapter 5), may be significant, as they possess properties that encourage the accumulation of peat, particularly since they produce slowly decaying litter (Couslon and Butterfield 1978).

However, the evidence from north of Lindholme (Chapter 5) suggests initial peat accumulation under conditions which have not previously been identified on Hatfield, at what is a very early date compared to basal contexts from much of the rest of the area (Chapter 5). It is also apparent that whilst the early wetland development at North Lindholme was in a poorly drained part of the landscape, it is not at all clear how paludification at this time was related to processes of rising watertables, or if these were even initially driven by changes in RSL. The earliest date for peat formation at north Lindholme of 5470–5220 cal. BC (SUERC-8952; 6355±35 BP) is recorded at HM21 (-0.11m OD).

According to current models of Holocene RSL (Long *et al.* 1998), sea levels rose from *c.* -9m OD at *c.* 5550 cal.

BC to OD by c. 2050 cal. BC, an average long term rate of c. 3.9 mm yr-1. Hence, around 5500 cal. BC, sea levels were over 8m *below* the base of peat formation at north Lindholme, and whilst lower lying areas closer to the rivers remained peat free (Chapters 5 and 6) floodplain wetlands across the Humber lowlands region were beginning to aggrade from around this time. The other basal dates on a rising altitudinal north–south transect at north Lindholme are HM18 (0.08m OD) 4905±35 BP (SUERC-8949; 3770–3630 cal. BC), HM2 (0.64m OD) 5430±35 BP (SUERC-8846; 4350–4230 cal. BC) and HM24 (0.93m OD) 4250±35 BP (SUERC-8877; 2920–2760 cal. BC). Peat accumulation had therefore already spread from -0.11m OD to 0.93m OD, thus over a vertical range of c. 1m, over a period of some 2000 years, *before* sea levels reached OD around 2050 cal. BC.

A complex relationship between base levels, local watertables and peat formation on Hatfield Moors therefore appears likely (*cf.* Foster and Glaser 1986, Foster and Wright 1990). Other factors which may have exacerbated paludification include climatic change, the effects of anthropogenic impact on the landscape, or a combination of these. The archaeological evidence for human activity (Chapter 2) near to the site of the earliest wetland development at Hatfield might be highlighted here, although there is no palaeoenvironmental evidence to indicate the clearance of woodland during the Mesolithic or Neolithic periods. Recent palynological study suggests only slight human impact on the vegetation during the Neolithic and the Bronze Age at the western edge of the Humberhead levels (Gearey *et al.* 2009). Whilst the presence of abundant charcoal in the basal deposits might reflect deliberate burning by prehistoric communities, the evidence for this is circumstantial at best.

The basal dates from the central area of the peatland at least cluster around 3300–3600 cal. BC and although generally from contexts above OD (see Chapter 5), paludification in advance of rising watertables driven by relative sea level rise may seem more likely on a landscape scale by the mid fourth millennium. It is unclear, however, how the mire was interacting as a system by this time. Presumably, the formation of peat, and possibly the development of ombrotrophic mire to the north of Lindholme, had already begun to impact on hydrology, but in the absence of more detailed study this remains somewhat speculative. It may be hypothesised that a range of autogenic and allogenic factors, or combinations thereof, were responsible for peat formation in different parts of the Moors across the relatively wide span of time now indicated. As Lawson *et al.* (2007, 23) have observed for blanket peat development in the Faroe Isles, "… the mechanisms underlying the change from limnic sediments to peat are likely to be substantially different to those underlying the paludification of mineral soils …" This would seem to be a particular issue for Hatfield given the identification of waterlain sediments to the north of Lindholme Island.

The subsequent development of ombrotrophic peat at Hatfield Moors occurs within a reasonably narrow time frame of 1000–1300 cal. BC. However, this is effectively based on only two securely dated sequences of HAT1 and HAT2 and there is no way of establishing how typical this age range may have been across the rest of the peatland. There is circumstantial evidence that earlier ombrotrophic peat formation began in the northern segment of the Moors (Chapter 5) and it may be significant that this is in the vicinity of the earliest known peat accumulation north of Lindholme.

The transition to ombrotrophic mire falls within the period for which there is evidence for an increase in waterlevels at sites in the lower Aire Valley as a result of rising sea levels (Kirby 2001) and, in BSW records, for a climatic deterioration around 1100 cal. BC (Charman 2010). Whether either of these might be implicated in the apparently rapid shift to raised mire is unclear. It has been suggested that the dense, highly humified peat produced by the *Calluna* and *Eriophorum* heath vegetation communities may have been instrumental in creating "… an almost impenetrable barrier between the mire and the underlying watertable, with the result that even a small increase in water supply or decrease in evaporation …[could have] facilitated raised watermound formation" (Whitehouse 2004, 87). Another feature of the early peat formation is the evidence for the burning of the mire surface indicating that the vegetation must have, seasonally at least, been relatively dry (see above, Chapter 5). A negative feedback loop may be postulated in that the burning of the peat surface would result in an increased tendency to waterlogging, since burnt peat has a low hydrological conductivity (Mallik *et al.* 1984).

On Thorne Moors, whilst early peat accumulation began at similar dates in the north, west and possibly central part of the Moors, ombrotrophic peat development was strongly time transgressive. Crowle Moor (CLM1) is the closest sampling site to the edge of the current Moors and recorded the latest shift to ombrotrophy. The silt and clay rich sediments and the dominance of *Alnus* in the pollen record at this site and also at Middle Moor indicate the proximity of both sites to groundwater influence, which probably resulted in the delayed development of ombrotrophic communities relative to sites such as Goole Moor (as outlined above). The lack of *Scheuchzeria–Sphagnum cuspidatum* communities at Crowle Moor, which precede full ombrotrophic conditions apparent at the other sampled locations, may also be a result of these different hydroedaphic conditions. Thus, although the pathway to ombrotrophy at Thorne Moors was ostensibly different to that at Hatfield Moors, there are certain similarities in the development of what may be described as a drier heath community, with *Eriophorum, Calluna* and other ericaceous shrubs, prior to the shift to true *Sphagnum* raised mire.

The evidence would seem to be somewhat equivocal with respect to the possible role of climate in the processes which led to paludification and peat inception, especially so given the fairly wide range for this event on both study areas. Other factors may have controlled or influenced pathways of mire development, including for example, changes in the fluvial network around the Moors. Although the results of the GIS modelling (Chapter 6) do not suggest that changes such as floodplain aggradation of the local rivers had a direct effect on initial peat inception, development might have been affected by fluvial processes in other ways. For example, Hughes and Barber (2003) hypothesised that the fen-bog transition at Tregaron Bog, south Wales, may have been affected by the nearby Teifi river channel. Similarly, Wheeler (1992) proposed that one of the major factors controlling the susceptibility of the Fenland mire systems of east England to acidification and their subsequent transition to ombrotrophy was their proximity to major rivers, especially where they drain through calcareous geology and were thus liable to flood wetland areas with base rich waters.

Waller *et al.* (1999) suggested that a similar mechanism may account for the character of developing mire systems at Walland Marsh, southern England; acidic communities are found at distance from the uplands, compared to eutrophic fen carr communities on the wetland edges, where the influence of base rich runoff is more pronounced. The identification of the possible palaeochannel feature across the centre of Thorne Moors at Middle Moor (see Chapters 2 and 6) may be highlighted here; as discussed above, peat accumulation began at a similar time at Rawcliffe Moor, Middle Moor and Crowle Moor, but ombrotrophic conditions did not develop at the latter two sites until much later than the former. The basal deposits infilling the palaeochannel feature at Middle Moor were described as silty organic sediment (Gearey 2005), suggesting perhaps a sluggish flow of groundwater, which may have delayed ombrotrophic mire growth until wider patterns of peat growth and channel aggradation eventually removed this influence.

Both the Old River Don and the River Went drain through the Magnesian Ridge to the west of Doncaster and would thus have provided base rich waters to vegetation communities proximal to their courses. It is difficult to invoke any one single mechanism to account for the similar sequences of mire growth that seem to have occurred over strongly diachronous time-scales at Thorne Moors, and this would appear to support the contention that "… peatlands may become ombrotrophic in a variety of watertable conditions and climatic regimes …" (Hughes and Barber 2003, 63).

8.5 Palaeoenvironmental evidence for human activity, the Humberhead Levels 'recurrence surfaces' and the identification of Holocene climatic change

Whilst there is generally a good correspondence between the timing and character of vegetation change attributed to human activity on an inter- and intra-site basis, there is evidence for spatial variability, with indications from the Goole Moor and Rawcliffe Moor sequences for anthropogenic disturbance to the woodland close to the northwestern corner of Thorne Moors in the Bronze Age. The chronological modelling has also allowed an investigation of the Humberhead levels 'recurrence surfaces' (Smith 2002; Chapter 2) which has demonstrated that of the five 'recurrence surfaces' only HHL IV (fourth–fifth centuries cal. BC) and HHL III (third–fourth centuries cal. AD) were likely to have been synchronous across both sites. The former occurred during a period which can be difficult to date with precision using radiocarbon, due to the presence of the 'Halstatt plateau' in the radiocarbon calibration curve between 760–420 cal. BC (Kilian *et al.* 1995, 2000).

There is a range of evidence from northern Britain, southern Scotland, Wales and Ireland for increased BSW in the middle of the first millennium AD (Charman 2010, Charman *et al.* 2006, Swindles *et al.* 2007, Blackford and Chambers 1991). Hence it could be posited that HHL/III may reflect allogenic forcing of mire hydrology related to climatic deterioration. HHL/IV would appear to be somewhat later than another well attested climatic change (The 'Sub-Boreal to Sub-Atlantic transition'), for which there is a range of palaeoenvironmental evidence for climatic deterioration recorded from peatland (e.g. Yeloff *et al.* 2007, Plunkett 2006, Swindles *et al.* 2007, Barber *et al.* 2004) and other records including from fluvial systems (e.g. Macklin *et al.* 2005) and lake levels (Magny 2004) across Europe around 850–550 cal. BC. This has been described as reflecting: "… the most profound climatic shift of the Holocene prior to the Little Ice Age." (Brown 2008b: 3). Solar forcing has been implicated in this event, as there is a close correlation between a marked reduction in the rate of radiocarbon production (a proxy for changing solar activity, as evidenced through the radiocarbon calibration curve) during the so called Homeric minimum (850–550 BC) (e.g. Swindle *et al.* 2007).

There is no evidence in the pollen record that any such climatic down-turn had a noticeable impact on later prehistoric human activity around the peatlands and the archaeological record is insufficient to draw any conclusions regarding the nature or extent of activity at this time. However, the later HHL III (cal. AD 200–500) is concurrent with evidence for a reduction in human activity in the form of falls in *Plantago lanceolata* in the pollen records from both Moors at estimated dates of *cal. AD 500–650, cal. AD 210–410* and *cal. AD 370–570* (all at 95%) in the HAT1, HAT2 and GLM1 sequences (Chapter 5) suggesting that the two events might be have been linked.

Gearey (2005) hypothesised that the pollen records might reflect the abandonment of low lying pastoral areas due to the effects of rising groundwater during the early Medieval period. Chiverrell's (2001) study of blanket peat sequences from the uplands of the North York Moors also demonstrated an increased BSW between cal. AD 260–540, with other indications of climatic deterioration in Europe in the fifth and sixth centuries AD (Charman 2010), providing possible evidence for regional climatic teleconnections. The relationship between autogenic processes and allogenic factors such as climate in mire development remains a matter of some debate (e.g. Tuittila *et al.* 2007, Anderson *et al.* 2003, Almquist-Jacobson and Foster 1995) which would clearly benefit from further study.

8.6 The regional archaeological context of Hatfield and Thorne Moors

Hatfield and Thorne Moors can be regarded as distinctive environments in the Humber lowlands during the later Holocene, as there are very few other sites in the region where extensive ombrotrophic mire developed. However, they formed part of a wider expanse of wetland environments that spanned both sides of the River Humber and included the lower reaches of the Rivers Aire, Ancholme, Derwent, Don, Foulness, Hull, Idle, Ouse, Torne, Trent, Went and Wharfe (Van de Noort and Davies 1993, Van de Noort 2004). This section contextualises Hatfield and Thorne Moors in relation to the evidence for human activity and the archaeology from this wider region. This chronological overview focuses on the prehistoric periods.

The Mesolithic period

Within the area surrounding Hatfield and Thorne Moors, sites and finds from the Mesolithic period have been identified in locations closely associated with watercourses such as the Rivers Went, Torne, Don and Old River Don (Figure 8.13) and this pattern seems to be the case regionally (Van de Noort *et al.* 1997, Van de Noort 2004). Whilst there are some notable areas such as the Vale of York where evidence for Mesolithic activity remains very limited (Van de Noort 2004), the landscape context of sites such as Misterton Carr to the south of Hatfield Moors (Head *et al.* 1997), Stone Carr in the Hull valley on the northern side of the Humber (Lillie and Chapman 2001) and Sutton Common on the southwestern edge of the Humber wetlands (Parker Pearson and Sydes 1997, Van de Noort *et al.* 2007) further suggest an association between human activity and watercourses during the prehistoric period. More specifically, Mesolithic finds from the Holderness region (summarised in Van de Noort 2004), alongside those such as Star Carr (Clark 1954, 1972, Rowley-Conwy 2010) and others within the Vale of Pickering in North Yorkshire (Conneller and Schadla-Hall 2003), demonstrate associations with open water and wetlands more broadly through the earlier Mesolithic period and later. This implies the potential, albeit on a smaller scale, for the survival of wet-preserved sites dating to the Mesolithic period on Hatfield Moors, perhaps in association with the early pools, such as the one associated with the Neolithic trackway and platform.

The earlier Neolithic period

Regionally, the pattern of activity for the earlier Neolithic period was focussed on the higher areas around the Humber wetlands, including the Lincolnshire Wolds and the Yorkshire Wolds. Occupation sites are ephemeral, consisting of concentrations of struck lithics and early ceramics, such as Grimston wares (Manby 1988). Although such sites are not well represented within the lowlands, this might be a reflection of archaeological visibility due to the 'masking effect' on earlier land-surfaces in this region by extensive alluvial deposits. Some indication of early Neolithic activity is demonstrated by the lithic concentrations identified during fieldwalking within the region (Van de Noort and Ellis 1997). Early monumental architecture is limited to the upland areas, with long barrows, mortuary enclosures and other structures recorded on the Yorkshire Wolds (e.g. Greenwell 1877, Ramm 1971, Manby 1976, Riley 1988, Stoertz 1997) and on the Lincolnshire Wolds (e.g. Jones 1998), as well as the adjacent areas. From the Humber lowlands the greatest concentration of earlier Neolithic activity appears to have been within the area of Holderness, although notable concentrations of lithic material from this period have been identified from the southern and western edges of the Humberhead levels area, although much of this material is difficult to date with any precision (*cf.* Head *et al.* 1997).

The later Neolithic period

The archaeology of the later Neolithic period is dominated by the wealth of sites from the upland area of the Yorkshire Wolds, and particularly around the areas of Rudston and Duggleby. At Rudston, a complex monumental landscape represents activities that continued from the earlier Neolithic through to the Bronze Age, including a cluster of at least four cursus monuments, later Neolithic "great barrows" and a henge (Dymond 1966, McInnes 1964, Manby 1988, Riley 1988, Chapman 2003, 2005). At Duggleby, the later Neolithic landscape was centred on a monumental "great barrow" (Mortimer 1905, Kinnes *et al.* 1983, Loveday 2002).

As outlined above, during the later Neolithic, there were significant changes across both Hatfield and Thorne Moors with the expansion of wetlands and the associated loss of woodland (Chapters 5 and 6). The more open nature of woodland on Hatfield Moors would perhaps have made this landscape regionally distinctive, and it is against this environmental backdrop that the Hatfield trackway and platform was constructed, around *2730–2450 cal. BC* (see Chapter 7). This date broadly equates to episodes of later Neolithic monument construction elsewhere, particularly hengiform structures such as Maiden's Grave near Rudston (McInnes 1964).

Whilst such structures are morphologically very different, some parallels can be identified. For example, the 'dog-leg' in the shape of the Hatfield trackway is reminiscent of the routes taken by some orthostat-lined avenues at sites such as Avebury (Burl 1979, Barrett 1994). Furthermore, the relationships between contemporaneous monuments such as henges and the natural environment are noteworthy, such as associations with water and arguably themes of environmental change (e.g. Tipping *et al.* 2004, Richards 1996). Within this context, the construction of a site such as the Hatfield trackway and platform perhaps makes sense as a local interpretation of a broader later Neolithic theme of monumentality.

Figure 8.13 Distribution of sites dating to the Mesolithic period from the local region indicating a focus on locations associated with watercourses and their floodplains (based on research conducted by the Humber Wetlands Project)

The Bronze Age period

The earlier Bronze Age within the region is characterised largely by round barrows which are recorded in particular density across the upland areas of the Yorkshire Wolds and the Lincolnshire Wolds (e.g. Stoertz 1997), and alternative approaches to the disposal of cremated human remains have been identified at Sutton Common during this period (Van de Noort et al. 2007). The distribution of artefactual evidence across the region reflects an increase in human activity (Van de Noort 2004).

There is also evidence for developments in water transport by the end of the third millennium BC, a theme that runs through the whole of the period. The earliest boat remains were discovered at North Ferriby on the River Humber (Wright 1990), with a series of vessels dating from 2030–1780 cal. BC (Wright et al. 2001) through to the early Iron Age (Switsur and Wright 1989). The remains of other Bronze Age vessels are also known from Brigg (Roberts 2007) and Kilnsea (Van de Noort et al. 1999), and probably represent both river-going and sea-going transport (cf. Chapman and Chapman 2005).

Themes of movement and access are also represented by aspects of the archaeological record for this period. In a different wetland context, two short trackways were excavated on the foreshore of the Humber near Melton, the first radiocarbon dated to 1440–1310 cal. BC, and the second producing two determinations of 1429–1120 cal. BC and 1100–840 cal. BC. These seem to have been constructed to cross creeks and channels, possibly as part of subsistence strategy involving seasonal grazing on the foreshore (Fletcher et al. 1999). The presence of trackways and boats arguably illustrates a focus on facilitating and aiding movement, perhaps in part related to the general expansion of wetland environments across this period on floodplains and river valleys.

The Iron Age period

The Iron Age appears to have been a period of significantly increased human activity around Hatfield Moors, with the clearance of *Tilia* dominated woodland implied in the pollen records at estimated dates of *390–40 cal. BC* and *210–20 cal. BC* (95%) at HAT1 and HAT2 respectively. The evidence from aerial photography may provide the archaeological signature of the felling of woodland, with ditch and enclosure systems interpreted as reflecting the expansion of settlement and mixed agriculture during the Iron Age (e.g. Riley 1980, Chapman 1998, 1999). Chapman (1997) has demonstrated that soil type in this region is directly related to cropmark visibility with poorly drained soils unable to produce distinct cropmarks, so such evidence must be used cautiously. In general, these crop marks are largely restricted to areas above c. 7m OD.

Various sites attributed to the Iron Age are located in the wider area around Hatfield Moors (Van de Noort et al. 1997). To the south and west of Hatfield village, 'brickwork' field plans have been recorded (Riley 1980), whilst around Sandtoft and the Isle of Axholme clear cropmarks of settlement and field systems with enclosures are visible. Other archaeological fieldwork has shown that these cropmarks tend to produce little pottery (e.g. Merrony 1993), further hindering the dating of sites and field systems. In a broader context, this would appear to correspond with archaeological evidence from elsewhere in Yorkshire, which has been interpreted as reflecting a high degree of "continuity and evolution ... in the [later] prehistoric and Roman periods" (Mackey 2003). Contemporaneous palaeoenvironmental data are generally sparse from this region, but demonstrate a very similar pattern of marked intensification of woodland clearance and indications of agricultural activity in both upland and lowland areas from the later first millennium BC (Gearey et al. 2009, Bartley et al. 1990, Tweddle 2001).

8.7 Summary

Few previous archaeological discoveries have been made on Hatfield and Thorne Moors in recent times (Chapter 2). Van de Noort and Ellis (1997) suggested that the absence of archaeological sites from the Moors were a product of archaeological visibility rather than a lack of human activity on these areas in the past. This hypothesis has been partially supported by the discovery of the Neolithic trackway and platform and associated lithics on Hatfield Moors. It is clear that other unreported archaeological finds have been made on the Hatfield and Thorne Moors in the past (Chapter 2), but the lack of contextual information reduces their archaeological value. This picture contrasts sharply with the rich archaeological record from many Irish raised mires (see Chapter 1). Nevertheless, the evidence discussed above suggests that the low concentration of archaeological finds from both areas may in fact reflect the loss of those deposits with the greatest potential to contain sites and material. There is potential for sites to be preserved within the surviving basal peats; the re-wetting of both areas may ensure the *in situ* survival of any such sites.

In terms of the survival and preservation *in situ* of any archaeological sites, it may be assumed that the future discovery of any post-Bronze Age wet-preserved archaeology on Hatfield Moors is unlikely due to the removal of much of the peat post-dating this period. Those deposits which do survive are relatively far from contemporary dryland–wetland interfaces and hence structures such as trackways and platforms would be unlikely to be found on the surviving core of the peatland at least. The investigation of Thorne Moors has drawn solely on data from previous studies (Chapter 2) but the generation of a model of the pre-peat landscape has permitted an understanding of the spatial and temporal patterns of mire growth and spread. This hopefully provides a platform for integrating these landscapes into broader narratives of Holocene landscape change, human activity and the archaeological record, as well as providing a foundation for their effective future management.

9. Conclusions: themes in the archaeo-environmental study of peatlands

with contributions by Peter Marshall

9.1 Introduction

This book has presented a 'landscape archaeology' of Hatfield and Thorne Moors, the two largest surviving areas of lowland raised mires in Britain (covering a combined area of over 33.5km^2). It has explored spatial and temporal patterns and processes of environmental change in relation to various themes including past human activity, the archaeological record and the future management of the peatlands. It was proposed in Chapter 1 that peatlands provided the potential for examining certain questions situated at the interface between palaeoecology and landscape archaeology and this chapter will present reflections and conclusions regarding these and related issues. Certain techniques used in landscape archaeology, such as aerial photography and geophysics, are of limited value in peatlands (see Chapter 1) and integrated archaeo-environmental investigations (see Chapter 2) are resource heavy. Hence, alternative approaches must be adopted and developed. One of the aims of the study was to explore how this might be achieved using more limited datasets, incorporating those derived from previous research as well as targeted fieldwork and analyses. As outlined in Chapter 2, mires present unique opportunities as well as challenges for archaeological research and this chapter will present proposals for a 'tool kit' for future integrated study.

The 'hidden landscapes' of the title of this book are both literal and figurative: the pre-peat land-surface may be regarded as the former, as this cannot be accessed or observed directly. The peat itself represents both a literal and figurative form of 'hidden landscape'; the peat layers represent a 'cumulative palimpsest' (*cf.* Bailey 2007; see Chapter 1) or vertical accumulations of former land-surfaces of the mire with any archaeological sequences overlying each other. The sub-fossil remains of pollen, insects and other material, provide evidence of the vegetation which grew both on the mire and on the dryland beyond, and also evidence of processes of past hydrological change which have been related to local processes of mire development or to regional 'forcing' by climate change (Chapter 2).

These landscapes are 'hidden' in the figurative sense that appropriate palaeoenvironmental data must first be obtained and the results collated to extrapolate and infer the chronological and spatial dimensions of landscape change. However, as outlined in Chapter 1, there is something of a disconnection between the intrinsically spatial character of the archaeological record and the essentially aspatial ouput provided by certain palaeoenvironmental analyses. This study has considered this problem by attempting to model processes of environmental change in four dimensions and by focussing on a case study investigating patterns of peat spread specific to the landscapes of the Humber peatlands. As also considered in Chapter 1, other broadly similar physical processes such as past sea level change represent critical events for contextualising the archaeological record and human activity, which may be well understood on a regional scale but present increasing uncertainty when finer grained spatial and temporal scales are concerned. Whilst it is clear that the operational and observational scales of past environmental changes will never be equivalent (see Chapter 1), this study has demonstrated that closer integration and enhanced understanding is possible. This final chapter will reflect on the methods and results of the study and outline possible lessons and directions for future research which seeks to integrate archaeological and palaeoenvironmental paradigms.

9.2 Hatfield and Thorne Moors: modelling mires

The Humber peatlands presented specific practical problems as peat has been extracted at an industrial scale from both, a process which was mechanised during the latter half of the twentieth century. More recently, both landscapes have been subject to re-wetting as part of peatland conservation measures (Figure 9.1). Whilst this can be viewed as a positive development following over a century of extraction, this process of flooding has rendered the vast majority of these landscapes physically inaccessible and

Figure 9.1 The landscape on the western side of Hatfield Moors in 2012

it is difficult to see how further archaeological fieldwork could now be carried out on these areas.

The specific objectives for the study of Hatfield and Thorne Moors focussed on their interpretation as archaeological landscapes and for providing data for their future management from a heritage perspective (Chapter 3). Hence, the first three objectives concerned the modelling of past environments through the investigation of their pre-peat landscapes, the patterns and processes of wetland inception and peat spread, and finally an assessment of peat deposits as a record of environmental change and human impact. The fourth objective focussed on the quantification of the archaeo-environmental resource in terms of the depths of surviving peat across these landscapes and the chronological range represented by these deposits. The approach adopted has relevance for the investigation of peatlands and other archaeological landscapes elsewhere (discussed below).

Access to the pre-peat landscapes could only be achieved through remote methods, primarily sampling through borehole survey, but also including datasets that had been captured initially for other purposes, such as the LIDAR and GPR data. These datasets needed to be validated prior to further analyses, and this was achieved through a combination of borehole transects and gridded borehole survey. This process also permitted the comparison of the relative accuracy of spatial datasets. Chronological modelling employed predominantly Bayesian methods implemented using the computer programme Ox-Cal, to assess the robustness of the radiocarbon dates associated with palaeoenvironmental studies from the peatlands and to generate *posterior density estimates* derived from Bayesian approaches (see Chapter 5 and Appendix 1) for specific 'events' interpreted from the pollen, insect and other proxy records and to assess the chronological relationship between certain of these events.

The combined modelling of spatial and chronological datasets to permit explicit analyses of the relationships between the two was addressed within the GIS and achieved through a variety of approaches including surface analyses and statistical comparisons. This process can be illustrated by the modelling of wetland inception (Chapter 6) where GIS models were generated to test previous hypotheses (Buckland and Smith 2003) regarding the processes underlying paludification leading to peat growth, including elevation, distance to watercourses, and surface flow-accumulation. The GIS was used to explore correlations between the spatial and chronological processes of peat inception implied by the hypotheses and the spatial and chronological pattern demonstrated by the available radiocarbon dates.

In the case of Hatfield Moors, the GIS was then used to derive a formula describing the statistical relationship between the calibrated radiocarbon dates and flow-accumulation values derived from surface analysis of the pre-peat landscape to generate a new GIS model of 'chronozones' representing the interpolated date of peat inception across the entire study area. A limited test of this procedure was performed using basal radiocarbon dates which had not been included in the initial process of modelling. However, it must be stressed that the models of peat inception are intended to be hypothetical representations of the currently available data, which may be tested by further work. As stated earlier, these models can perhaps be best regarded as *data visualisations*, rather than as *representative visualisations* (see McCoy and Ladefoged 2009).

Exploring and understanding pre-peat landscapes

The modelling of the pre-peat landscapes integrated a variety of datasets including elements of LIDAR data, GPR data and peat-depth data from previous surveys, alongside new borehole data from both transects and gridded surveys (Chapter 3). The process involved an initial phase of 'ground-truthing' the legacy datasets to establish their validity, followed by their integration into DEMs using methods of interpolation to generate continuous land-surfaces. The process highlighted issues regarding the applicability of legacy datasets, the potential use of approaches such as GPR, and issues relating to the resolution of data required (see below).

The pre-peat landscape models shed significant light on the variations in the pattern and process of Holocene environmental change described by past research (Chapter 2). In the case of Hatfield Moors, the current extent of Lindholme Island represents the uppermost part of a larger morainic ridge which extends in a somewhat subdued form beneath the surviving peat to the north and south. These morainic deposits of Lindholme are partially overlain and surrounded by coversands (see Chapter 2), forming relatively flat relief, but with low ridges of more pronounced sand dunes in the west and south (see Figure 4.1).

The pre-peat landscape of Thorne Moors showed greater topographic variation with pronounced depressions in the 'Lake Humber' silts and clays in the north and east of the study area. There is some indication that the Rawcliffe depression in the north, one of the earliest areas of peat growth, extends and deepens below the reclaimed areas of peatland. A possible palaeochannel feature was identified running west–east in the central part of the Moors, presumably incised during the early Holocene (see Chapter 2). The morphology of the pre-peat landscape of both areas appears to have been highly significant in terms of the subsequent timing and patterns of peat inception and spread, with the palaeochannel possibly permitting the influence of groundwater on the vegetation communities in the Middle Moor area and delaying the onset of ombrotrophic conditions.

This process has thus demonstrated the significance of the pre-peat context for subsequent peatland development (see below) and it is also clear that without this understanding it would have been very difficult to fully contextualise the results of the detailed previous research (Chapter 2: Smith 1985, 2002, Whitehouse, Buckland, Boswijk *et al.* 2001, Whitehouse, Buckland, Wagner *et al.* 2001, Gearey 2005). A combination of these previous data and the modelling of peat growth and spread, based on the pre-peat landscape model, underpins the narrative of landscape change described in Chapter 8. The modelling of the pre-peat landscapes of Hatfield and Thorne Moors provides the first comprehensive three-dimensional understanding of these environments and also represents the first attempt to model a land-surface obscured beneath peat on a landscape scale for specifically archaeological purposes. This has implications for future research into peatland growth and development as well as for the acquisition of data with relevance to the future management of these peatlands (see below).

Modelling wetland inception and peat spread

The modelling of wetland inception and peat spread provided reconstructions of the mires at different periods (Chapter 6), identifying the *foci* for wetland inception and the spatial pattern of peat spread across both landscapes. The specific 'time-slices' represented were chosen on the basis of their relationships to cultural periods and, for periods of significant environmental change, they were represented in higher resolution. Hence, for Hatfield Moors the rendered divisions related to the later Mesolithic (latter part of the sixth millennium BC), the early Neolithic period (*c.* 4000–3300 BC), the later Neolithic period (*c.* 3000–2500 BC), and the early Bronze Age period (*c.* 2500–2000 BC), by which time wetland inception and peat spread had covered the vast majority of the Moors. For Thorne Moors the rendered divisions related to the early Neolithic period (*c.* 3500 BC), the later Neolithic period (*c.* 3000–2500 BC), the early Bronze Age (*c.* 2500–2000 BC), and the middle Bronze Age (*c.* 1500 BC).

As outlined above, the modelling of wetland inception and peat spread was achieved by testing the existing hypotheses for wetland inception and peat spread using the GIS. In the case of Hatfield Moors, it was identified that there was a close correlation between early dates for peat inception and high values for flow-accumulation. In other words, micro-topographic context appears initially to have been more significant than gross elevation in terms of peat inception, with the lower southeastern segment of Hatfield Moors demonstrating later dates for this process in comparison to the relatively higher areas of relief in the north. The radiocarbon dating programme also demonstrated that wetland inception began earlier in the northern half of the study area. Early organic sediment accumulation began in a shallow pool, or perhaps series of pools, in slight hollows in the pre-peat substrate during the sixth millennium BC (later Mesolithic period). At this time, much of the rest of the study area was dryland, although the likely vegetation of *Pinus* dominated heath probably made the area somewhat distinctive within the context of the dense deciduous woodland across much of the Humber lowlands area.

This evidence for sediment accumulation within shallow water, contrasts with the previous palaeoenvironmental evidence for peat growth in a relatively dry heath environment across much of this area from the third to fourth millennium BC (Chapter 5). The possible causes of paludification are unclear and, given the available data, it is probably easier to reject, rather than propose, hypotheses. Certainly, the impact on local hydrology by vegetation clearance or disturbance by human communities seems very unlikely (see also Smith 2002: 89) during the Mesolithic and it is difficult to envisage a direct link to flooding from local rivers given the relative position of base level at this time (see Chapter 3). Rising watertables associated with climate change may be regarded as the most probable mechanism with the impact of channel aggradation perhaps also important; although there is a shortage of comparative palaeoclimate records for this part of the Holocene (see Charman 2010).

The analyses therefore suggest that peat inception at Hatfield was multi-focal and probably began in areas of poor drainage across the pre-peat landscape irrespective of their position relative to OD. The flow-accumulation modelling identified a number of such 'nodes' across the Moors and it is possible that future investigation and radiocarbon dating of sediments from these locations might push the date for peat inception back further still. The available range of radiocarbon dates also indicates that the growth of peat took place across a longer period of time than previously recognised (e.g. Smith 2002, Whitehouse 2004; Chapter 2) and that ombrotrophic peat growth might have commenced significantly earlier (Chapter 5; see below). The early pattern of mire development at Hatfield Moors is reminiscent of that identified in palaeoecological studies of the development of the Flow Country peatlands in northern Scotland where '... Early changes [in mire development] must be considered extremely localised, probably applicable to only a small area metres or tens of metres across at the most.' (Charman 1994, 296; see also Charman 1992, 1995).

For Thorne Moors, hypothesis testing was less conclusive, although the available data indicates that wetland inception commenced in the areas of Rawcliffe Moor, Crowle Moor and the central area of Middle Moor at some time during the earlier Neolithic, and in the case of the latter two areas, was possibly associated with the infilling of palaeochannels which are now concealed beneath the peat. The reliability of the Middle Moor radiocarbon chronology is unclear, but a basal date of 3345–2929 cal. BC (NZA-18334) is available from this location (see Chapter 2 and Appendix 1). The *posterior density estimates* for peat accumulation (Chapter 5) range from the earliest recorded date of *3545–3460 (6%) or 3425–2930 (89%) cal. BC* at Rawcliffe to *1500–1155 cal. BC* at Thorne Moors Site 1 (Figure 9.2; Table 9.1). Later dates for inception tend to be recorded on the higher and presumably better drained parts of the landscape and by the end of the Bronze Age wetland had spread across the entire landscape.

Despite over 30 years of palaeoecological research into peatlands, various aspects of pattern and process in peat growth and spread remain poorly understood and there is a need for more detailed site based studies to assess the relative role of biotic, geomorphological, topographic and climatic factors. Tipping (2008) has recently summarised the importance of understanding the spread of peat on a landscape scale with respect to establishing the context of prehistoric human activity in the Scottish uplands, concluding that even with a comprehensive sampling strategy and an understanding of the morphology of the pre-peat landscape '... it is still difficult from these data to describe the process of blanket peat 'spread'.' (Tipping 2008, 2110).

The research presented here, alongside other studies (e.g. Charman 1992, 1994, 1995, Foster and Fritz 1987), confirms that an understanding of the morphology of the pre-peat landscape is a clearly essential aspect of any investigation of landscape evolution. It has long been recognised that landscape-scale processes and patterns of change cannot be understood from 'single-age estimates' (Edwards and Hirons 1982), but exploring past environmental change in four dimensions will probably always be challenging, if only in terms of the resources required to investigate multiple sequences from a given site (e.g. Smith and Cloutman 1988). The modelling approach using GIS represents a method for manipulating and analysing spatial and chronological datasets and for generating testable models, adding to other recent applications of this technology in the interpretation of palaeoenvironmental data (e.g. Fyfe 2007).

Reconstructing raised mire development: spatial variations in the transition to ombrotrophy

Previous research on Hatfield and Thorne Moors recognised that the different recorded pathways to ombrotrophic mire growth were probably related in part to the contrasting character of the substrates beneath each peatland (e.g. Whitehouse, Boswijk and Buckland 2001). The processes of paludification and the subsequent ombrotrophic mire development might have been fairly complex both spatially and temporally. Once peat accumulation had commenced the pattern and timing of subsequent spread was probably related to a complex of variables, such as autogenic changes in peat forming vegetation, alterations to the local drainage patterns as the spread of peat progressed, and perhaps even the effects of human disturbance. The transition to ombrotrophy occurred during the Bronze Age at *1115–840 cal. BC* (HAT1) and *1665–1380 cal. BC* (HAT2) (Figure 9.2 and Table 9.1; see below). Patterns of mire development have implications for archaeological interpretation and understanding (see Chapters 2 and 7) but for Hatfield, this

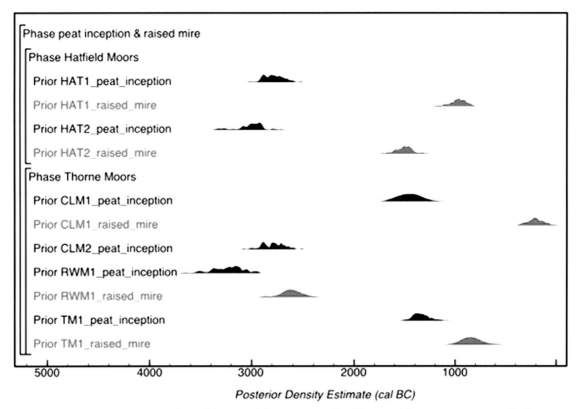

Figure 9.2 Posterior density estimates (derived from modelling described in Chapter 5) for peat inception and subsequent raised (ombrotrophic) mire development on Hatfield and Thorne Moors. Note: CLM2 record does not include the fen-mire transition

Table 9.1 Posterior density estimates BC (derived from chronological modelling described in Chapter 5) for peat inception and raised (ombrotrophic) mire development on Hatfield and Thorne Moors.

Parameter	Depth (m)	68% probability	95% probability
HAT1 peat inception	2.00	2905–2845 (22%) or 2835–2705 (46%)	2930–2600
HAT1 raised mire	1.46	1025–890	1115–840
HAT2 peat inception	2.54	3095–3070 (6%) or 3035–2885 (62%)	3350–3220 (9%) or 3195–3160 (1%) or 3135–2865 (83%) or 2810–2760 (2%)
HAT2 raised mire	1.42	1565–1425	1665–1380
CLM1 peat inception	1.55	1575–1345	1670–1245
CLM1 raised mire	0.95	270–120	340–65
CLM2 peat inception	4.15	2915–2855 (22%) or 2815–2675 (46%)	3015–2980 (2%) or 2965–2580 (93%)
RWM1 peat inception	2.12	3390–3100	3545–3460 (6%) or 3425–2930 (89%)
RWM1 raised mire	1.86	2700–2520	2790–2415
TM1 peat inception	0.89	1425–1275	1500–1155
TM1 raised mire	0.60	940–755	1025–655

is based on analyses of very few sequences given the size of the peatlands and it is not possible to establish the exact spatial or chronological patterning of this shift.

The locations with the earliest dates for wetland inception do not equate to the earliest transition to ombrotrophic peat, the development of which seems to have been strongly time transgressive on Thorne Moors, ranging from *2790–3100 cal. BC* at Rawcliffe Moors to *340–65 cal. BC* at Crowle Moors Site 1 (see Figure 9.2 and Table 9.1). There is evidence for a similar contrast between peat inception and raised mire development on Hatfield Moors but the greater availability of data concerning this process for Thorne Moors hampers the meaningful comparison of the spatial and chronological variability of the process between the two peatlands. Certainly, despite the considerable body of multiple sequence, multi-proxy studies carried out on Hatfield and Thorne Moors, and the detailed knowledge of changes in specific locations (Chapter 2), it is difficult to reconstruct these processes of change on a landscape scale.

These data were hence insufficient to reconstruct the evolution of the mire surface itself in four dimensions in the

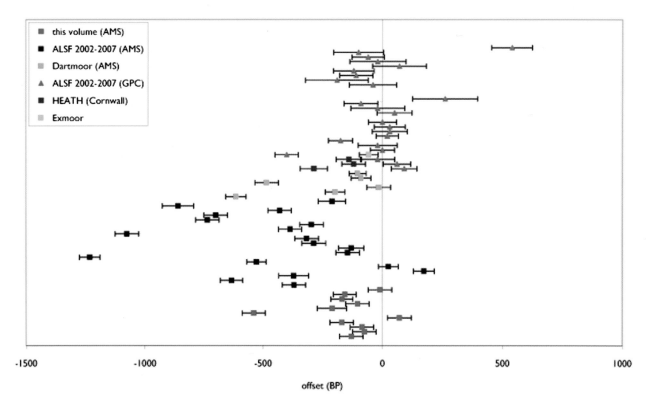

Figure 9.3 Offsets between radiocarbon measurements of the humic acid and humin fractions of bulk sediment samples, from ALSF 2002–2007 (AMS) (Bayliss et al. 2007, 2008), Dartmoor (Fyfe et al. 2008), ALSF 2004–2007 (GPC) (Bayliss et al. 2007, 2008), Heath (Cornwall) (English Heritage pers. comm.) and Exmoor (Matthews 2008) (ALSF = Aggregates Levy Sustainability Fund)

manner of that undertaken by The Lisheen Archaeological Project at Derryville (see Chapter 2). However, the former project focussed on an area of peatland of around 200ha, compared with the 3350ha of Hatfield and Thorne Moors. The density of archaeological remains and level of detail achieved in the interpretation of the raised mire landscape of Derryville was exceptional. At one level this study cannot be compared to that described in this book, but the work at Hatfield and Thorne Moors has demonstrated what is possible using somewhat fewer resources and by 'recycling' data.

Radiocarbon dating and Bayesian modelling: exploring spatial and temporal dimensions of palaeoenvironmental records

Robust chronological control is central to palaeoenvironmental interpretation and, as outlined in Chapter 1, recent developments in Bayesian modelling now provide a methodology to assess this aspect of analysis. The approach has permitted the chronological robustness of the palaeoenvironmental sequences from Hatfield and Thorne Moors to be determined, allowing an assessment of the precision of the associated interpretations and the synchronicity of palaeoenvironmental 'events'. The results (Chapter 5) indicated that all of the palaeoenvironmental analyses carried out by Smith (2002) display a good degree of chronological robustness, but there are problems with the radiocarbon dates from HAT4 (Whitehouse 2004) and Porters Drain (Gearey 2005) on Hatfield Moors and Middle Moor on Thorne Moors (Gearey 2005).

It is probably significant that these sequences which relied on AMS dating, particularly of bulk sediment, demonstrated less robust chronologies compared to those based on standard bulk radiocarbon dates. The dating of the multiple fractions discussed in Chapter 5 indicates that the basal peats from Hatfield are inhomogeneous and consist of material of a range of different ages (e.g. see Shore *et al.* 1995), probably as a result of the formation processes of the peat deposits. Recent studies of the radiocarbon dates of multiple peat fractions from floodplain deposits (Howard *et al.* 2009) and archaeological sites (Brock *et al.* 2011) indicates a complex relationship between the relative age of different fractions and the character of deposit accumulation.

The AMS dating of bulk sediment samples from a variety of environments appears to show a systematic tendency for the humic fraction to be younger than the humin fraction, while such an offset is not so readily apparent in the use of 'large' bulk radiometric size samples (Figure 9.3). This is undoubtedly because a small amount of contamination by younger or older carbon can severely affect the results (Shore *et al.* 1995, Nilsson *et al.* 2001, Wolhfarth *et al.* 1998). While the use of bulk conventional dates in such circumstances may, on the other hand, mean that the subsequent radiocarbon age is less prone to the influence of small amounts of older or younger carbon. Further investigation of the formation processes of peatlands and the implications for precise and accurate dating of the various fractions of sediment is required.

The Bayesian methodology also facilitated the formal estimation and comparison of the significant palaeoenvironmental 'events' which had been identified by previous research (The 'Humberhead Levels 'recurrence surfaces' Chapter 2). This aspect focussed on the timing of the evidence for potential late Holocene climate change and the palynological record, with particular reference to the timing and character of human impact on the dryland vegetation around the two sites (Smith 1985). The comparison of the palynological data demonstrates the spatial variation in human impact during early prehistory in particular (Chapter 5). This assessment of the HHL 'recurrence surfaces' is significant with respect to the arguably poorly understood relationship between past climate change, peatland dynamics and the palaeohydrological record (see Chapter 8) as there are relatively few, well dated, multi-proxy, multi-sequence studies of raised mire sequences with proven chronological robustness.

The results imply that perhaps only HHL III and IV (see Chapter 2) may be regarded as synchronous (Chapters 5 and 8), implying potential evidence for palaeohydrological changes, possibly reflecting climatic deterioration in east England during the late Bronze Age–early Iron Age and the early Medieval periods, although identifying meaningful causality between cultural and environmental change is problematic (see e.g. Coombes and Barber 2005). However, it is difficult to see how it would be possible to disentangle the modelled chronological patterns in terms of which of the BSW shifts might be regarded as a primarily 'climatic' signal, and which not, other than through a determination of their synchroneity in the manner presented in this and other studies (see Chapter 2). This process arguably runs the risk of generating something of a circular argument.

The spatial and chronological variations observed in the BSW records across Hatfield and Thorne Moors were quite probably also controlled by localised differences in the morphology of the mire surface and autogenic processes of change (e.g. cyclical regeneration of hummock-hollow complexes, Barber, 1981, changes in drainage patterns or even catastrophic threshold shifts, such as 'bog bursts', see Caseldine and Gearey 2005). Disentangling such autogenic signals from allogenic forcing rather relies on the availability of a single independent dataset which is regarded as unequivocally demonstrating a climate 'signal' rather than 'noise' related to internal system dynamics (*sensu* Blaauw and Maquoy 2012). Ultimately, all palaeoclimate data are, by definition, derived from proxy records so it is hard to see how 'signal' can could be entirely separated from 'noise'. However, the identification of palaeohydrological shifts within a mire system may be significant in terms of contextualising the wetland archaeological record of a specific system (see Chapter 2; Gearey and Caseldine 2006) irrespective of the precise mechanisms involved.

Investigating the chronological relationship between inferred periods of climatic change and human activity in the wider landscape may be approached in a formal manner using the Bayesian methodology. For example, the pollen records do not indicate any significant reduction in human activity following the possible climate deterioration of HHL IV 'recurrence surface', whilst there is evidence for this following HHL V in the form of reductions in 'anthropogenic indicators' in the pollen record (see Chapter 5). The latter may reflect relatively localised changes in land-use and agriculture, and need not necessarily be interpreted as demonstrating 'negative' or 'catastrophic' impacts wrought by any episode of climate change on human communities in the area. The potential climatic change during the later Bronze Age possibly represented by HHLV is actually associated with an increase in evidence for human activity in the pollen records from Hatfield and Thorne Moors (Chapter 5). This may be compared to Dark's (2006) meta-analysis of pollen records from across Britain which concludes that there is little evidence for significant changes in land-use around 850 cal. BC, with some areas even apparently showing increased agricultural activity. Plunkett (2009) has also demonstrated how palaeoenvironmental and archaeological interpretation may also be linked to generate new perspectives on both records.

Finally, it was proposed (Chapter 2) that the Humber peatlands provided a good case study for investigating spatial and temporal patterns of Holocene environmental change due to the availability of detailed, multi-core and multi-proxy analyses. The time and expense of analyzing multiple sequences generally prohibits such study and few comparative analyses from the same peatland system are available. The results demonstrate the potential complexity of patterns and process in terms of Holocene peatland development but also indicate the dangers of relying upon interpretation from a single palaeoenvironmental sequence. It is clear that any one of the detailed palaeoenvironmental sequences provides a record of environmental changes at a specific point in space only, but any subsequent interpretations regarding climate change and human impact on a wider scale might potentially be highly misleading. For example, the timing of initial human impact on the vegetation varied significantly between the different sequences (Chapter 5).

This may perhaps be less surprising as regards the pollen record from sampling sites from the same peatland system but several hundreds of metres apart, reflecting for example the proximity of a given sampling site to different vegetation communities in the past (see for example Bunting *et al.* 2004). However, recent detailed assessment of chronological and spatial variations in multiple palaeoenvironmental sequences from a raised mire system in the Netherlands have demonstrated clear evidence for variations in the fluctuations of pollen taxa, assumed to be of 'regional' vegetation taxa, in cores sampled only tens of metres apart, with the proxy records demonstrating differences: '… not only in the timing but also in the shapes of the features [i.e. the curves in the pollen diagrams] themselves' (Blaauw and Mauquoy 2012, 9). These conclusions may have significant implications both for the future practice of palaeoecology and for subsequent archaeological interpretation based on such data.

Landscape archaeology and the palaeoenvironmental record

The approach adopted in this book has sought to highlight both the quantitative value and what might be described as the essentially qualitative worth of palaeoenvironmental data. For example, the manner in which palynological data were used to describe the timing and extent of environmental change (Chapter 5) was effectively quantitative, using pollen percentages and chronological control to describe broad patterns of change. This certainly provided a broad environmental context for different periods (Chapter 8), perhaps encapsulating one of the commonest applications of such data in landscape archaeology.

However, information from palaeoenvironmental studies may also be interpreted in a more qualitative manner, for example, providing information regarding 'functional' aspects of past landscapes: Ombrotrophic mires might be characterised by deep, waterlogged and unstable peat, hazardous to access on foot and with associated implications for the perception of and relationship to these environments by past peoples (see Van de Noort 2004), and subsequent implications for archaeological interpretation (e.g. Bond 2010).

These and other related questions of the relationship between past people and environments rely on both a quantitative and also qualitative appreciation of landscape change. The raised mire environments would have been unlikely to have provided wood for fuel or building, but reeds for basket making or thatching for example, would have been available. It is overly simplistic to conceptualise processes of environmental change in a uni-directional manner with respect to processes of enculturation: a change from fen woodland to mire might have led to impeded movement and the loss of certain resources, but the mire environment would have provided greater openness and visibility, new forms of vegetation and different textures in the landscape (cf. Evans 2003). The apparent low biological productivity and treacherous nature of raised mires may not necessarily have been equated to the perception of these as wilderness or luminal and numinuous environments (cf. Van de Noort 2004; see also Midgley 2010: 62).

In this sense, palaeoenvironmental data may extend beyond the quantitative limits of the data to assist interpretation in a qualitative manner, as proposed for the late Neolithic trackway and platform, where an enriched interpretation of the possible cultural resonance of materials used in construction is derived in part from understanding of the environmental context of the site and wider landscape. Likewise, other processes such as shifts in the relative wetness of a mire surface have potential cultural implications depending upon the implied spatial and temporal resolution of this process. Localised hydrological changes might have affected access to and across mire surfaces, but synchronous change to a wetter or drier mire across an entire system may reflect climatic change (see Chapter 2) and hence indicate an allogenic process of change with a regional spatial component.

Arguably, it is not the longer term patterns of large scale environmental change, such as the Holocene relative sea level rise (cf. Van de Noort 2011), but these somewhat enigmatic 'sub-Milankovich' fluctuations, implied by peatland palaeoclimate records (e.g. such as the mid-ninth century BC 'downturn' outlined above), which present the greatest potential, as well as the greatest challenge in terms of investigating the potentially complex and recursive relationship between climate change, human activity and the archaeological record during the Holocene. More problematic still is the devising of meaningful methodologies for investigating the relationship between climatic and cultural change (e.g. Brown 2008b, 13).

However, understanding the rate, pattern and character of peatland development and subsequent palaeohydrological changes are not only key to understanding processes of wetland development (see Chapter 2), but also to accessing the wider environmental context which would have been an integral part of the lifeways of the prehistoric peoples who occupied this part of eastern England. As outlined above, robust and precise palaeoenvironmental chronologies therefore have inherent importance in terms of providing context to cultural interpretation. However, a nuanced approach to the archaeological record requires palaeoenvironmental and other related data, whilst palaeoenvironmentalists must recognise that interpretation may fruitfully extend beyond the quantitative description of past environments and contribute to aspects of social and cultural interpretation.

9.3 Peatlands, space and time: reflections

How much data are enough? Resolution and deposit modelling

This work has highlighted several issues regarding data requirements for the modelling of 'hidden' land-surfaces in particular. One of the problems centres on the resolution of data required for the appropriate sampling of the 'hidden' pre-peat landscapes. The GPR data from Hatfield Moors (Chapter 3) had been collected at intervals of around 50m along transects, and up to 200m between them. In addition, peat depth data collected by Natural England was also potentially a very useful source of information, although it had been collected at an unknown resolution. The gridded borehole survey was implemented to determine the accuracy of these data, using resolutions of between 20m and 50m. The results indicated the degree of topographic variation within the pre-peat landscape and the potential contrast between data collected at different resolutions (cf. Fletcher and Spicer 1988). This was demonstrated clearly within the area of Packards South where local topographic variations associated with a dune system were not apparent in the lower resolution datasets.

As has been observed when using a sampling approach, the resolution of the input data (e.g. that from gridded borehole survey) required to accurately represent areas of comparatively low topographic variability, contrasts sharply with that needed to represent complex three-dimensional surfaces (e.g. Fletcher and Spicer 1988). The density of sample points required is determined by the topographic complexity of the landscape, but defining an appropriate resolution for the creation of interpolated surfaces is challenging enough when the form of the landscape can be observed (e.g. Chapman and Van de Noort 2001).

Clearly, for a 'hidden' landscape such as that sealed beneath peat, it is not possible to collect data subjectively by selecting sampling locations and resolutions, an issue which is compounded when working at a landscape scale. Further research on sampling resolution could be achieved by comparing different densities of input data, although ultimately it will always be difficult to establish what resolution is required until at least some knowledge of the situation is available. Even then, choosing an appropriate resolution will be determined by both the resources available and by the types of modelling to be performed. All resolutions of sampling will result in a model of an unknown land-surface, but whether such a model is appropriate will be determined by the intended aims of the study.

Whilst there is no absolute quantity of data that is required for the study of the archaeo-environmental resource of a raised mire landscape, it is clear that quality does matter, as demonstrated by the variations in the datasets for Hatfield and Thorne Moors. In the case of Hatfield Moors, the availability of GPR and LIDAR had a considerable impact on the ability to generate accurate models of both the pre-peat landscape and the morphology of the current surface of the peatlands. Similarly, the greater number of radiocarbon dates for both the base and the top of the peat from Hatfield Moors facilitated subsequent modelling and interpretation at the landscape scale. Together, these datasets enabled the modelling of the pre-peat land-surface, of wetland inception and peat spread, and the generation of models quantifying the surviving peatland resource. For such integration, spatial datasets must first be conformed from specific to continuous formats. For example, the three-dimensional points generated by sampling through borehole excavation, GPR or LIDAR survey had to be to be interpolated to generate a continuous surface. Similarly, data relating to environmental change including wetland inception and peat spread were obtained from samples taken from specific points in space across these landscapes and through the modelling process, used to generate continuous surfaces representing an interpolated chronozone of wetland inception. Through the creation of such 'continuous' models of peatlands it is possible to better visualise the data as a land-surface and thus to interpret sites within this context.

Quantifying the archaeo-environmental record of peatlands for cultural resource management

The fourth objective of the study was to quantify the surviving archaeo-environmental resource of the peatlands (Chapter 6). For Hatfield Moors three outputs were generated to address this objective. Firstly, a model of peat depths across the Moors was generated, which demonstrated that deposits of up to a depth of 3.5m survived, but for over 50% of the landscape less than 1m of peat remains and less than 1% of it with more than 3m. Secondly, from the analysis of wetland inception (see above), a model was generated showing the probable earliest date of peat and hence the potential for the survival of the archaeo-environmental resource. Finally, a model of the likely dates of the surface peat across the study area was created. This showed that over 90 percent had been cut down to Bronze Age levels, with only limited areas of surface peat dating to the Iron Age or later. Hence, from these three models, it was possible to determine the likely depth of peat for any location across the Moors and to estimate the probable chronological range represented by the deposits.

There is of course a direct relationship between peat survival and the potential for wet-preservation of archaeological remains (*cf.* Coles and Coles 1996). The youngest surviving peat on Hatfield dates to the Bronze Age, indicating that it is unlikely that sites of later periods will remain *in situ*, except perhaps at a few discrete locations around the fringes of the study area where more recent peat survives. However, in many areas there is the potential for the preservation of wet-preserved sites, dating to as early as the later Mesolithic.

For Thorne Moors the process of quantifying the archaeo-environmental resource was limited due to the lower number of available datasets, with the resulting models less precise due to the lower resolution of data. Despite this, two outputs were generated. The first consisted of a model of peat depths from across Thorne Moors which demonstrated that a greater quantity of peat survives than for Hatfield Moors, with less than 15% of the landscape having less than 1m of organic deposits. A range of between 0.02m and 2.22m of peat was shown to survive overall, with most areas having between 1m and 2m. Within the topographic low points, such as the area of Rawcliffe Moor to the north and the channel feature at Middle Moor towards the centre of Thorne Moors, a greater depth of peat was identified with up to 4.13m of organic deposits surviving.

As for Hatfield Moors, the second output was a model of the estimated date of wetland inception across the landscape. Unlike for Hatfield Moors, the third output relating to the dates of the surface peat in across Thorne Moors was not possible due to insufficient three-dimensional surface data. This was exacerbated by there being only one recent radiocarbon date available for the recent peat surface which, for Middle Moor at least, provided a date of cal. AD 350–540 (Gearey 2005). The available data indicate that deposits survive on the Moors dating potentially from the earlier Neolithic through to the early post-Roman period. It is therefore feasible that any archaeology associated with human activity on the Moors may also be preserved *in situ*.

It can be hypothesised that the most likely locations for the preservation and survival of archaeological material on the remaining core of peatland would be in the very earliest stages of peat growth (Mesolithic–Neolithic), before the early centres of wetland had coalesced and hence movement and access were hindered but not prevented. The local context of the Hatfield trackway and platform, the lithic finds (Chapter 7) and the Thorne Moors Bronze Age trackway may support this inference, indicating that human communities were present on both Moors during the earlier stages of peat growth, but at a time before movement across and through the landscape became impracticable. The Hatfield trackway and platform site was located at the later Neolithic wetland–dryland interface north of Lindholme Island, often a key context for the location of archaeological sites in prehistory (e.g. Gowen *et al.* 2005; see also Chapter 2).

Only these early centres and interfaces of wetland development now survive *in situ*, although the preservation of the trackway site itself was partial. Any further sites associated with early centres of wetland development are probably within or beneath the peat matrix and hence effectively preserved *in situ*, although the effects of the on-going re-wetting of the peatlands on any associated archaeological sites or deposits are unclear.

It is difficult to identify or model the wetland–dryland interface for periods later than perhaps the early Bronze Age (Chapter 6), but for both peatlands it has probably largely been cut away and/or lies under reclaimed agricultural land. Despite this, the modelling of peat development and survival and the associated palaeoenvironmental analyses may provide some clues to the possible character of post-Bronze Age archaeological sites. Spatial variation in the character of the mire surface appears to be one key at least to archaeological site location in raised mires (Gowen *et al.* 2005), but given the lack of surviving ombrotrophic peat on either area (Chapter 6), this could not be meaningfully modelled for either study area. Little research has been carried out to investigate the practical aspects of prehistoric trackway construction in terms of the technological limitations of different techniques of build with respect to specific hydrological conditions (but see Casparie and Moloney 1994); it could be argued that the raised mire deposits may have been too wet for such any form of trackway construction through much of later prehistory.

Alternatively, if such sites had been built they are likely to have been extensive linear structures built to cross the mires, which it could be hypothesised, would be more likely to have been identified by the various fieldwalking projects (Chapter 2), even if partially destroyed, compared to shorter structures across discrete wet hollows or pools on the mire surface. If this is accepted, then the failure of past survey to locate any archaeology, may relate to an actual absence of surviving sites on the main industrially extracted area of the peatlands at least. The current and likely future management regime for the peatlands means that it is very unlikely that these hypotheses regarding the possible *in situ* preservation of archaeological sites will ever be tested.

9.4 The heritage management of peatlands in the twenty-first century

Models of the surviving peat following centuries of extraction are useful data for the cultural resource management of the Humber peatlands, but just as importantly demonstrate a methodology for assessing other extracted peatlands. Models of peat depth and extent are also essential for determining carbon storage in these environments (JNCC 2011). Many of the raised mire landscapes in the UK are designated as Sites of Special Scientific Interest (SSSI) on ecological grounds and it has been estimated that around 10% of the lowland bogs in Great Britain and Northern Ireland have been designated as SSSIs (Lindsay and Immirzi 1996) or Areas of Special Scientific Interest (Northern Ireland Habitat Action Plan 2003) respectively. Until recently, 83% of the total area under peat extraction takes place on areas that have been designated as SSSIs or have otherwise been identified for their nature conservation value (DETR 1998, Van de Noort *et al.* 2002).

A total of 109 sites within lowland peatlands have been afforded protection as Scheduled Ancient Monuments by English Heritage, although the majority of archaeological sites thus protected are located in the Somerset Levels (see Chapter 2). Some raised mires have been classified as Special Areas of Conservation (SAC) under the European Habitats Directive (European Commission 1992, Alexander *et al.* 2008), although this has arguably not resulted in the effective protection of the archaeo-environmental resource. For example, in 1994 when ownership of Hatfield and Thorne Moors passed to English Nature (now Natural England), peat cutting rights were retained with the agreement that extraction could continue except for the lower 0.5m of peat.

The restoration of degraded and damaged peatlands is of increasing significance in the UK. Recent restoration work on Hatfield and Thorne Moors is part of the International Union for the Conservation of Natures Peatlands Programme, which has identified a goal of bringing one million hectares of peatlands in the UK into good condition or 'restorative management' by 2020 (Bain *et al.* 2011). Restoration usually involves techniques to stabilise eroding surfaces, re-establish peatland vegetation and raise the watertable, and hence encourage waterlogged conditions that will enable peat to grow again (Worrall *et al.* 2010). In addition to the general conservation and biodiversity value of healthy peatlands, restoration of degraded peatlands may reduce carbon losses to both the atmosphere (e.g. Tuittila *et al.* 1999) and the aqueous environment (e.g. Waddington *et al.* 2009, Holden *et al.* 2007). Although the precise processes are complex and in many ways poorly understood (see Worrall *et al.* 2010), restoration of damaged and cut-over peatlands look set become an increasing imperative part of climate change mitigation in the twenty-first century.

It is essential that these developing agendas benefit from archaeological input to ensure the promotion and protection of the archaeo-environmental resource within the context of 'Ecosystem Service' frameworks which encapsulate the under-valued contributions which peatlands make to society (see Gearey *et al.* in press). Resource assessments of the form described in this book represent an attempt to demonstrate that, as discussed below, even extracted and damaged mires can still provide significant heritage 'service value'. Developing methods to assess the archaeological potential of peatlands and to assist in their future management is of enhanced importance against a backdrop of management imperatives which stress re-wetting and restoration, and bring potential additional threats as well as opportunities for sites such as the Humber peatlands and other such landscapes.

9.5 Landscape archaeology and peatlands: developing a 'tool-kit' for future archaeo-environmental investigations

Wetland archaeology requires an interdisciplinary approach which combines archaeological with palaeoenvironmental methodologies and expertise, and the synergies which can arise from such collaborative work have been well demonstrated by previous projects (e.g. Pryor *et al.* 1985, Pryor 2001, Van de Noort and Davies 1993, Van de Noort 2002, Menotti 2012, Leah *et al.* 1998). The results from this work underline the importance of an integrated approach and provide new avenues for understanding the evolution of peatland landscapes and for interpreting the archaeo-environmental resource. The archaeological record from mires may be remarkable, but sites and finds are often relatively poorly understood in relation to their contemporary environmental and landscape context. For example, despite in-depth analyses of the Iron Age bog body 'Lindow Man' from northwest England (Stead *et al.* 1986), there was no attempt to establish the wider landscape context such as its proximity to contemporary dryland for example, arguably a key aspect of any attempt to interpret this find. The modelling approach outlined in this book has considerable potential for providing a landscape context to such finds and a method for understanding patterns of environmental stasis and change.

Whilst this book has not aimed to provide a 'tool kit' for future integrated archaeo-environmental investigations of extensive peatlands, it is possible to propose a broad framework for such work. The significance of the morphology of the pre-peat landscape for contextualising palaeoenvironmental research and for devising associated radiocarbon dating programmes has emerged as a key theme. The mapping of such landscapes in three dimensions is difficult but GPR provides a robust methodology for this process, with the excavation of grids of cores providing a means of 'ground truthing' if required. Three dimensional modelling can of course be carried out using accurately georeferenced grids of cores (Chapter 3; Chapman and Gearey 2002), but this process is potentially more time consuming compared to geophysical survey, although this of course depends on the relative complexity of the hidden landscape and adopted sampling resolutions. Given the problems of 'sampling blind' using any form of remote intervention (discussed above), especially for sites with little pre-existing information, it may be necessary to adopt a 'nested approach' to survey, beginning with relatively coarse intervals, subsequently refining and targeting specific areas as data accrues.

Remote sensing technologies clearly have important roles to play in archaeological research in peatlands and wetlands. The models of the peat depth and associated archaeo-environmental resource were derived from a combination of GPR and LIDAR. The value of the latter for the remote prospection of archaeological landscapes is now well established (e.g. Challis 2005, Crutchley 2006, Doneus *et al.* 2008, Hesse 2010, Challis *et al.* 2011). Although GPR cannot be used to locate archaeological sites in wetlands (but see Utsi 2004), it can be employed to map the interface between organic sediments and the underlying geology, as described above, providing critical data regarding pre-peat land-surfaces. Such data may also be used in a 'predictive' capacity, identifying locations where deposits and sites are more or less likely to be preserved *in situ* (see Carey *et al.* 2006, Howard *et al.* 2008 for examples of this application in alluvial landscapes).

Knowledge of peat depth with respect to concealed land-surfaces allows sampling of cores for subsequent palaeoenvironmental analyses to be carried out in an informed and targeted manner, taking account of the potential implications of the morphology of the pre-peat surface for formation processes. Interpretation of subsequent data can thus be carried out with a greater appreciation of context. Certainly, understanding archaeological sites and human activity in peatlands requires knowledge of patterns and processes of environmental change on a range of chronological and spatial scales.

Multi sequence, multi-proxy studies are necessary for this, with associated comprehensive chronological control. The Bayesian modelling has demonstrated that the chronological relationship between 'events' should be assessed in a formal manner, to prevent erroneous assumptions regarding 'synchroneity' and hence inferences concerning past patterns and processes of change. The programme has also indirectly stressed the long known problems regarding the radiocarbon dating of different fractions of peat. Detailed programmes of palaeoenvironmental work can be compromised by an inadequate radiocarbon dating programme, or even by the nature of the deposit formation processes which might complicate the selection of the 'correct' fraction for submission. Given the relatively recent advent of Bayesian methodologies, further re-assessment of other palaeoenvironmental studies are also warranted, as flawed chronological control may significantly affect conclusions drawn from otherwise detailed palaeoenvironmental analyses.

9.6 Peatlands, palaeoecology and archaeology

This book began with a brief discussion of aspects of space and time (chronology) within archaeological and palaeoenvironmental research. It was proposed in Chapter 1 that wetland and peatland environments in particular presented specific opportunities for investigating spatial and temporal patterns and processes of environmental change (see also Blaauw *et al.* 2010) as well as the potential for enriching the interpretation of the archaeological record. The results have demonstrated something of the potential complexity of and variability between palaeoenvironmental records from the 'same' peatland systems and the potential for incomplete or misleading interpretations of spatial patterns of change derived from single sequences. The importance and difficulties of robust chronological control and the value of formal chronological modelling methodologies have been demonstrated. Methods for accessing and modelling 'hidden land-surfaces' have been developed, and the importance of topographic data for contextualising patterns of mire inception has been highlighted. Likewise, an appreciation of the spatial patterns of landscape change across a range of spatial scales has been illustrated through the interpretation of a rather unusual late Neolithic site.

This study has demonstrated that the reconstruction of spatial patterns of past environmental change is complex, even where the physical remains of the process being investigated (i.e. in this case, peat growth and spread) survive in the present, providing material for analyses. Other processes of relevance and interest to both palaeoenvironmental and archaeological debate, such as Holocene climatic change, provide challenges in terms of extrapolating the spatial extent of the process from proxy data such as those from BSW records.

As stated in Chapter 1, despite the considerable body of palaeoecological research into peatlands over the last three decades or so, there have been few attempts to assess the results or methodological advances in this area from a specifically archaeological perspective (Gearey and Chapman 2004). This may be related in part to the different paradigms of research; much palaeoecological research into raised mires has aimed at assessing changes in Holocene palaeoclimate (*cf.* Charman 2002). The time and resources required for analyses of this kind means that the general trend in sampling strategies has been for single core studies, which are regarded as 'representative' of the entire mire system under consideration. Such approaches have been underpinned by prevailing theories regarding the growth of mires (e.g. Barber 1981) alongside investigations of multiple cores which tend to indicate that, assuming the appropriate sampling location is selected (e.g. Barber 1994), the results of the subsequent analyses will be representative of the 'response' of the entire system to past hydrological changes. The degree of coherence between Holocene palaeoclimatic records from mire sequences and an associated correspondence with other proxies may validate this approach (Charman 2010; but see also Swindles *et al.* 2012).

However, archaeological investigations of peatlands, although drawing explicitly on palaeoenvironmental analyses and data (Chapter 2), arguably represent a very different paradigm compared to palaeoecology. Archaeological focus begins on the local scale where '… stratigraphy reveals, delimits and contextualises artefacts and provides the basic data for archaeological interpretation.' (Brown 2008a, 279). For peatland sites, this generally entails a detailed study of the on-site stratigraphy and palaeoenvironmental analyses of associated deposits, carried out first and foremost to investigate the local context of a site, and being less concerned, initially at least, with explicitly identifying the signal of allogenic processes, such as climatic change. However, interpreting the archaeological record also requires an understanding of the dynamics of the landscape on a wider spatial and temporal scale, with all the methodological difficulties and interpretative uncertainties associated with such a nested approach of this kind. Single core investigations of the form undertaken in many palaeoclimatic studies, no matter how detailed, are arguably of less use for understanding mires as archaeological landscapes.

Moreover, the data generated by certain archaeological investigations of peatlands, especially that including detailed descriptions and recordings of peat stratigraphy and associated palaeoenvironmental analyses (e.g. Gowen *et al.* 2005, Bermingham and Delaney 2005; see Chapter 2), arguably provides valuable, but to date somewhat neglected, information for investigating the link between mire formation processes and the palaeohydrological record. Closer integration between essentially 'archaeological' and 'palaeoenvironmental' approaches to peatlands may hold the key to better understanding the relationship between allogenic and autogenic processes in particular.

The GIS approach presented here is one which permits the storage, analysis and manipulation of these diverse sources of information, but this is really just a means to an end and not the end itself. Integrated archaeo-environmental approaches to past landscapes have always been plagued by methodological problems concerning the meaningful integration of the three dimensions of space and the fourth dimension of time (*cf.* Wachowicz and Healey 1994). Advances in Bayesian modelling now allow formal testing and correlation of palaeoenvironmental and archaeological chronologies which, alongside recent progress in the interpretation of pollen data in a spatially meaningful manner (e.g. Fyfe 2006), present new possibilities in terms of 'bridging the gap' between the spatial and temporal uncertainties in the records. Hopefully this will lead to a new paradigm in our practical and theoretical approaches to, and understanding of, past landscapes, human activity and the archaeological record.

Whilst wetland landscapes may present unique opportunities with the preservation of a four dimensional archive of landscape absent from dryland contexts, the cost and technical problems of investigating the 'hidden landscapes' of peatlands and the current emphasis on restoration in some areas and continuing peat extraction in

Figure 9.4 Sunset on Hatfield Moors

others means that the full archaeological potential of such environments may never be fully realised (*cf.* Gearey and Chapman 2004). The extensive areas of prehistoric and later palaeolandscape preserved beneath upland and lowland peatlands remain, to all intents and purposes, largely uninvestigated, except where destructive processes such as drainage and peat cutting expose deposits and buried surfaces. Out of sight should not, in this case, necessarily mean out of mind.

The approach developed for Hatfield and Thorne Moors demonstrates one approach to the problem as well as the potential of these hidden landscapes. Not all peatlands benefit from the extensive past research carried out on Hatfield and Thorne Moors. However, data which may not have been originally collected for specifically archaeological purposes, such as the LIDAR, peat depth and geophysical data from Hatfield Moors, may be available for other areas. This may be 'recycled', and used alongside targeted palaeoenvironmental study and additional radiocarbon dating, to generate robust and testable models useful for archaeological interpretation and management.

From a disciplinary perspective, the work presented here has drawn on a range of theoretical and methodological frameworks, from the 'archaeological sciences' including scientific dating, GIS modelling and palaeoenvironmental analyses and interpretation, through to various themes in 'cultural' archaeology. All approaches to the archaeological record and past landscapes ultimately involve the investigation of various physical features of landscape and observable, and in some measure quantifiable, correlations with archaeological sites and the resulting implications for past human lives. Whether the aim is to map the physical distribution of lithics scatters across a region, for example, or an investigation of 'subjective, lived environments', the focus is on the inter-relationship between past people, sites, monuments and features of the physical environment which may be observed, recorded and mapped, whether by conventional methods used in landscape archaeology, or by 'embodied' means exemplified by 'phenomenological' or 'experiential-locational' approaches. It is, by definition, impossible to investigate a landscape which cannot be observed or recorded directly and different methodologies and technologies present different perspectives and opportunities. As Sturt (2006, 123) has observed, with reference to investigating the early Holocene 'hidden landscapes' of the Fenlands of east England, this 'encourages an approach to landscape which moves beyond seeing engagements with space today as a direct correlate to understanding life in the past.'

This need not mean a straight-jacketed, representational approach to aims and methodologies for understanding past landscapes. Hacigüzeller (2012, 257) has stated that 'by interacting with digital or non-digital technology and a series of animate and inanimate materials, we create narratives about past lives as well.' Developing approaches for the study of hidden landscapes such as peatlands is potentially complex due to the wide range of disciplinary approaches involved, and the spatial and temporal precision and accuracy of different datasets. This book has detailed one approach to this which has focussed on raised mires, but there should be considerable value in applying the lessons learned and the methods used to other landscapes.

Appendix 1. Radiocarbon dating protocol and Bayesian chronological modelling methodology

Radiocarbon dating programme

The 40 samples from both the base and the top of peat from Hatfield Moors (Chapter 5) consisted of short-lived charcoal, fragments of waterlogged wood, plant macrofossils and bulk peat samples and were submitted for dating by Accelerator Mass Spectrometry (AMS) at the Scottish Universities Environmental Research Centre (SUERC), East Kilbride and the Oxford Radiocarbon Accelerator (OxA). The preference for identifiable plant macrofossils for radiocarbon dating is related to two main considerations:

- Sample size: improvements in the size of sample that can be measured (*c.* 2mg of organic carbon), thus increasing the degree of stratigraphic resolution.
- Taphonomic: the belief that plant macrofossils are more reliable than "bulk" samples of sediment matrix as the source of carbon in the former is known and they are not made up of heterogeneous material that could be of different ages (Walker *et al.* 2001, Lowe and Walker 2000).

The samples were pretreated by the acid-base-acid protocol (Stenhouse and Baxter 1983), converted to carbon dioxide in pre-cleaned sealed quartz tube (Vandeputte *et al.* 1996), graphitised as described by Slota *et al.* (1987), and were measured by AMS (Xu *et al.* 2004). The measurements obtained from the Oxford Radiocarbon Accelerator were obtained from samples of short-lived charcoal, "bulk" sediment and waterlogged plant macrofossils. These were prepared following the procedures described in Hedges *et al.* (1989) and Bronk Ramsey *et al.* (2002) and measured by Accelerator Mass Spectrometry as described by Bronk Ramsey *et al.* (2004). The conventional radiocarbon ages (Stuiver and Polach 1977) are quoted in accordance with the international standard known as the Trondheim convention (Stuiver and Kra 1986). The results of the Hatfield dating programme are discussed in Chapter 5.

Calibration

All radiocarbon calibrations have been calculated using the calibration curve of Reimer *et al.* (2004) and the computer program OxCal v4.0.5 (Bronk Ramsey 1995, 1998, 2001, 2008). The calibrated date ranges cited are those for 95% confidence and quoted in the form recommended by Mook (1986), with the end points rounded outwards to 10 years. The ranges in plain type in Tables A1–A18 have been calculated according to the maximum intercept method (Stuiver and Reimer 1986), and those quoted in italics are *posterior density estimates* derived from mathematical modelling (see below). All other ranges are derived from the probability method (Stuiver and Reimer 1993).

Chronological modelling of the legacy palaeoenvironmental datasets

Although the calibrated dates may be accurate estimates of the sample in question, It is the dates of the 'events' represented in the palaeoenvironmental sequences, which are represented by those samples which are of interest. The dates of these events can be estimated by using the absolute dating information from the radiocarbon measurements, and the stratigraphic relationships between samples. There has been much discussion over the years about the most 'accurate' way of producing such "age-depth" models to estimate the date of "events" in a given palaeoenvironmental record, and a range of methods have been utilised, for example linear interpolation, splines and linear regression (Bennett 1994), mixed-effect models (Heegard *et al.* 2005) and fuzzy regression (Boreux *et al.* 1997). Methodology is now available which allows the combination of these different types of information explicitly, to produce robust estimates of the dates of interest (see for example Bronk Ramsey 2008, Blockley *et al.* 2008). It should be emphasised that the *posterior density estimates* produced by this process are interpretative *estimates*, which could change as further data become available and as other researchers choose to model the data from different perspectives.

The technique used was a form of Markov Chain Monte Carlo sampling, applied using the program OxCal v4.0.5 (http://c14.arch.ox.ac.uk/). Details of the algorithms employed by this program are available from the on-line manual or in Bronk Ramsey (1995, 1998, 2001, 2008). Initially uniform deposition, or *U-Sequence,* models were implemented using the computer programme OxCal (Bronk Ramsey 2008) in which the sediment accumulation rate is unknown but assumed to have been constant (Christen *et al.* 1995), an assumption often made in palaeoecological research (e.g. Barber *et al.* 2003, Chiverrell 2001).

In all but one case (see below), the *U-Sequence* models for individual sequences showed poor overall agreement between the radiocarbon measurements and this assumption. Constructing *U-Sequence* models allowing for changes in the deposition rate to be specified at boundaries (Blaauw and Christen 2005) identified in the plant macrofossil or stratigraphic records (see Chapter 2) also showed poor overall agreement; perhaps because such models are essentially uniform deposition models, broken up into sections defined by changes in the sequence (Bronk Ramsey 2008). An alternative approach to that encapsulated in uniform deposition models is to assume that deposition is random or based on a "Poisson" process (Bronk Ramsey 2008). Such *P-Sequence* models were run ($k = 10$ cm^{-1}, i.e. 10 depositional events per cm) incorporating the boundaries identified through changes in the plant macrofossil records or in the stratigraphic record of each sequence.

For the models presented below, each plotted radiocarbon distribution represents the relative probability that an event occurred at a particular time. The *P-Sequence* models show probability distributions of dates from the modelled sequences (see chapter 5). For each of the radiocarbon dates in these models, two distributions have been plotted, one in outline, which is the result of simple radiocarbon calibration and a solid one, which is based on the chronological model used (*P-Sequence*). The percentage value in brackets after the radiocarbon measurements are the individual index of agreement (Bronk Ramsey 1995). This index provides a measure of how well the *posterior distribution* agrees with the *prior distribution* (for further details see Bayliss *et al.* 2007, Blockley *et al.* 2008).

Hatfield Moors

Hatfield Moors Site 1 (HAT1)

Ten measurements (Table A1.1) were obtained on bulk peat samples from a vertical peat face dug into the side of a drainage ditch which encircles Lindholme (Smith 2002). The *P-Sequence* model (Figure A1.1) shows good overall agreement between the model and data ($A_{overall}$= 104.3%).

Hatfield Moors Site 2 (HAT2)

Thirteen measurements (Table A1.2) were obtained on bulk peat samples taken from monoliths taken from the sides of a pit dug next to Porters Drain (Smith 2002). The *P-Sequence* model (Figure A1.2) shows good overall agreement between the model and data ($A_{overall}$= 108.3%).

Hatfield Moors Site 3 (HAT3)

A single measurement (Table A1.3 and Figure A1.3) was obtained on a sample of bulk *Pinus* twigs from the peat base (Whitehouse 2004).

Hatfield Moors Site 4 (HAT4)

Eight measurements (Table A1.4 and Figure A1.4) were obtained on bulk samples of *Pinus/Betula* twigs from a vertical section on the southern edge of Hatfield Moors (Whitehouse 2004). The results do not provide an accurate chronology. The six samples from the middle of the sequence (SRR-6121-6126) are statistically consistent (T'=8.5; ν=5; T'(5%)=3.8; Ward and Wilson, 1978) and could therefore be of the same actual age. This suggest that the submission of bulk samples of twigs from 5cm sections of peat has probably incorporated material of varying ages that has essentially produced an average age.

Table A1.1 Hatfield Moors Site 1 radiocarbon results

Laboratory Number	Material/context	Radiocarbon Age (BP)	Calibrated date (95% confidence)	Posterior Density Estimate (95% probability)
CAR-168	Peat; 2.00m	4180±70	2920–2500 cal. BC	*2910–2580 cal. BC*
CAR-169	Peat; 1.72m	3570±70	2140–1730 cal. BC	*1960–1730 cal. BC*
CAR-170	Peat; 1.46m	2775±65	1120–800 cal. BC	*1120–840 cal. BC*
CAR-171	Peat; 1.34m	2695±65	1000–780 cal. BC	*980–780 cal. BC*
CAR-172	Peat; 1.13m	2345±65	750–210 cal. BC	*750–410 cal. BC*
CAR-173	Peat; 1.12m	2470±65	800–390 cal. BC	*730–390 cal. BC*
CAR-174	Peat; 0.92m	2145±65	390–10 cal. BC	*380–40 cal. BC*
CAR-176	Peat; 0.60m	1475±55	cal. AD 430–660	*cal. AD 440–480 (1%) or 510–660 (94%)*
CAR-177	Peat; 0.59m	1460±75	cal. AD 420–680	*cal. AD 470–490 (1%) or 530–680 (94%)*
CAR-179	Peat; 0.43m	1175±75	cal. AD 660–1020	*cal. AD 680–980*

Lindholme Bank Road (LIND_B)

Two measurements (Table A1.5 and Figure A1.5) were obtained on bulk samples of *Calluna* seeds from a section in the vicinity of Lindholme Island (Whitehouse 2004).

Porters Drain

Six measurements (Table A1.6 and Figure A1.6) were obtained on AMS bulk samples of peat from the Porters Drain (Gearey 2005). Both *P-Sequence* and *U-Sequence* models from this sequence showed poor overall agreement between the model and data ($A_{overall}$ = <60.0%). This is probably the result of the sequence containing a number of outliers, however, identifying these is somewhat problematic given all the samples were AMS sized bulk peat samples.

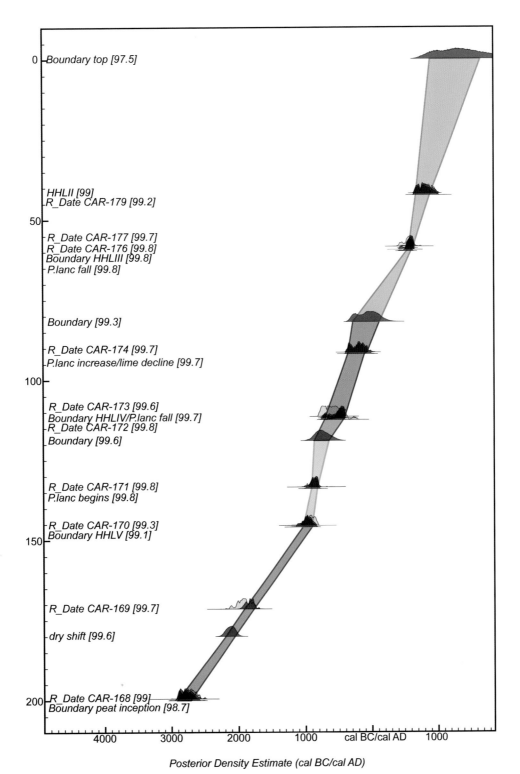

Figure A1.1 Probability distributions of dates from Hatfield Moors Site 1 (HAT1) P-Sequence model

170 *Modelling archaeology and paleoenvironments in wetlands*

Hatfield Moors 'recurrence surfaces'

Five measurements (Table A1.7) were obtained on bulk peat samples intended to date three consecutive 'recurrence surfaces' A, B, and C (Turner 1962).

Thorne Moors

Crowle Moor Site 1 (CLM1)

Thirteen measurements (Table A1.8) from Crowle Moor Site 1 ('The Crown Peat Works Site') were obtained on bulk

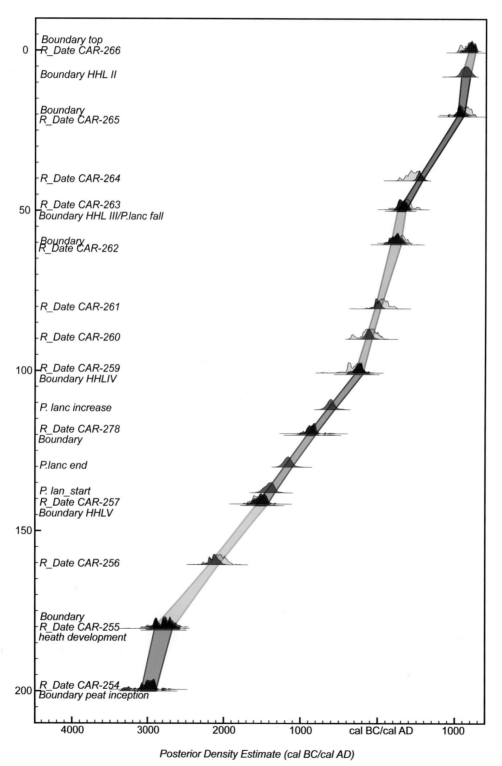

Figure A1.2 Probability distributions of dates from Hatfield Moors site 2 (HAT2) P-Sequence model

Appendix 1. Radiocarbon dating protocol and Bayesian chronological modelling methodology 171

Table A1.2 Hatfield Moors Site 2 radiocarbon results

Laboratory Number	Material/context	Radiocarbon Age (BP)	Calibrated date (95% confidence)	Posterior Density Estimate (95% probability)
CAR-254	Peat; 2.00m	4335±75	3320–2780 cal. BC	*3340–3210 (9%) or 3180–3150 (1%) or 3130–2860 (83%) or 2800–2760 (2%) cal. BC*
CAR-255	Peat; 1.81m	4255±75	3080–2630 cal. BC	2950–2580 cal. BC
CAR-256	Peat; 1.61m	3685±65	2280–1895 cal. BC	2270–2010 cal. BC
CAR-257	Peat; 1.42m	3240±70	1690–1390 cal. BC	1670–1380 cal. BC
CAR-278	Peat; 1.20m	2675±70	980–760 cal. BC	1020–750 (93%) or 690–660 (1%) or 640–590 (1%)
CAR-259	Peat; 1.01m	2265±70	420–160 cal. BC	390–110 cal. BC
CAR-260	Peat; 0.90m	2085±70	360 cal. BC–cal. AD 70	210–30 cal. BC
CAR-261	Peat; 0.81m	1945±65	100 cal. BC–cal. AD 230	90 cal. BC–cal. AD 70
CAR-262	Peat; 0.61m	1775±65	cal. AD 80–420	cal. AD 110–350
CAR-263	Peat; 0.50m	1700±65	cal. AD 140–540	cal. AD 220–420
CAR-264	Peat; 0.41m	1616±65	cal. AD 250–600	cal. AD 460–620
CAR-265	Peat; 0.21m	910±65	cal. AD 990–1270	cal. AD 900–920 (2%) or 970–1170 (93%)
CAR-266	Peat; 0.00m	865±60	cal. AD 1020–1280	cal. AD 1090–1280

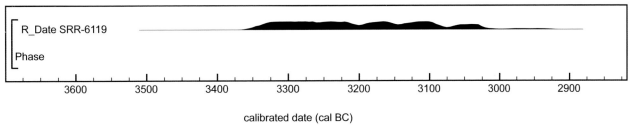

Figure A1.3 Probability distribution of dates from Hatfield Moors Site 3 (HAT3). The distribution represents the relative probability that an event occurred at a particular time. The distribution is the result of simple radiocarbon calibration (Stuiver and Reimer 1993)

Table A1.3 Hatfield Moors Site 3 radiocarbon result

Laboratory Number	Material/context	Radiocarbon Age (BP)	Calibrated date (95% confidence)
SRR-6119	Bulk *Pinus* twigs: 0.95–1.03m	4480±45	3360–3010 cal. BC

Table A1.4 Hatfield Moors Site 4 radiocarbon results

Laboratory Number	Material/context	Radiocarbon Age (BP)	Calibrated date (95% confidence)
SRR-6127	Bulk *Pinus/Betula* twigs: 1.50–1.55m	3164±40	1520–1320 cal. BC
SRR-6126	Bulk *Pinus/Betula* twigs: 1.20–1.25m	2965±45	1380–1020 cal. BC
SRR-6125	Bulk *Pinus/Betula* twigs: 1.05–1.10m	2890±40	1260–930 cal. BC
SRR-6124	Bulk *Pinus/Betula* twigs: 0.95–1.00m	2960±45	1380–1010 cal. BC
SRR-6123	Bulk *Pinus/Betula* twigs: 0.85–0.90m	3020±45	1410–1120 cal. BC
SRR-6122	Bulk *Pinus/Betula* twigs: 0.75–0.80m	2960±40	1370–1040 cal. BC
SRR-6121	Bulk *Pinus/Betula* twigs: 0.40–0.45m	2870±40	1200–920 cal. BC
SRR-6120	Bulk *Pinus/Betula* twigs: 0.10–0.15m	1405±45	cal. AD 570–680

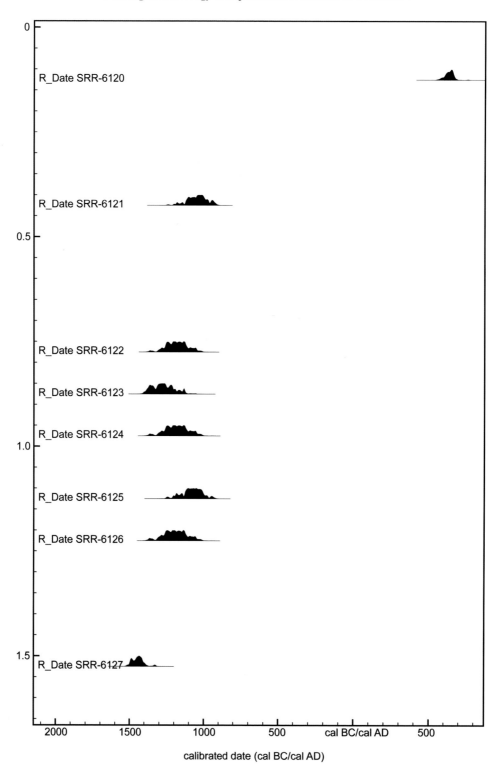

Figure A1.4 Probability distribution of dates from Hatfield Moors site 4 (HAT4)

peat samples from a peat monolith and excavated blocks of peat, taken from the section of a hand-dug pit (Smith 2002). The *P-Sequence* model (Figure A1.7) shows good overall agreement between the model and data ($A_{overall}$= 66.9%).

Crowle Moor Site 2 (CLM2)

Four measurements (Table A1.9) were obtained from Crowle Moor Site 2 (Crowle Moor Nature Reserve) on bulk peat samples taken with a large Russian-type sampler. Additionally two measurements (Figure A8) were obtained from a *Pinus sylvestris* that had been burnt and chopped (CAR-313) and a horizon in the peat at the site that contained significant quantities of charcoal (CAR-247). The *P-Sequence* model (Figure A1.8) shows good overall agreement between the model and data ($A_{overall}$= 85.6%).

Appendix 1. Radiocarbon dating protocol and Bayesian chronological modelling methodology

Table A1.5 Lindholme Bank Road (LIND_B) radiocarbon results

Laboratory Number	Material/context	Radiocarbon Age (BP)	Calibrated date (95% confidence)
Beta-91800	*Calluna* seeds: 1.60–1.72m	3990±60	2840–2300 cal. BC
Beta-91799	*Calluna* seeds: 1.15–1.20m	2828±50	1130–840 cal. BC

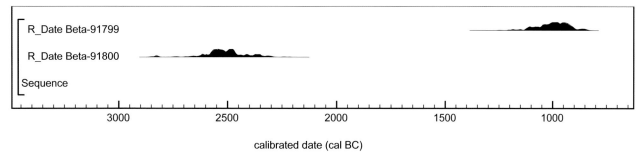

Figure A1.5 Probability distribution of dates from Lindholme Bank Road (LIND_B). Each distribution represents the relative probability that an event occurred at a particular time. These distributions are the result of simple radiocarbon calibration (Stuiver and Reimer 1993)

Goole Moor Site 1 (GLM1 area)

Eleven measurements (Table A1.10) were obtained from bulk peat samples taken from a section excavated from the Blackwater dyke side (Smith 2002). The *P-Sequence* model (Figure A1.9) shows good overall agreement between the model and data ($A_{overall}$= 62.0%).

Goole Moor Site 2 (GLM1 area)

Four measurements (Table A1.11) were obtained on bulk peat samples taken from a 1m monolith collected from a section of a pit dug next to the sloping edge of the Blackwater dyke. The *P-Sequence* model (Figure A1.10) shows good overall agreement between the model and data ($A_{overall}$= 125.1%).

Goole Moor Site 3 (GLM1 area)

Two measurements (Table A1.12) were obtained on bulk peat samples taken from a 1m monolith collected from a section of a pit dug next to the sloping edge of the Blackwater dyke (Smith 2002). The *U-Sequence* model (Figure A1.11) shows good overall agreement between the model and data ($A_{overall}$= 133.0%).

Rawcliffe Moor Site 1 (RWM1)

Twelve measurements (Table A1.13) were obtained on bulk peat samples from a pit dug into the side of a dyke (Smith 2002). The *P-Sequence* model (Figure A1.12) shows good overall agreement between the model and data ($A_{overall}$= 104.2%).

Thorne Moor Site 1 (TM1 area)

Six measurements (Table A1.14) were obtained on bulk peat samples from a pit dug adjacent to the trackway site excavated by Buckland (1979) (Smith 2002).

Thorne Moor Site 2 (TM1 area)

Two measurements (Table A1.15) were obtained on bulk peat samples from a pit dug adjacent to the trackway site (Buckland 1979, Smith 2002) and TM1 from a very obvious 'recurrence surface' which could be observed in all peat cuttings in the area.

The *P-Sequence* model (Figure A1.13) shows good overall agreement between the model and data ($A_{overall}$= 100.8%), although one measurement CAR-188 had to be removed.

Thorne Moor Trackway site (TM1 area)

Three measurements (Table A1.16) were obtained on wood washed from peat overlying clayey silt and an associated timber trackway (Buckland 1979). Given the small number of samples and uncertainty about the absolute relationship between the samples and the sequence was modelled as a simple *Sequence* (Bronk Ramsey 1995). The model (Figure A1.14) shows good agreement between the radiocarbon results and stratigraphy ($A_{overall}$=112.6%).

Thorne Waste Site 1

A single measurement (Table A1.17) was obtained on a bulk peat sample (Turner 1962) from a site *c.* 800m south of TM1 (Smith 2002).

Table A1.6 Porters Drain radiocarbon results

Laboratory Number	Material/context	Radiocarbon Age (BP)	Calibrated date (95% confidence)
NZA-18212	Peat; 2.40m	3828±50	2470–2130 cal. BC
NZA-18211	Peat; 1.60m	3265±40	1640–1440 cal. BC
NZA-18210	Peat; 0.96m	2286±40	410–200 cal. BC
NZA-18209	Peat; 0.76m	2116±45	360–10 cal. BC
NZA-18208	Peat; 0.48m	2012±45	170 cal. BC–cal. AD 80
NZA-18207	Peat; 0.24m	1686±40	cal. AD 240–430

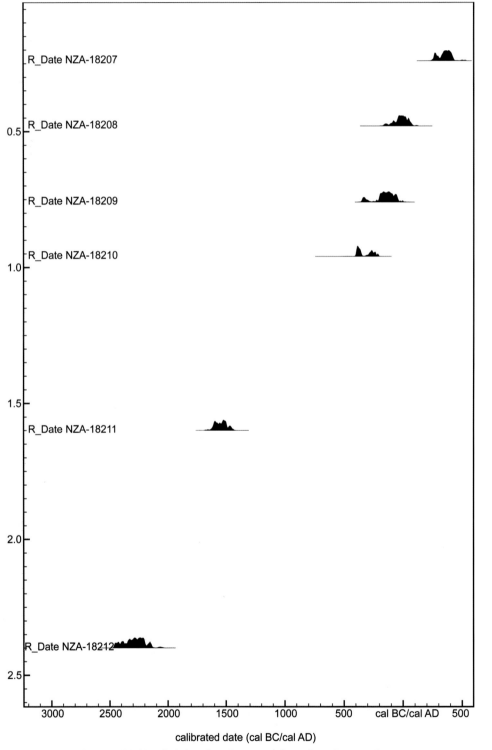

Figure A1.6 Probability distribution of dates from Porters Drain

Appendix 1. Radiocarbon dating protocol and Bayesian chronological modelling methodology

Table A1.7 Hatfield Moors 'recurrence surfaces' radiocarbon results

Laboratory Number	Material/context	Radiocarbon Age (BP)	Calibrated date (95% confidence)
Q-487	Fresh *Sphagnum* peat; 0.82–0.83m	2215±110	520 cal. BC–cal. AD 10
Q-486	Humified *Sphagnum* peat; 0.82–0.83m	1381±110	cal. AD 420–890
Q-485	Fresh *Sphagnum* peat; 0.53–0.54m	1381±110	cal. AD 420–890
Q-484	Humified *Sphagnum–Calluna* peat; 0.32–0.33m	1392±110	cal. AD 420–890
Q-483	Fresh *Sphagnum* peat; 0.31–0.32m	1384±110	cal. AD 420–890

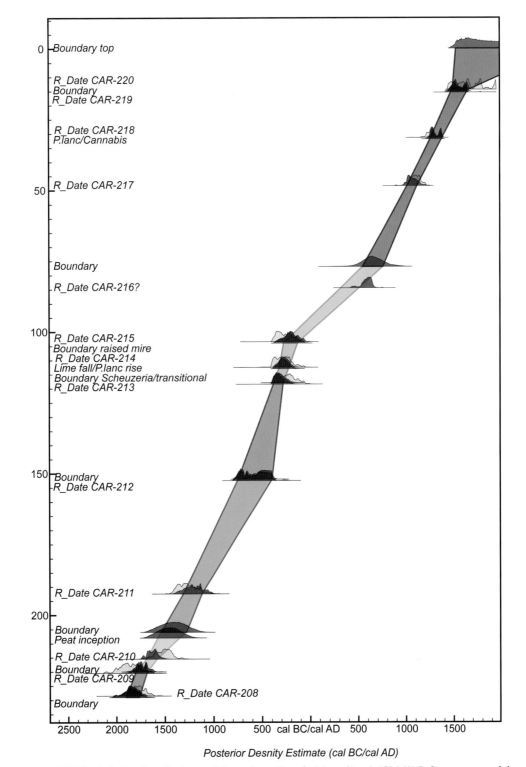

Figure A1.7 Probability distributions of dates from Crowle Moor Site 1 (CLM1) P-Sequence model

Table A1.8 Crowle Moor Site 1 radiocarbon results

Laboratory Number	Material/context	Radiocarbon Age (BP)	Calibrated date (95% confidence)	Posterior Density Estimate (95% probability)
CAR-208	Peat; 2.29m	3475±65	1960–1620 cal. BC	1970–1690 cal. BC
CAR-209	Peat; 2.21m	3540±70	2120–1690 cal. BC	1860–1620 cal. BC
CAR-210	Peat; 2.16m	3230±75	1690–1320 cal. BC	1750–1520 cal. BC
CAR-211	Peat; 1.93m	3025±65	1440–1050 cal. BC	1380–1020 cal. BC
CAR-212	Peat; 1.53m	2405±70	780–370 cal. BC	770–390 cal. BC
CAR-213	Peat; 1.18m	2185±65	400–40 cal. BC	400–200 cal. BC
CAR-214	Peat; 1.13m	2225±70	410–90 cal. BC	370–170 cal. BC
CAR-215	Peat; 1.04m	2175±60	400–40 cal. BC	330–50 cal. BC
CAR-216	Peat; 0.85m	1445±60	cal. AD 440–680	cal. AD 430–680
CAR-217	Peat; 0.49m	950±55	cal. AD 980–1220	cal. AD 980–1210
CAR-218	Peat; 0.32m	725±60	cal. AD 1210–1400	cal. AD 1250–1400
CAR-219	Peat; 0.16m	360±55	cal. AD 1430–1660	cal. AD 1450–1650
CAR-220	Peat; 0.15m	225±55	cal. AD 1520–1955*	cal. AD 1470–1680

Table A1.9 Crowle Moor Site 2 radiocarbon results

Laboratory Number	Material/context	Radiocarbon Age (BP)	Calibrated date (95% confidence)	Posterior Density Estimate (95% probability)
CAR-309	Peat; 4.15m	4230±70	3010–2620 cal. BC	3020–2580 cal. BC
CAR-311	Peat; 3.00m	3620±75	2000–1760	2210–1770 cal. BC
CAR-310	Peat; 1.00m	1825±75	cal. AD 20–400	cal. AD 130–400
CAR-312	Peat; 0.80m	1480±65	cal. AD 420–670	cal. AD 410–650
CAR-313	Burnt and chopped *Pinus sylvestris* trunk	3435±70	1930–1530 cal. BC	–
CAR-247	Charcoal horizon from peat, associated with burning & disappearance of *Pinus*	3620±60	2200–1770 cal. BC	–

Table A1.10 Goole Moor Site 1 radiocarbon results

Laboratory Number	Material/context	Radiocarbon Age (BP)	Calibrated date (95% confidence)	Posterior Density Estimate (95% probability)
CAR-232	Peat; 1.95m	4515±70	3500–2930 cal. BC	3500–3470 (2%) or 3380–3010 (90%) or 2980–2930 (3%) cal. BC
CAR-233	Peat; 1.56m	3715±70	2340–1920 cal. BC	2290–1920 cal. BC
CAR-293	Peat; 1.50m	3415±70	1900–1520 cal. BC	1980–1720 cal. BC
CAR-234	Peat; 1.36m	3560±65	2130–1730 cal. BC	1920–1680 cal. BC
CAR-235	Peat; 0.97m	2350±60	740–230 cal. BC	760–690 (7%) or 660–380 (88%)
CAR-236	Peat; 0.96m	2380±60	760–370 cal. BC	730–650 (7%) or 550–350 (88%) cal. BC
CAR-294	Peat; 0.58m	1745±60	cal. AD 120–430	cal. AD 130–410
CAR-237	Peat; 0.47m	1675±60	cal. AD 230–540	cal. AD 360–550
CAR-238	Peat; 0.46m	1525±60	cal. AD 400–650	cal. AD 380–570
CAR-492	Peat; 0.17m	1200±60	cal. AD 670–990	cal. AD 840–1020
CAR-491	Peat; 0.16m	1005±60	cal. AD 890–1170	cal. AD 860–1040

Thorne Waste Site 2

Three measurements (Table A1.18) were obtained on bulk peat samples (Turner 1962) from a site *c.* 800m south of TM1 (Smith 2002). Given the small number of samples and uncertainty about the absolute relationship between the samples the sequence was modelled as a simple *Sequence* (Bronk Ramsey 1995). The model (Figure A1.15) shows good agreement between the radiocarbon results and stratigraphy ($A_{overall}$=66.0%).

Appendix 1. Radiocarbon dating protocol and Bayesian chronological modelling methodology 177

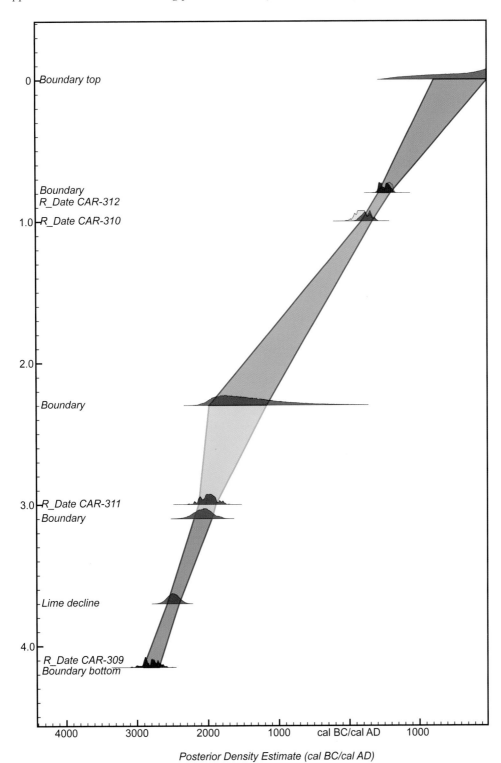

Figure A1.8 Probability distributions of dates from Crowle Moor Site 2 (CLM2) P-Sequence *model*

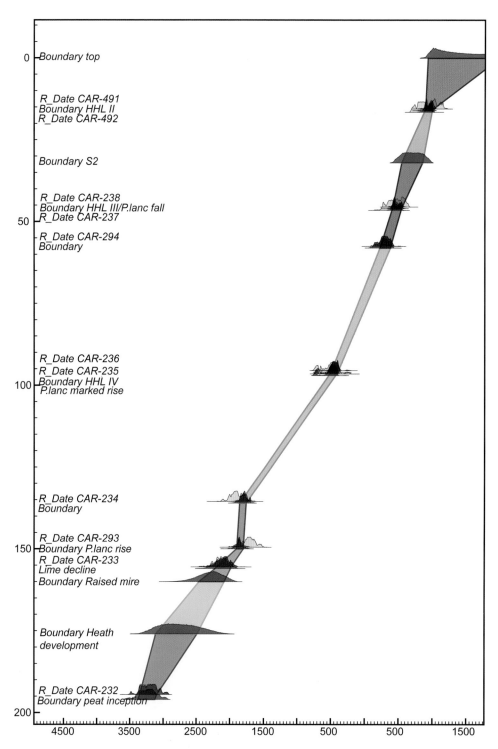

Figure A1.9 Probability distributions of dates from Goole Moor Site 1 (GLM1) P-Sequence *model*

Appendix 1. Radiocarbon dating protocol and Bayesian chronological modelling methodology 179

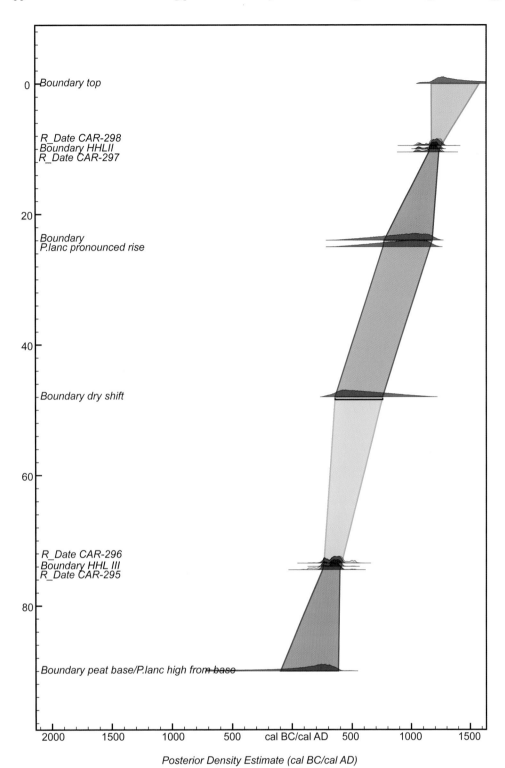

Figure A1.10 Probability distributions of dates from Goole Moor Site 2 (GLM2) P-Sequence model

Table A1.11 Goole Moor Site 2 radiocarbon results

Laboratory Number	Material/context	Radiocarbon Age (BP)	Calibrated date (95% confidence)	Posterior Density Estimate (95% probability)
CAR-295	Peat; 0.75m	1730±60	cal. AD 130–430	*cal. AD 230–420*
CAR-296	Peat; 0.74m	1675±60	cal. AD 230–540	*cal. AD 240–430*
CAR-297	Peat; 0.11m	860±55	cal. AD 1020–1280	*cal. AD 1030–1090 (10%) or 1110–1260 (85%)*
CAR-298	Peat; 0.10m	825±60	cal. AD 1040–1290	*cal. AD 1050–1100 (9%) or 1120–1280 (85%*

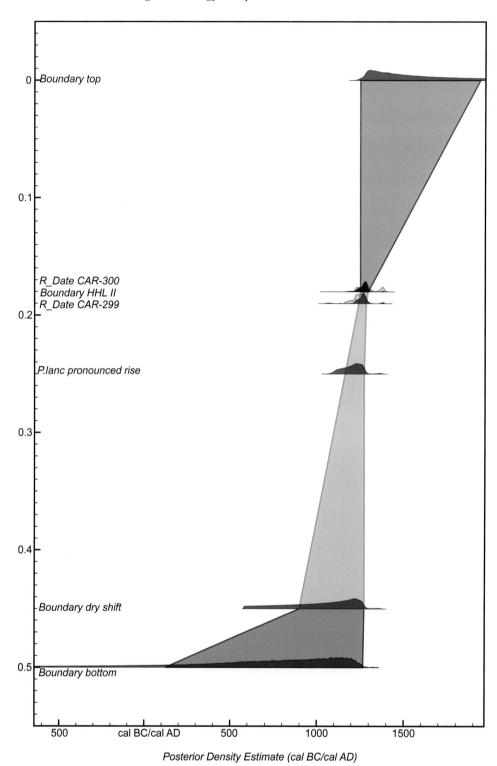

Figure A1.11 Probability distributions of dates from Goole Moor Site 3 (GLM3) U-Sequence model

Table A1.12 Goole Moor Site 3 radiocarbon results

Laboratory Number	Material/context	Radiocarbon Age (BP)	Calibrated date (95% confidence)	Posterior Density Estimate (95% probability)
CAR-299	Peat; 0.19m	770±55	cal. AD 1160–1300	cal. AD 1200–1300
CAR-300	Peat; 0.18m	725±55	cal. AD 1210–1390	cal. AD 1210–1310

Appendix 1. Radiocarbon dating protocol and Bayesian chronological modelling methodology 181

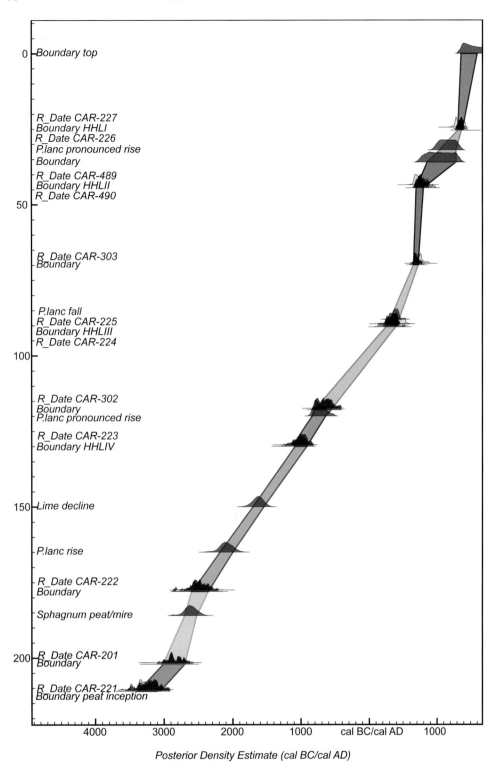

Figure A1.12 Probability distributions of dates from Rawcliffe Moor Site 1 (RAW1) P-Sequence *model*

182 *Modelling archaeology and paleoenvironments in wetlands*

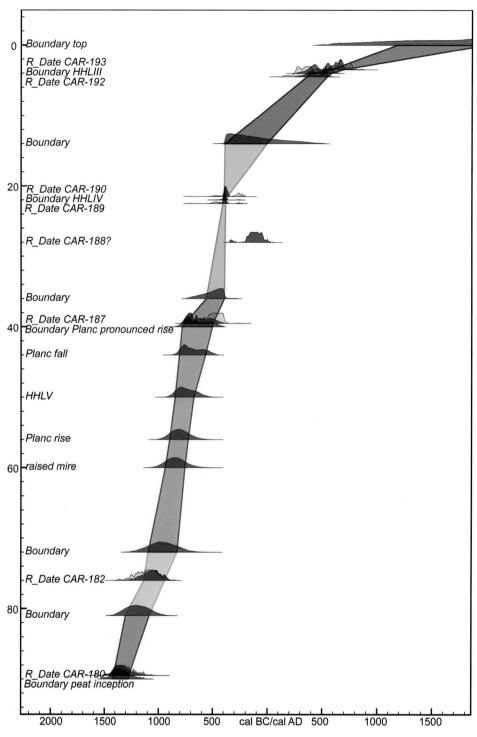

Posterior Density Estimae (cal BC/cal AD)

Appendix 1. Radiocarbon dating protocol and Bayesian chronological modelling methodology 183

Table A1.13 Rawcliffe Moor Site 1 radiocarbon results

Laboratory Number	Material/context	Radiocarbon Age (BP)	Calibrated date (95% confidence)	Posterior Density Estimate (95% probability)
CAR-221	Peat; 2.11m	4545±75	3520–3010 cal. BC	*3500–3430 (6%) or 3380–2920 (89%) cal. BC*
CAR-201	Peat; 2.02m	4255±70	3030–2630 cal. BC	*3030–2630 cal. BC*
CAR-222	Peat; 1.78m	3975±75	2840–2230 cal. BC	*2830–2200 cal. BC*
CAR-223	Peat; 1.30m	2825±75	1200–830 cal. BC	*1200–830 cal. BC*
CAR-302	Peat; 1.18m	2525±60	810–400 cal. BC	*810–420 cal. BC*
CAR-224	Peat; 0.91m	1725±60	cal. AD 130–430	*cal. AD 220–420*
CAR-225	Peat; 0.90m	1665±70	cal. AD 230–550	*cal. AD 240–450*
CAR-303	Peat; 0.70m	1350±55	cal. AD 600–780	*cal. AD 620–780*
CAR-490	Peat; 0.45m	1235±60	cal. AD 650–970	*cal. AD 670–870*
CAR-489	Peat; 0.44m	1295±60	cal. AD 640–890	*cal. AD 690–890*
CAR-226	Peat; 0.26m	555±60	cal. AD 1280–1450	*cal. AD 1280–1400*
CAR-227	Peat; 0.25m	675±60	cal. AD 1250–1410	*cal. AD 1290–1410*

Table A1.14 Thorne Moor Site 1 radiocarbon results

Laboratory Number	Material/context	Radiocarbon Age (BP)	Calibrated date (95% confidence)	Posterior Density Estimate (95% probability)
CAR-180	Peat; 0.90m	3060±65	1460–1120 cal. BC	*1500–1130 cal. BC*
CAR-182	Peat; 0.76m	2900±65	1310–900 cal. BC	*1250–910 cal. BC*
CAR-187	Peat; 0.40m	2415±65	780–380 cal. BC	*780–440 cal. BC*
CAR-188	Peat; 0.28m	2095±45	350 cal. BC–cal. AD 10	–
CAR-189	Peat; 0.23m	2335±40	490–360 cal. BC	*480–250 cal. BC*
CAR-190	Peat; 0.22m	2310±45	420–230 cal. BC	*480–220 cal. BC*

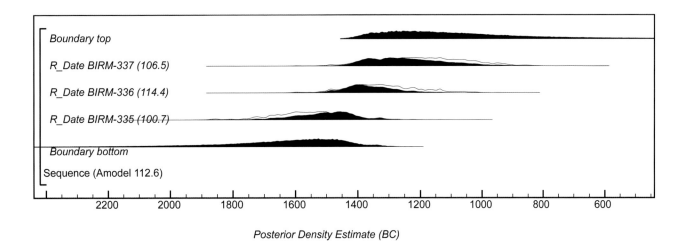

Figure A1.13 Probability distributions of dates from Thorne Moor Sites 1 and 2 (TM1 and TM2) P-Sequence model

Table A1.15 Thorne Moor Site 2 radiocarbon results

Laboratory Number	Material/context	Radiocarbon Age (BP)	Calibrated date (95% confidence)	Posterior Density Estimate (95% probability)
CAR-192	Peat;	1705±60	cal. AD 170–530	*cal. AD 280–560*
CAR-193	Peat;	1350±60	cal. AD 590–780	*cal. AD 430–460 (1%) or 540–770 (94%)*

184 *Modelling archaeology and paleoenvironments in wetlands*

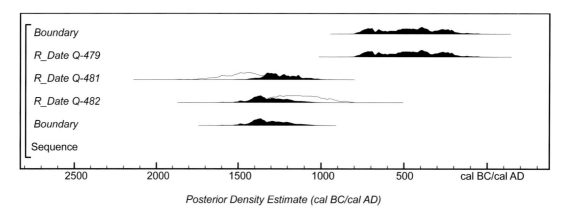

Figure A1.14 Probability distributions of dates from Thorne Moor trackway site: each distribution represents the relative probability that an event occurs at a particular time. For each of the radiocarbon dates two distributions have been plotted. The outline plots shows the results of radiocarbon calibration, and the solid plots show the results of chronological modelling

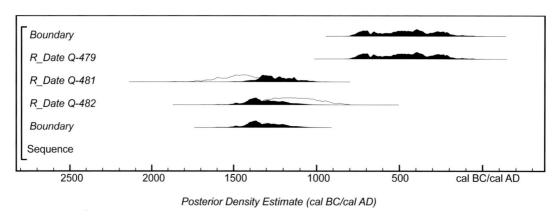

Figure A1.15 Probability distributions of dates from Thorne Waste Site 2: each distribution represents the relative probability that an event occurs at a particular time. For each of the radiocarbon dates two distributions have been plotted. The outline plots shows the results of radiocarbon calibration, and the solid plots show the results of chronological modelling

Table A1.16 Thorne Moor trackway site radiocarbon results

Laboratory Number	Material/context	Radiocarbon Age (BP)	$\delta^{13}C$ (‰)	Calibrated date (95% confidence)	Posterior Density Estimate (95% probability)
BIRM-335	Wood, *Betula*; 1.10m	3260±100		1760–1310 cal. BC	1740–1310 cal. BC
BIRM-336	Burnt wood Pine; 1.00m	3080±90	-21.8	1530–1050 cal. BC	1520–1180 cal. BC
BIRM-358	*Pinus* bark used in trackway; 0.95m	2980±110		1500–900 cal. BC	1450–990 cal. BC

Table A1.17 Thorne Waste Site 1 radiocarbon result

Laboratory Number	Material/context	Radiocarbon Age (BP)	Calibrated date (95% confidence)
Q-477	Highly humified *Sphagnum-Calluna-Eriophorum* peat; 0.18-0.19m	1855±120	160 cal. BC–cal. AD 430

Table A1.18 Thorne Waste Site 2 radiocarbon results

Laboratory Number	Material/context	Radiocarbon Age (BP)	Calibrated date (95% confidence)	Posterior Density Estimate (95% probability)
Q-482	Humified *Sphagnum* peat; 0.96–0.97m	2942±115	1440–830 cal. BC	1520–1070 cal. BC
Q-481	Humified *Sphagnum* peat; 0.95–0.96m	3170±115	1740–1130 cal. BC	1440–1020 cal. BC
Q-479	Humified *Sphagnum* peat; 0.83–0.84m	2329±110	780–120 cal. BC	790–170 cal. BC

Appendix 2. Coleoptera from the Hatfield trackway and platform site

Table A2.1 Coleoptera from the Hatfield and trackway and platform site

Sample number	S1	S2	S3	S4	S5	S6
Carabidae						
Dyschirius globosus (Hbst.)	2	1	–	8	9	–
Trechus sp.	–	1	–	–	–	–
Bembidion lampros (Hbst.)	–	–	–	1	–	–
Bradycellus harpalinus (Serv.)	–	–	–	1	–	–
Bradycellus sp.	–	–	–	–	1	–
Pterostichus diligens (Sturm)	1	2	5	14	17	2
Pterostichus nigrita/rhaeticus (Payk.)/Heer	–	3	–	1	2	–
Pterostichus minor (Gyll.)	–	–	–	1	2	–
Pterostichus sp.	1	1	–	2	1	–
Abax parallelopipedus (Pill. & Mitt.)	–	1	–	–	–	–
Olisthopus rotundatus (Payk.)	–	–	–	2	–	–
Agonum fuliginosum (Panz.)	–	–	–	4	–	1
Oxypselaphus obscurus (Hbst.)	–	1	–	1	9	–
Cymindis vaporariorum (L.)	–	–	–	–	–	1
Haliplidae						
Haliplus sp.	–	1	1	–	–	–
Dytiscidae						
Hydroporus scalesianus Steph.	–	–	–	–	–	1
Hydroporus angustatus Sturm	–	–	1	–	–	–
Hydroporus obscurus Sturm	1	–	–	–	–	2
Hydroporus pubescens (Gyll.)	1	2	–	2	–	–
Hydroporus discretus Fairm. & Bris.	–	–	–	–	–	2
Hydroporus nigrita (F.)	–	16	2	11	11	–
Hydroporus spp.	–	6	1	32	20	2
Agabus guttatus (Payk.)	–	–	–	1	–	–
Agabus bipustulatus (L.)	–	–	–	1	–	–
Ilybius guttiger (Gyll.)	–	1	–	3	–	–
Ilybius aenescens Thoms.	–	–	–	–	1	–

Sample number	S1	S2	S3	S4	S5	S6
Hydraenidae						
Ochthebius sp.	–	–	1	–	–	–
Hydrophilidae						
Cercyon sp.	–	–	–	1	–	–
Enochrus affinis (Thun.)	–	6	1	2	–	–
Hydrophilidae indet.	1	–	–	–	–	–
Leiodidae						
Agathidium sp.	–	1	–	–	–	–
Scydmenidae						
Stenichnus collaris (Müll. & Kunze)	–	–	–	1	1	–
Staphylinidae						
Olophrum piceum (Gyll.)	1	–	–	5	5	–
Acidota crenata (F.)	–	2	1	1	–	1
Stenus sp.	1	–	–	3	4	1
Scopaeus sp.	–	–	–	–	1	–
Lathrobium elongatum (L.)	–	2	–	–	–	–
Lathrobium (s.l.) sp.	2	1	–	4	3	–
Xantholinus linearis (Ol.)	–	–	–	–	–	1
Philonthus sp.	–	1	–	–	1	–
Staphylinus sp.	–	–	1	–	–	–
Quedius sp.	–	–	–	–	1	–
Aleocharinae indet.	1	2	–	2	2	1
Pselaphidae						
Pselaphidae indet.	2	–	–	–	–	–
Bryaxis bulbifer (Reich.)	–	–	–	1	5	1
Reichenbachia juncorum Leach	–	–	–	–	1	–
Elateridae						
Actinicerus sjaelandicus (Müll.)	–	1	–	–	–	–
Scirtidae						
Microcara testacea (L.)	–	1	–	–	–	–
Cyphon spp.	2	3	–	1	–	2
Nitidulidae						
Nitidulidae indet.	1	–	–	–	–	–
Brachypteridae						
Brachypterus urticae (F.)	–	–	–	1	–	–

Appendix 2. Coleoptera from the Hatfield trackway and platform site

Sample number	S1	S2	S3	S4	S5	S6
Lathridiidae						
Corticaria/Corticarina sp	–	–	–	–	–	1
Latridiidae indet.	–	1	–	–	–	–
Colydiidae						
Cerylon histeroides (F.)	–	–	–	–	1	–
Sphindidae						
Sphindus dubius (Gyll.)	1	–	–	–	–	–
Anobidae						
Anobium sp.	–	–	–	–	1	–
Scarabaeidae						
Aphodius sphacelatus (Panz.)	1	–	–	–	–	–
Phyllopertha horticola (L.)	–	–	–	1	–	–
Chrysomelidae						
Donacia sp.	1	–	–	–	–	–
Plateumaris discolor (Panz.)	–	22	–	3	–	4
Plateumaris discolor/sericea (Panz.)/(L.)	–	–	2	–	6	–
Lochmaea suturalis (Thoms.)	–	1	–	–	–	–
Altica sp.	–	–	–	–	–	1
Scolytidae						
Tomicus piniperda (L.)	1	–	–	–	1	–
Leperisinus fraxini (Panz.)	–	–	–	–	2	–
Dryocoetinus alni (Georg)	1	–	–	–	–	–
Pityophthorus pubescens (Marsham)	3	–	–	–	–	–
Pityogenes bidentatus (Hbst.)	1	–	–	–	–	–
Curculionidae						
Apion (s.l.) sp.	–	–	–	–	1	–
Dryophthorus corticalis (Payk.)	–	–	1	–	1	–
Rhyncolus sculpturatus Waltl	1	–	–	–	–	1
Rhyncolus spp.	–	–	–	–	2	–
Micrelus ericae (Gyll.)	1	4	2	10	10	5
Ceutorhynchus sp.	–	–	–	–	–	1
Rhynchaenus quercus (L.)	2	–	–	1	–	–
Rhynchaenus fagi (L.)	–	1	–	–	–	–
Totals MNI	**30**	**85**	**19**	**122**	**122**	**31**
Total Species	**23**	**31**	**12**	**39**	**31**	**20**

Bibliography

Aaby, B. 1976. Cyclic variations in climate over the past 5500 years, reflected in raised bogs. *Nature* 263, 281–284

Aldenderfer, M. 1996. Introduction, in M. Aldenderfer and H. D. G. Maschner (ed.) *Anthropology, space, and geographic information systems*, 3–18. Oxford: Oxford University Press

Alexander, D., N. C. Bragg, R. Meade, G. Padelopoulos and O. Watts 2008. Peat in horticulture and conservation: the UK response to a changing world. *Mires and Peat* 3, 1–10

Allen, T. F. H. and T. W. Hoekstra 1991. Role of heterogeneity in scaling of ecological systems under analysis, in J. Kolass and S. T. A. Pickett (ed.) *Ecological heterogeneity*, 47–68. New York: Springer

Almquist-Jacobson, H. and D. R. Foster 1995. Toward an integrated model of raised-bog development: theory and field evidence. *Ecology* 76(8), 2503–2516

Altschul, J. H. 1990. Red flag models: the use of modelling in management contexts, in K. M. S. Allen, S. W. Green and E. B. W. Zubrow (ed.) *Interpreting space: GIS and archaeology*: 226–38. London: Taylor and Francis

Amesbury, M. J., D. J. Charman, R. M. Fyfe, P. G. Langdon and S. West 2008. Bronze Age upland settlement decline in southwest England: testing the climate change hypothesis. *Journal of Archaeological Science* 35, 87–98

Anderson, D. E., H. A. Binney and M. A. Smith 1998. Evidence for abrupt climatic change in northern Scotland between 3900 and 3500 calendar years BP. *The Holocene* 8, 97–103

Anderson, E. F., L. McLoughlin, F. Liarokapis, C. Peters, P. Petridis and S. de Freitas 2009. Serious games in cultural heritage, in M. Ashley and F. Liarokapis (ed.) *10th International Symposium on virtual reality, archaeology and cultural heritage (VAST '09), VAST-STAR, Short and Project Proceedings*, 29–48

Anderson, R. L., D. R. Foster and G. Motzkin 2003 Integrating lateral expansion into models of peatland development in temperate New England. *Journal of Ecology* 91, 68–76

Appleton, J. 1975. *The experience of landscape*. London: John Wiley

Aston, M. 1985. *Interpreting the landscape: landscape archaeology in local studies*. London: Batsford

Aston, M. and T. Rowley 1974. *Landscape archaeology: an introduction to fieldwork techniques on post-Roman landscapes*. Newton Abbot: David and Charles

Baillie, M. G. L. 1991. Suck in and smear: two related chronological problems for the 90s. *Journal of Theoretical Archaeology* 2, 12–16

Baillie, M. G. L. 1992. Great oaks from little acorns… precision and accuracy in Irish dendrochronology, in F. Chambers (ed.) *Climate change and human impact on the landscape*, 33–46. London: Chapman and Hall

Baillie, M. G. L. and D. M. Brown 1996. Dendrochronology of Irish bog trackways, in B. Raftery (ed.) *Trackway excavations in the Mountdillon bogs, Co. Longford, 1985–1991*, 395–402. Dublin: Irish Archaeological Wetland Unit Transactions 3, Crannóg Publication

Baillie, M. G. L. and D. M. Brown 2002. Oak dendrochronology: some recent archaeological developments from an Irish perspective. *Antiquity* 76, 497–505

Baillie, M. G. L. and M. A. R. Munroe, 1988. Irish tree rings, Santorini and volcanic dust veils. *Nature* 332, 344–346

Bailey, G. 1981. Concepts, time-scales and explanation in economic prehistory, in A. Sheridan and G. N. Bailey (ed.) *Economic archaeology*, 97–117. Oxford, British Archaeological Report International Series 96

Bailey, G. N. 2007. Time perspectives, palimpsests and the archaeology of time. *Journal of Anthropological Archaeology* 26, 198–223.

Bain, C. G., A. Bonn, R. Stoneman, S. Chapman, A. Coupar, M. Evans, B. Gearey, M. Howat, H. Joosten, C. Keenleyside, J. Labadz, R. Lindsay, N. Littlewood, P. Lunt, C. J. Miller, A. Moxey, H. Orr, M. Reed, P. Smith, V. Swales, D. B. A. Thompson, P. S. Thompson, R. Van de Noort, J. D. Wilson and F. Worrall 2011. *IUCN UK commission of inquiry on peatlands*. Edinburgh: IUCN Peatland Programme

Bakewell, C. 1833. *An introduction to geology*. London

Barber, K. E. 1981. *Peat stratigraphy and climate change: a palaeoecological test of the theory of cyclic peat bog regeneration*. Rotterdam: A.A. Balkema

Barber, K. E. 1982. Peat-bog stratigraphy as a proxy climate record, in A. F. Harding (ed.) *Climatic change in later prehistory*, 103–13. Edinburgh: University Press

Barber, K. E. 1994. Deriving Holocene palaeoclimates from peat stratigraphy: some misconceptions regarding the sensitivity and continuity of the record. *Quaternary Newsletter* 72, 1–10

Barber, K. E., F. M. Chambers, D. Maddy, R. E. Stoneman and J. S. Brew 1994. A sensitive high resolution record of Late-Holocene climatic change from a raised bog in northern England. *The Holocene* 4, 198–205

Barber, K. E., L. Dumayne-Peaty, P. Hughes, D. Mauquoy and R. Scaife 1998. Replicability and variability of the recent macrofossil and proxy-climate record from raised bogs: field stratigraphy and macrofossil data from Bolton Fell Moss and Walton Moss, Cumbria, England. *Journal of Quaternary Science* 13, 15–28

Barber, K. E., F. M. Chambers and D. Maddy 2003. Holocene palaeoclimates from peat stratigraphy macrofossil proxy climate records from three oceanic raised bogs in England and Ireland. *Quaternary Science Reviews* 22, 521–539

Barber, K. E., Chambers, F. M. and Maddy, D. 2004. Holocene climatic history of northern Germany and Denmark. Peat macrofossil investigations at Dosenmoor, Schleswig-Holstein and Svanemose, Jutland. *Boreas* 33, 132–144.

Barrett, J. C. 1994. *Fragments from antiquity*. Oxford: Blackwell

Bartley, D. D. and C. Chambers 1992. A pollen diagram, radiocarbon ages and evidence of agriculture of Entwistle Moor, Lancashire. *New Phytologist* 121, 311–320

Bateman, M. D. 1998. The origin and age of coversand in north Lincolnshire. *Permafrost and periglacial processes* 9, 313–325

Bateman, M. D., Buckland, P. C., Chase, B., Frederick, C. D. and G. D. Gaunt 2008. The Late Devensian proglacial Lake Humber: new evidence from littoral deposits at Ferrybridge, Yorkshire, England. *Boreas* 37, 195–210

Bayliss, A. 2007. Introduction: scientific dating and the Aggregates Levy Sustainability Fund 2002–6, in A. Bayliss, G. Cook, C. Bronk Ramsey and J. van der Plicht (ed.) *Radiocarbon Dates from samples funded by English Heritage under the Aggregates Levy Sustainability Fund 2002–6*, vii–xiv. Swindon: English Heritage

Bayliss, A. 2008. Introduction: scientific dating and the Aggregates Levy Sustainability Fund 2004–7, in A Bayliss, G. Cook, C. Bronk Ramsey, J. van der Plicht and G. McCormac (ed.) *Radiocarbon Dates from samples funded by English Heritage under the Aggregates Levy Sustainability Fund 2004–7*, vii–xvii. Swindon: English Heritage

Bayliss, A., C. Bronk Ramsey, J. van der Plicht, and A. Whittle 2007. Bradshaw and Bayes: towards a timetable for the Neolithic. *Cambridge Archaeological Journal* 17, 1–28

Behre, K. -E. 1981. The interpretation of anthropogenic indicators in pollen diagrams. *Pollen et Spores* 23, 225–243

Bennett, K. D. and H. J. B. Birks 1990. Post-glacial history of alder (*Alnus glutinosa* (L.) Gaertn.) in the British Isles. *Journal of Quaternary Science* 5, 123–133

Berglund, B. E. 2003. Human impact and climate changes – synchronous events and a causal link? *Quaternary International* 105, 7–12

Bermingham, N. 2005. *Reconstructing the evolution of a raised mire system: Kilnagharnagh Bog, Co. Offaly, Ireland*. Hull: Unpublished PhD, University of Hull

Bermingham, N. and M. Delaney 2005. *The bog body from Tumbeagh*. Bray: Wordwell

Bernick, K. (ed.) 1998. *Hidden dimensions. The cultural significance of wetland archaeology*. Vancouver: University of British Columbia Press

Binford, L. 1981. Behavioural archaeology and the *Pompeii* premise. *Journal of Anthropological Research* 35, 255–273

Blaauw, M, and J. A. Christen 2005. Radiocarbon peat chronologies and environmental change. *Applied Statistics* 54, 805–816

Blaauw, M., R. Bakker, J. A. Christen, V. A. Hall and J. van der Plicht 2007. A Bayesian framework for age modelling of radiocarbon-dated peat deposits: case studies from the Netherlands. *Radiocarbon* 49, 357–368

Blaauw, M., J. A. Christen and D. Mauquoy 2010. Peatlands as a model system for exploring and reconciling Quaternary chronologies. *PAGES News* 18(1), 9–10

Blaauw, M., J. A. Christen, D. Mauquoy, J. van der Plicht and K. D. Bennett 2007. Testing the timing of radiocarbon-dated events between proxy archives. *The Holocene* 17, 283–288

Blaauw, M. and D. Mauquoy 2012. Signal and variability within a Holocene peat bog – chronological uncertainties of pollen, macrofossil and fungal proxies. *Review of Palaeobotany and Palynology* 186, 5–15

Blaauw, M., B. van Geel and J. van der Plicht 2004. Solar forcing of climate change during the mid-Holocene: indications from raised bog in the Netherlands. *The Holocene* 14, 35–44

Blackford, J. J. and F. M. Chambers 1991. Proxy records of climate from blanket mires: evidence for a Dark Age (1400BP) climatic deterioration in the British Isles. *The Holocene* 1, 63–67

Blockley, S. P. E., M. Blaauw, C. Bronk Ramsey and J. van der Plicht 2007. Building and testing age models for radiocarbon dates in Lateglacial and Early Holocene sediments. *Quaternary Science Reviews* 26, 1915–1926

Blundell, A. and K. E. Barber 2005. A 2800-year palaeoclimatic record from Tore Hill Moss, Strathspey, Scotland: the need for a multi-proxy approach to peat-based climate reconstructions. *Quaternary Science Reviews* 24, 1261–1277

Bond, C. 2004. The Sweet Track, Somerset: a place mediating culture and spirituality? in T. Insoll (ed.) *Belief in the past. The proceedings of the 2002 Manchester conference on archaeology and religion*, 37–50. Oxford: British Archaeological Report International Series 1212

Bond, C. 2007. Walking the track and believing: the Sweet Track as a means of accessing earlier Neolithic spirituality, in D. A. Barrowclough and C. Malone (ed.) *Cult in Context: reconsidering ritual in archaeology*, 158–166. Oxford: Oxbow Books

Bond, C. 2010. The 'God-dolly' wooden figurine from the Somerset Levels, Britian: context, place and meaning in A. Cyphers and D. Gheorghiu (ed.) *Anthropomorphic and zoomorphic miniature figures in Eurasia, Africa and Meso-America: morphology, materiality, technology, function and context*, 43–54. Oxford: British Archaeological Report International Series 1238

Bonsall, C., M. G. Macklin, D. E. Anderson and R. W. Payton 2002. Climate change and the adoption of agriculture in north-west Europe. *European Journal of Archaeology* 5, 9–23

Boreux. J. –J., G. Pesti, L. Duckstein and J. Nacolas 1997. Age model estimation in paleoclimatic research: fuzzy regression and radiocarbon uncertainties. *Palaeogeography, Palaeoclimatology, Palaeoecology* 128, 29–37

Borren, W., W. Bleurten and E. D. Lapshina 2004. Holocene peat and carbon accumulation rates in the southern Taiga of western Siberia. *Quaternary Research* 61, 42–51

Boswiijk, I. G. 1998. *A dendrochronological study of oak and pine from the raised mire of the Humberhead Levels, England, UK*. Sheffield: Unpublished PhD thesis, University of Sheffield

Boswijk, G. and N. J. Whitehouse 2002. *Pinus* and *Prostomis*: a dendrochronological and palaeoentomological study of a study of mid-Holocene woodland in Eastern England. *The Holocene*, 12, 585–596

Boswijik, G., N. J. Whiteshouse, B. M. Smith and P. C. Buckland 2001. Thorne Moors (SE7316), in M. D. Bateman, P. C. Buckland, C. D. Frederick and N. J. Whitehouse (ed.) *The Quaternary of east Yorkshire and north Lincolnshire field guide*, 169–177. London: Quaternary Research Association

Bowden, M. 1999. *Unravelling the landscape: an inquisitive approach to archaeology*. Stroud: Tempus

Bradley, R. 1991. Ritual, time and history. *World Archaeology* 23, 209–219

Bradley, R. 1993. *Altering the earth. The origins of monuments* in Britain and continental Europe. Edinburgh: Society of Antiquaries of Scotland

Bradley, R. 1998. *The significance of monuments. On the shaping of human experience in Neolithic and Bronze Age Europe*. London and New York: Routledge

Bradley, R. 2003. A life less ordinary: the ritualisation of the domestic sphere in later prehistoric Europe. *Cambridge Archaeological Journal* 13, 5–23

Bradley, R. 2007. *The prehistory of Britain and Ireland*. Cambridge: University Press

Brayshay, B. A. and M. H. Dinnin 1999. Integrated palaeoecological evidence for biodiversity at the floodplain-forest margin. *Journal of Biogeography* 26, 115–131

Briggs, C. S. and R. C. Turner 1986. The bog burials of Britain and Ireland, plus Gazetteer, in I. M. Stead, J. B. Bourke and D. Brothwell (ed.) *Lindow Man, the body in the bog*, 144–62 and 181–96. London: British Museum

Brock, F., S. Lee, R. Housley and C. Bronk Ramsey 2011. Variation in the radiocarbon age of different fractions of peat: a case study from Ahrenshöft, northern Germany. *Quaternary Geochronology* 6, 505–55

Brockmeier, J. 2000. Autobiographical time. *Narrative Inquiry* 10, 51–73

Bronk Ramsey, C. 1995. Radiocarbon Calibration and Analysis of Stratigraphy: The OxCal Program. *Radiocarbon* 37, 425–30

Bronk Ramsey, C. 1998. Probability and dating. *Radiocarbon* 40, 461–74

Bronk Ramsey, C. 2001. Development of the Radiocarbon Program OxCal. *Radiocarbon* 43, 355–363

Bronk Ramsey, C. 2008. Deposition models for chronological records. *Quaternary Science Reviews* 27, 42–60

Bronk Ramsey, C., T. Higham, and P. Leach 2004. Towards high precision AMS: progress and limitations. *Radiocarbon* 46, 1, 17–24

Brown, A. G. 2008a. Geoarchaeology, the four dimensional (4D) fluvial matrix and climatic causality. *Geomorphology* 101, 278–297

Brown, A. G. 2008b. The Bronze Age climate and environment of Britain. *Bronze Age Review* 1, 7–22

Brück, J. 2005. Experiencing the past? The development of phenomenological archaeology in British prehistory. *Archaeological Dialogues* 12, 45–72

Brunning, R. 2001. A wet example: an assessment of wetland archaeology in Somerset, England, in B. Raftery and J. Hickey (ed.) *Recent developments in wetland research*, 243–250. Dublin: WARP Occasional Paper 14

Buck, C. E., W. G. Cavanagh, and C. D. Litton 1996. *Bayesian approach to interpreting archaeological data*. Chichester: Wiley

Buckland, P. C. 1979. *Thorne Moors: a palaeoecological study of a Bronze Age site*. Birmingham: Department of Geography, University of Birmingham

Buckland, P. C. 1993. Peatland archaeology: a conservation resource on the edge of extinction. *Biodiversity and Conservation* 2, 513–517

Buckland, P. C. and M. H. Dinnin 1997. The rise and fall of a wetland habitat: recent palaeoecological research on Thorne and Hatfield Moors. *Thorne and Hatfield Moors Papers* 4, 1–18

Buckland, P. C. and M. J. Dolby 1973. Mesolithic and later material from Misterton Carr, Notts – an interim report. *Transactions of the Thoroton Society of Nottingham* 17, 5–33

Buckland, P. C. and H. K. Kenward 1973. Thorne Moors: the palaeoecological implications of a Late Bronze Age site. *Nature* 241, 405

Buckland, P. C. and J. Sadler 1985. The nature of late Flandrian alluviation in the Humberhead levels. *East Midlands Geographer* 8, 239–251

Buckland, P. C. and B. Smith 2003. Equifinality, conservation and the origins of lowland raised mires: the case study of Thorne and Hatfield Moors. *Thorne and Hatfield Moors Papers* 6, 30–51

Bulleid, A. and H. St G. Gray 1948. *The Meare lake Village, volume 1*. Taunton: Taunton Castle

Bullock, J. A. 1993. Host plants of British beetles: a list of recorded associations. *Amateur Entomologist* 11, 1–24

Bunting, M. J., M-J. Gaillard, S. Sugita, R. Middleton and A. Broström 2004. Vegetation structure and pollen source area. *The Holocene* 14, 651–660

Burl, A. 1979. *Prehistoric Avebury*. London: Yale

Burrough, P. A. 1986. *Principles of geographical information systems for land resources assessment*. Oxford: Clarendon Press

Buteux, S. and H. Chapman 2009. *Where rivers meet. The archaeology of Catholme and the Trent-Tame confluence*. York: Council for British Archaeology Research Report 161

Cao, C. and N. Lam 1997. Understanding the scale and resolution effects in remote sensing and GIS, in D. A. Quattrochi and M. F. Goodchild (ed.) *Scale in remote sensing and GIS*, 57–72. Boca Raton, Florida: CRC Press

Carey, C. J., T. G. Brown, K. C. Challis, A. J. Howard and L. Cooper 2006. Predictive modelling of multiperiod geoarchaeological resources at a river confluence: a case study from the Trent-Soar, UK. *Archaeological Prospection* 13, 241–250

Caseldine, C. 2012. Conceptions of time in (palaeo)climate science and some implications. *Wiley Interdisciplinary Reviews: Climate Change* 3 (4), 329–338

Caseldine, C. J. and B. R. Gearey 2005. Evaluation of a multi-proxy approach to reconstructing surface wetness changes on a complex raised mire system at Derryville Bog, Co. Tipperary, Ireland: identification of responses to a series of prehistoric bog bursts. *The Holocene* 15, 585–601

Caseldine, C. J. and B. R. Gearey 2007. Multiproxy approaches to palaeohydrological investigations of raised bogs in Ireland: a case study from Deryville, Co. Tipperary, in N. J. Whitehouse and E. M. Murphy (ed.) *Environmental archaeology in Ireland*, 259–276. Oxford: Oxbow Books

Caseldine, C. J., B. R. Gearey, J. Hatton, E. Reilly, I. Stuijts and W. Casparie 2001. From the 'Wet' to the 'Dry' – palaeoecological studies at Derryville, Co. Tipperary, in B. Raftery and J. Hickey (ed.) *Recent developments in wetland research*, 99–115. Dublin: University College Dublin Department of Archaeology/WARP Occasional Paper 14

Caseldine, C., R. Fyfe, C. Langdon and G. Thompson 2007. Simulating the nature of vegetation communities at the opening of the Neolithic on Achill Island, Co. Mayo, Ireland – the potential role of models of pollen dispersal and deposition. *Review of Palaeobotany and Palynology* 144, 135–144

Casparie, W. A. 1972. *Bog development in southeastern Drenthe (the Netherlands)*. The Hague: Junk

Casparie, W. A. 1982. The Neolithic wooden trackway XXI (Bou) in the raised bog at Nieuw-Dordrecht (the Netherlands). *Palaeohistoria* 24, 115–164

Casparie, W. A. 1984. The three Bronze Age footpaths XVI (Bou), XVII (Bou) and XVIII (Bou) in the raised bog of southeast Drenthe (the Netherlands). *Palaeohistoria* 26, 41–94

Casparie, W. A. 1986. The two Iron Age trackways XIV (Bou) and XV (Bou) in the raised bog of southeast Drenthe (the Netherlands). *Palaeohistoria* 28, 169–210

Casparie, W. A. 1987. Bog trackways in the Netherlands. *Palaeohistoria* 29, 35–65

Casparie, W. A. 1993. The Bourtanger Moor: endurance and

vulnerability of a raised bog system. *Hydrobiologica* 265, 203–215

Casparie, W. A. and M. Gowen 1998. Lisheen Archaeological Project, Co. Tipperary, Ireland. *NewsWARP* 23, 29–38

Casparie, W. A. and A. Moloney 1992. Niederschlagsklima und Bautechnik neolithischer hölzerner Moorwege. Moorarchäologie in Nordwest-Europa. Gedenkschrift für Dr h.c. Hajo Hayen. *Archäologische Mitteilungen aus Nordwestdeutschland* 15, 69–88

Casparie, W. A. and A. Moloney 1994. Neolithic wooden trackways and bog hydrology. *Journal of Paleolimnology* 12, 49–64

Casparie, W. A. and A. Moloney 1996. Corlea 1. Palaeoenvironmental aspects of the trackway, in B. Raftery (ed.) *Trackway excavations in the Mountdillon Bogs, Co. Longford 1985–1991*, 367–377. Dublin: Crannog Publications

Casparie, W. A. and J. G. Streefkerk 1992. Climatological, stratigraphic and palaeoecological aspects of mire development, in J. T. A. Veerhoeven (ed.) *Fens and bogs in the Netherlands: vegetation, history, nutrient dynamics and conservation*, 81–129. Dordrecht: Kluwer Academic Press

Cebecauer, T., J. Hofierka and M. Šúri 2002. Processing digital terrain models by regularized spline with tension: tuning interpolation parameters for different input datasets. *Proceedings of the open source GIS – GRASS users conference 2002*, 123–134

Challis, K. 2005. Airborne laser altimetry in alluviated landscapes. *Archaeological Prospection* 13(2), 103–127

Challis, K., C. Carey, M. Kincey and A. J. Howard 2011. Airborne lidar intensity and geoarchaeological prospection in river valley floors. *Archaeological Prospection* 18(1), 1–13

Chambers, F. M., J. G. A. Lageard, G. Boswijk, P. A. Thomas, K. J. Edwards and J. Hillam 1997. Dating prehistoric bog fires in northern England to calendar years by long-distance cross-matching of pine chronologies. *Journal of Quaternary Science* 12, 253–256

Chapman, H. P. 2000. *Comparative and introspective GIS – an analysis of cell-based GIS methods and their applications within landscape archaeology*. Unpublished PhD thesis, University of Hull

Chapman, H. P. 2001. Understanding and using archaeological surveys – the 'error conspiracy', in Z. Stančič and T. Veljanovski (ed.) *Computing archaeology for understanding the past – CAA200. Computer applications and quantitative methods in archaeology*, 19–23. Oxford: British Archaeological Report International Series 931

Chapman, H. P. 2003. Rudston 'Cursus A' – engaging with a Neolithic monument in its landscape setting using GIS. *Oxford Journal of Archaeology* 22, 345–356

Chapman, H. P. 2005. Re-thinking the 'cursus problem' – investigating the Neolithic landscape of Rudston, East Yorkshire, UK using GIS. *Proceedings of the Prehistoric Society* 71, 159–170

Chapman, H. 2006. *Landscape archaeology and GIS*. Stroud: Tempus

Chapman, H. P. and P. R. Chapman 2005. Seascapes and landscapes – the siting of the Ferriby Boat finds in the context of prehistoric pilotage. *International Journal of Nautical Archaeology* 34, 43–50

Chapman, H. P. and J. L. Cheetham 2002. Monitoring and modelling saturation as a proxy indicator for in situ preservation in wetlands: a GIS-based approach. *Journal of Archaeological Science* 29, 277–289

Chapman, H. P. and B. R. Gearey 2000. Palaeoecology and the perception of prehistoric landscapes: some comments on visual approaches to phenomenology. *Antiquity* 74, 316–319

Chapman, H. P. and B. R. Gearey 2003. Archaeological predictive modelling in raised mires – concerns and approaches for their interpretation and future management. *Journal of Wetland Archaeology* 2, 77–88

Chapman, H. P. and M. C. Lillie 2004. Investigating 'Doggerland' through analogy. The example of Holderness, East Yorkshire (UK), in N. Fleming (ed.) *Submarine prehistoric archaeology of the North Sea*, 65–69. York: Council for British Archaeology Research Report 141

Chapman, H. P. and R. Van de Noort 2001. High-resolution wetland prospection, using GPS and GIS: landscape studies at Sutton Common (South Yorkshire), and Meare Village East (Somerset). *Journal of Archaeological Science* 28, 365–375

Charman, D. J. 1992. Blanket mire formation at the Cross Lochs, Sutherland, northern Scotland. *Boreas* 21, 53–72

Charman, D. J. 1994. Patterned fen development in northern Scotland: developing a hypothesis from palaeoecological data. *Journal of Quaternary Science* 9, 285–297

Charman, D. J. 1995. Patterned fen development in Northern Scotland: hypothesis testing and comparison with ombrotrophic blanket peats. *Journal of Quaternary Science* 10(4), 327–342

Charman, D. J. 2002. *Peatlands and Environmental Change*. Chichester: Wiley

Charman, D. J. 2010. Centennial climate variability in the British Isles during the mid–late Holocene. *Quaternary Science Reviews* 29, 1539–1554

Charman, D. J., K. E. Barber, M. Blaauw, P. G. Langdon, D. Mauquoy, T. J. Daley, P. D. M. Hughes and E. Karofeld 2001. Climate drivers for peatland palaeoclimate records. *Quaternary Science Reviews* 28, 18111–1819

Charman, D. J., A. Blundell, R. C. Chiverrell, D. Hendon and P. C. Langdon 2006. Compilation of non-annually resolved Holocene proxy climate records: stacked Holocene peatland palaeo-watertable reconstructions from northern Britain. *Quaternary Science Reviews* 35, 336–350

Charman, D. J., C. J. Caseldine, A. Baker, B. R. Gearey and J. Hatton 2001. Palaeohydrological records from peat profiles and speleothems in Sutherland, Northwest Scotland. *Quaternary Research* 55, 223–234

Charman, D. J., D. Hendon and S. Packman 1999. Multiproxy surface wetness records from replicate cores on an ombrotrophic mire: implications for Holocene palaeoclimate records. *Journal of Quaternary Science* 14, 451–463

Charman, D. J., D. Hendon and W. Woodland 2000. *The identification of testate amoebae (Protozoa:Rhizopoda) in peat*. London: Quaternary Research Association Technical Guide 9

Chiverell, R. C. 2001. Proxy record of late Holocene climate change from May Moss, northeast England. *Journal of Quaternary Science* 16, 9–29

Christen, J., R. Clymo and C. Litton 1995. A Bayesian approach to the use of C-14 dates in the estimation of the age of peat. *Radiocarbon* 37, 431–441

Clark, J. G. D. 1954. *Excavations at Star Carr. An early Mesolithic site at Seamer near Scarborough, Yorkshire*. Cambridge: Cambridge University Press

Clark, J. G. D. 1972. *Star Carr: a case study in bioarchaeology*. Reading: Addison-Wesley

Coles, B. 1995. *Wetland management: a survey for English Heritage*. Exeter: Wetland Archaeological Research Project

Coles, B. J. 1998. Doggerland: a speculative survey. *Proceedings of the Prehistoric Society* 64, 45–81

Coles, J. M. 1979. The Abbot's Way. Taunton: Somerset Levels Papers 6, 16–49

Coles, J. M. 1987. *Meare Village East. The excavations of A. Bulleid and H. St. George Gray 1932–1956*. Taunton: Somerset Levels Papers 13

Coles, B and J. M. Coles 1986. *Sweet Track to Glastonbury: the prehistory of the Somerset Levels*. London: Thames and Hudson

Coles, J. and B. Coles 1989. *People of the wetlands. Bogs, bodies and lake-dwellers*. London: Thames and Hudson

Coles, J. M. and B. Coles 1996. *Enlarging the past – the contribution of wetland archaeology*. Exeter: Wetland Archaeology Research Project and the Society of Antiquaries of Scotland

Coles, J. M. and D. Hall. 1997. The Fenland Project: from survey to management and beyond. *Antiquity* 71, 831–844

Coles, J. M. and S. C. Minnitt 1995. *Industrious and fairly civilised: Glastonbury Lake Village*. Exeter: Somerset Levels Project

Coles, J. M. and B. J. Orme 1976. The Sweet Track, railway site. *Somerset Levels Papers* 2, 32–65

Coles, J. M. and B. J. Orme 1980. *Prehistory of the Somerset Levels*. Exeter: Somerset Levels Project

Conneller, C. 2003. Star Carr recontextualised, in J. Moore and L. Bevan (ed.) *Peopling the Mesolithic in a northern environment*, 81–86. Oxford: British Archaeological Report International Series 955

Conolly, J. and M. Lake 2006. *Geographical Information Systems in archaeology*. Cambridge: Cambridge University Press

Conneller, C. and T. Schadla-Hall 2003. Beyond Star Carr: the Vale of Pickering in the 10th millennium BP. *Proceedings of the Prehistoric Society* 69, 85–106

Cook, G., A. Dugmore and J. S. Shore 1998. The influence of pretreatment on humic acid yield and ^{14}C age of *Carex* peat. *Radiocarbon* 40, 21–7

Coombes, P. V. and K. E. Barber 2005. Environmental determinism in Holocene research: causality or coincidence? *Area* 37, 303–311.

Cox, M., J. Chandler, C. Cox, J. Jones and H. Tinsley 2001. The archaeological significance of patterns of anomalous vegetation on a raised mire in the Solway Estuary and the processes involved in their formation. *Journal of Archaeological Science* 28, 1–18

Crawford, O. G. S. 1953. *Archaeology in the field*. London: Phoenix House

Cris, R., Buckmaster, S., Bain, C. and A. Bonn 2011. *UK Peatland Restoration: Demonstrating Success*. Edinburgh: IUCN UK National Committee Peatland Programme

Crockett, A. D., M. J. Allen and R. G. Scaife 2002. A Neolithic trackway within peat deposits at Silvertown, London. *Proceedings of the Prehistoric Society* 68, 185–213

Cross, S., C. Murray, J. Ó Neill and P. Stevens 1999. *The Lisheen Archaeological Project. Catalogue of sites*. M. Gowen Ltd. unpublished report

Cross, S., C. Murray, J. Ó Neill and P. Stevens 2001. Derryville Bog: A vernacular landscape in the Irish Midlands, in B. Raftery and J. Hickey (ed.) *Recent developments in wetland research*, 87–99. Dublin: Department of Archaeology, University College Dublin/WARP Occasional Paper 14

Cross May, S., C. Murray, J. Ó Néill and P. Stevens 2005. Chronology, in M. Gowen (ed.) *The Lisheen Mine Archaeological Project 1996–8*, 55–77. Co. Wicklow: Wordwell Ltd

Crutchley, S. 2006. Light detection and ranging (lidar) in the Witham Valley, Lincolnshire: an assessment of new remote sensing techniques. *Archaeological Prospection* 13 (4), 251–257

Dark, P. 2006. Climate deterioration and land-use change in the first millennium BC: perspectives from the British palynological record. *Journal of Archaeological Science* 33, 1381–1395

De la Pryme, A. nd. Ms. Ephemeris Vitae Abraham Pryme, in C. Jackson (ed.) 1889. *Publications of the Surtees Society*

De la Pryme, A. 1701. Part of a letter from the Rev. Mr Abraham de la Pryme concerning trees found underground on Hatfield Chase. *Philosophical Transactions of the Royal Society of London* 22, 980–992

Department of the Environment 1990. *Planning policy guidance: archaeology and planning (PPG16)*. London: Department of the Environment

DETR 1998. *Peatland issues: report on the working group on peat extraction and related matters*. London: Department of the Environment, Transport and the Regions

Dinnin, M. H. 1993. *Thorne and Hatfield Moors outline management plan*. Wakefield: English Nature

Dinnin, M. H. 1994. *An archaeological and palaeoecological investigation of Thorne Moors, with particular reference to the buried prehistoric forest*. Sheffield: ARCUS Report 148

Dinnin, M. 1997a. The drainage history of the Humberhead Levels, in R. Van de Noort and S. Ellis (ed.) *Wetland heritage of the Humberhead Levels, an archaeological survey*, 19–30. Hull: Humber Wetlands Project, University of Hull

Dinnin, M. 1997b. The palaeoenvironmental survey of West, Thorne and Hatfield Moors, in R. Van de Noort and S. Ellis (ed.) *Wetland heritage of the Humberhead Levels, an archaeological survey*, 157–190. Hull: Humber Wetlands Project, University of Hull

Dinnin, M. 1997c. The palaeoenvironmental survey of the Rivers Idle, Torne and Old River Don, in R. Van de Noort and S. Ellis (ed.) *Wetland heritage of the Humberhead Levels, an archaeological survey*, 81–155. Hull: Humber Wetlands Project, University of Hull

Dinnin, M. H. and R. Van de Noort 1999. Wetland habitats, their resource potential and exploitation, in B. Coles, J. Coles and M. Schou Jørgensen (ed.) *Bog bodies, sacred sites and wetland archaeology*, 69–78. Exeter: Wetland Archaeology Research Project (WARP)

Doneus, M., C. Brieses, M. Fera and M. Janner 2008. Archaeological prospection of forested areas using full-waveform airborne laser scanning. *Journal of Archaeological Science* 35, 882–893

Donisthorpe, H. St J. K. 1939. *The Coleoptera of Windsor Forest*. London: Published privately

Dresser, P. Q. 1970. *A study of sampling and pretreatment for radiocarbon dating*. Belfast: Unpublished PhD thesis, Queens University Belfast

Dudley, H. E. 1949. *Early days in north-west Lincolnshire*. Scunthorpe: Caldicott

Dunston, G. 1909. *The rivers of Axholme with a history of navigable rivers and canals of the district*. Hull: A. Brown and Sons

Dymond, D. P. 1966. Ritual monuments at Rudston, East Yorkshire, England. *Proceedings of the Prehistoric Society* 32, 86–95

Edmonds, M. 1993. Interpreting causewayed enclosures in the past and the present. *Interpretative Archaeology* 99–142

Edwards, K. J. and K. R. Hirons 1982. Cereal pollen grains in pre-elm decline deposits: implications for the earliest agriculture in Britain and Ireland. *Journal of Archaeological Science* 11, 71–80

Ellis, C., A. Crone, E. Reilly and P. Hughes 2002. Excavation of a Neolithic wooden platform, Stirlingshire. *Proceedings of the Prehistoric Society* 68, 247–56

English Heritage 2007. *3D laser scanning for heritage: advice and guidance to users on laser scanning in archaeology and architecture*. Luton: English Heritage

Environment Agency 2006. *LIDAR quality control report for polygon P_3513 (R. Torne)*. Bath: Unpublished Report

European Commission 1992. *The Habitats Directive*. Brussels

Erdtman, G. 1928. Studies in the post-arctic history of the forests of north-west Europe: investigations in the British Isles. *Geölogisk Forenings Stockholm Förhandlingar* 50, 123–192

Evans, J. G. 2003. *Environmental archaeology and the social order*. London: Routledge

Eversham, B. 1991. Thorne and Hatfield Moors – implications of land use change for nature conservation. *Thorne and Hatfield Moors Papers* 2, 3–18

Eversham, B. C., P. C. Buckland and M. H. Dinnin 1995. Lowland raised mires: conservation, palaeoecology and archaeology in the Humberhead Levels, in M. Cox, V. Straker and D. Taylor (ed.) *Wetlands: archaeology and nature conservation*, 75–85. London: HMSO

Fairclough, G. J. 1999. *Historic landscape characterisation: "The state of the art"*. London: English Heritage

Fairclough, G. and S. Rippon (ed.) 2002. *Europe's cultural landscape: archaeologists and the management of change*. Brussels: Europae Archaeologieae Consilium

Feehan, J. and G. O'Donovan 1996. *Bogs of Ireland: an introduction to the natural, cultural and industrial heritage of Irish peatlands*. Dublin: University College Dublin Environmental Institute

Fletcher, M. and D. Spicer 1988. Clonehenge: an experiment with gridded and non-gridded survey data, in S. P. Q. Rahtz (ed.) *Computer and quantitative methods in archaeology*, 309–324. Oxford: British Archaeological Report International Series 446(ii)

Fletcher, W., H. Chapman, R. Head, H. Fenwick, R. Van de Noort and M. Lillie 1999. The archaeological survey of the Humber estuary, in R. Van de Noort and S. Ellis (ed.) *Wetland heritage of the Vale of York: an archaeological survey*, 205–241. Hull: Humber Wetlands Project, University of Hull

Foley, R. A. 1981. A model of regional archaeological structure. *Proceedings of the Prehistoric Society* 47, 1–17

Foster, D. R. and P. Glaser, P. 1986. The raised bogs of south eastern Labrador, Canada: classification, distribution, vegetation and recent dynamics. *Journal of Ecology* 74, 47–71

Foster, D. R. and D. H. Wright. 1990. Role of ecosystem and climate change in bog formation in central Sweden. *Ecology* 71, 450–463

Fraser, D. 1983. *Land and society in Neolithic Orkney*. Oxford: British Archaeological Report 117

Friday, L. E. 1988. *A key to the adults of British water beetles. Field Studies 7*. Taunton: Field Studies Council

Fyfe, R. 2006. GIS and the application of a model of pollen deposition and dispersal: a new approach to testing landscape hypotheses using the POLLANDCAL models. *Journal of Archaeological Science* 33, 483–493

Fyfe, R. 2007. The importance of local-scale openness within regions dominated by closed woodland. *Journal of Quaternary Science* 22(6), 571–578

Fyfe, R. M., J. Brück, R. Johnston, H. Lewis, T. Roland and H. Wickstead 2008. Historical context and chronology of Bronze Age enclosure on Dartmoor, UK, *Journal of Archaeological Science* 35, 2250–2261

Fyfe, R. and J. Woodbridge 2012. Differences in time and space in vegetation patterning: analysis of pollen data from Dartmoor, UK. *Landscape Ecology* 27, 745–760

Gaffney, V. and M. van Leusen 1995. Postscript – GIS, environmental determinism and archaeology: a parallel text, in G. Lock and Z. Stančič (ed.) *Archaeology and geographical information systems: a European perspective*: 367–382. London: Taylor and Francis

Gaillard, M-J, S. Sugita, M. J. Bunting, R. Middleton, A. Broström, C. Caseldine, T. Giesecke, S. E. V. Hellman, S. Hicks, K. Hjelle, C. Langdon, A-B Nielsen, A. Poska, H. von Stedingk and S. Veski 2008. The use of modelling and simulation approach in reconstructing past landscapes from fossil pollen data: a review and results from the POLLANDCAL network. *Vegetation History and Archaeobotany* 17, 419–443

Gaunt, G. 1976. The Devensian maximum ice limit in the Vale of York. *Proceedings of the Yorkshire Geological Society* 40, 631–637

Gaunt, G. 1994. *Geology of the country around Goole, Doncaster and the Isle of Axholme. Memoir for one inch sheets 79 and 88 (England and Wales)*. London: HMSO

Gaunt, G. 1987. The geology and landscape development of the region around Thorne Moors. *Thorne Moors Papers* 1, 5–28

Gear, A. J. and B. Huntley 1991. Rapid changes in the range limits of Scots pine 4000 years ago. *Science* 251, 544–517

Gearey, B. R. 2005. *Palaeoenvironmental investigations at Thorne and Hatfield Moors: research in mitigation of continued peat extraction*. Hull: WAERC, University of Hull report EN-SCO/01-05

Gearey, B. R., N. Bermingham, C. Moore and R. Van de Noort 2012. *Review of archaeological survey and mitigation policy relating to Bord na Móma peatlands since 1990*. Unpublished report for the National Monuments Service, Department of the Arts, Heritage and the Gaeltacht

Gearey, B. R. and C. J. Caseldine 2006. Archaeological applications of testate amoebae analyses: a case study from Derryville, Co. Tipperary, Ireland. *Journal of Archaeological Science* 33, 49–55

Gearey, B. R. and H. P. Chapman 2004. Towards realising the full archaeoenvironmental potential of raised (ombrotrophic) mires in the British Isles. *Oxford Journal of Archaeology* 23, 199–208

Gearey, B. R. and H. P. Chapman 2006. Digital gardening: a method for simulating elements of palaeovegetation and its implications for the interpretation of prehistoric landscapes, in T. L. Evans and P. Daly (ed.) *Digital archaeology: bridging method and theory*, 171–191. London: Routledge

Gearey, B. R. and H. P. Chapman, H.P. 2011. 'Surprising but not always informative'? Archaeological investigations of the Hatfield trackway and platform. *Thorne and Hatfield Moors Paper* 8, 17–31

Gearey, B. R., S. Fitch, E-J. Hopla, S. Fitch and D. Tappin in press. Methodological and interpretative issues for integrated palaeoenvironmental and archaeological investigations of submerged landscapes: a case study from the southern North Sea. In V. Gaffney (ed.) *Between the Salt Water and the Sea Strand: Comparative Methodologies in Submerged Landscape Research*. Oxford: Oxbow

Gearey, B. R. and M. C. Lillie 1999. Aspects of Holocene vegetational change in the Vale of York, in R. Van de Noort and S. Ellis (ed.) *Wetland Heritage of the Vale of York. Humber Wetlands Project, University of Hull*, 109–125. Hull: Humber Wetlands Project, University of Hull

Gearey, B., N. Bermingham, H. Chapman, D. Charman, W. Fletcher, R. Fyfe, J. Quartermaine and R. Van de Noort 2010. *IUCN Review: Peatlands and the historic environment*. International Union for the Conservation of Nature

Gearey, B. R, Marshall, P. and Hamilton, D. 2009. Correlating archaeological and palaeoenvironmental records using a Bayesian approach: a case study from Sutton Common, south Yorkshire, England. *Journal of Archaeological Science* 35, 1477–1487

Gibson, J. J. 1977. The theory of affordances, in R. E. Shaw and J. Bransford (ed.) *Perceiving, acting and knowing: toward an ecological psychology*, 67–82. Hillsdale, NJ: Lawrence Erlbaum Associates

Gibson, J. J. 1979. *The ecological approach to visual perception.* Boston: Houghton Mifflin

Giddens, A. 1984. *The constitution of society.* Cambridge: Polity Press

Gillings, M. and G. T. Goodrick 1996. Sensuous and reflexive GIS: exploring visualization and VRML. *Internet Archaeology* 1, 5.1

Gillings, M. and A. Wise 1999. *GIS guide to good practice.* Oxford: Oxbow Books/Archaeology Data Service

Girling, M. 1984. Investigations of a second insect assemblage from the Sweet Track. *Somerset Levels Papers* 10, 79–91

Glob, P. V. 1998. *The bog people.* London: Faber and Faber

Godwin, H. 1940. Pollen analysis and forest history in England and Wales. *New Phytologist* 39, 370–400

Godwin, H. 1960. Prehistoric wooden trackways of the Somerset Levels: their construction, age and relation to climate change. *Proceedings of the Prehistoric Society* 26, 1–36

Gowen, M., J. Ó Neill and M. Philips (ed.) 2005. *The Lisheen Mine Archaeological Project 1996–8.* Co. Wicklow: Wordwell

Greenwell, Canon W. 1877. *British barrows.* Oxford: Clarendon

Hacigüzeller, P. 2012. GIS, critique representation and beyond. *Journal of Social Archaeology* 12(2), 245–263

Hayen, H. 1987. Peat bog archaeology in Lower Saxony, West Germany, in J. M. Coles and A. J. Lawson (ed.) *European wetlands in prehistory*, 117–136. Oxford: Clarendon

Head, R., H. Chapman, H. Fenwick, R. Van de Noort and M. Lillie 1997. The archaeological survey of the Rivers Aire, Went, former Turnbridge Dike (Don north branch) and the Hampole Beck, in R. Van de Noort and S. Ellis (ed.) *Wetland heritage of the Humberhead Levels: an archaeological survey*, 229–264. Hull: Humber Wetlands Project, University of Hull

Heathwaite, A. L., K. Gottlich, E.-G. Burmeister, G. Kaule and T. Grospietsch 1993. Mires: definitions and form, in A. L. Heathwaite (ed.) *Mires. Process, exploitation and conservation*, 1–76. Chichester: John Wiley

Hedges, R. E. M., I. A. Law, C. R. Bronk and R. A. Housley 1989. The Oxford Accelerator Mass Spectrometry facility: technical developments in routine dating. *Archaeometry* 31, 99–113

Heegard, E., H. J. B. Birks and R. Telford 2005. Relationship between calibrated ages and depth in stratigraphical sequences: an estimation procedure by mixed-effect regression. *The Holocene* 15, 612–618

Hendon, D. and D. J. Charman 1997. The preparation of testate amoebae (Protozoa:Rhizopoda) samples from peat. *The Holocene* 7, 199–205

Hendon, D., D. J. Charman and M. Kent 2001. Comparison of the palaeohydrological records derived from testate amoebae from peatlands in northern England: within site variability, between site comparability and palaeoclimatic implications. *The Holocene* 11, 127–148

Hesse, R. 2010. LiDAR-derived local relief models – a new tool for archaeological prospection. *Archaeological Prospection* 17(2), 67–72

Hayen, H. 1987. Peat bog archaeology in Lower Saxony, West Germany, in J. M. Coles and A. J. Lawson (ed.) *European wetlands in prehistory*, 117–136. Oxford: Clarendon

Higuchi, T. 1983. *Visual and spatial structure of landscapes.* Massachusetts: MIT

Hillam, J., C. M. Groves, D. M. Brown, M. G. L. Baillie, J. M. Coles and B. J. Coles 1990. Dendrochronology of the English Neolithic. *Antiquity* 64, 210–220

Hodgkinson, D., E. Huckerby, R. Middleton and C. E. Wells 2001. *The lowland wetlands of Cumbria.* Lancaster: Lancaster Imprints

Holden, J. and T. P. Burt. 2003. Hydrological studies on blanket peat: the significance of the acrotelm-catotelm model. *Journal of Ecology* 91, 86–102

Holden, J., L. Shotbolt, A. Bonn, T. P. Burt, P. J. Chapman, A. J. Dougill, E. D. G. Fraser, K. Hubacek, B. Irvine, M. J. Kirby, M. S. Reed, C. Prell, S. Stagl, L. C. Stringer, A. Turner and F. Worrall 2007. Environmental change in moorland landscapes. *Earth-Science Reviews* 82, 75–100

Hoskins, W. G. 1955. *The making of the English landscape.* London: Hodder and Stoughton

Howard, A. J., A. G. Brown, C. J. Carey, K. Challis, L. P. Cooper, M. Kincey and P. Toms 2008. Archaeological resource modelling in temperate river valleys: a case study from the Trent Valley, UK. *Antiquity* 82, 1040–1054

Hughes, P. D. M. 2000. A reappraisal of the mechanisms leading to ombrotrophy in British raised mires. *Ecology Letters* 3, 7–9

Hughes, P. D. M and K. E. Barber 2003. Mire development across the fen bog transition on the Teifi floodplain at Tregaron Bog, Ceredigion, Wales and a comparison with 13 other raised bogs. *Journal of Ecology* 91, 253–264

Hughes, P. D. M., D. Mauquoy, K. E. Barber and P. G. Langdon 2000. Mire development pathways and palaeoclimatic records from a full Holocene peat archive at Walton Moss, Cumbria, England. *The Holocene* 10, 465–479

Hunter, J. 1828. *South Yorkshire: the history and topography of the deanery of Doncaster in the diocese and county of York. Volume 2.* London

Hyman, P. S. 1992. *A review of the scarce and threatened Coleoptera of Great Britain; Part 1.* (revised and updated by M. S. Parsons). Peterborough: UK Joint Nature Conservation Committee

ICOMOS 1996. *The ICOMOS charter on the protection and management of underwater cultural heritage.* Sofia (http://www.international.icomos.org/under_e.htm)

Ingold, T. 1992. Culture and the perception of the environment, in E. Croll and D. Parkin (ed.) *Bush Base: forest farm*, 39–56. London: Routledge

Ingold, T. 1993. The temporality of landscape. *World Archaeology* 25, 152–174

Ingram, H. A. P. 1983 Hydrology, in A. J. P. Gore (ed.) *Mires, swamps, bog, fen and moor. General studies in ecosystems of the world* 4a, 67–158. Amsterdam: Elsevier

Ingram, M. J., G. Farmer and T. M. L. Wigley 1981. Past climates and their impact on man: a review, in T. M. L. Wigley, M. J. Ingram and G. Farmer (ed.) *Climate and history: studies in past climates and their impacts on man*, 3–50. Cambridge: Cambridge University Press

JNCC 2011. *Towards an assessment of the state of UK peatlands.* Joint Nature Conservancy Council Report 445

Johnson, M. 2007. *Ideas of landscape.* Oxford: Blackwell

Jones, D. 1998. Long barrows and Neolithic elongated enclosures in Lincolnshire: an analysis of the air photographic evidence. *Proceedings of the Prehistoric Society* 64, 83–114

Jones, J., H. Tinsley and R. Brunning 2007. Methodologies for assessment of the state of preservation of pollen and plant macrofossil remains in waterlogged deposits. *Environmental Archaeology* 12, 71–86

Kamermans, H., J. Deeben, D. Hallewas, M. van Leusen, P. Verhagen and P. Zoetbrood 2002. *Predictive modelling for archaeological heritage management.* Leiden: Leiden University Press

Kaufman, L. and F. Clement 2007. How culture comes to mind: from social affordances to cultural analogies. *Intellectia* 2, 1–30

Kenward, H. K., A. R. Hall and A. K. Jones 1980. A tested set of techniques for the extraction of plant and animal macrofossils from waterlogged archaeological deposits. *Science and Archaeology* 22, 3–15

Kenward, H. K., C. Engleman, C. Robertson and F. Large 1986. Rapid scanning of urban archaeological deposits for insect remains. *Circaea* 3, 163–172

Kilian, M. R., J. van der Plicht and B. van Geel 1995. Dating raised bogs: new aspects of AMS ^{14}C wiggle matching, a reservoir effect and climate change. *Quaternary Science Reviews* 14, 959–966

Kilian, M. R., B. van Geel and J. van der Plicht 2000. ^{14}C AMS wiggle matching of raised bog deposits and models of peat accumulation. *Quaternary Science Reviews* 19, 1011–1033

Kinnes, I., T. Schadla-Hall, P. Chadwick and P. Dean 1983. Duggleby Howe reconsidered. *Archaeological Journal* 140, 83–108

Kirby, J. R. 2001. Regional late Quaternary marine and perimarine record, in M. D. Bateman, P. C. Buckland, C. D. Frederick, N. J. Whitehouse (ed.) *The Quaternary of east Yorkshire and north Lincolnshire field guide*, 25–34. London: Quaternary Research Association

Koch, K. 1989. *Die Käfer Mitteleuropas. Ökologie*, 1 and 2. Krefeld: Goecke and Evers

Koch, K. 1992. *Die Käfer Mitteleuropas. Ökologie* 3. Krefeld: Goecke and Evers

Kohler, T. A. and S. C. Parker 1986. Predictive models for archaeological resource location, in M. B. Schiffer (ed.) *Advances in archaeological method and theory*, volume 9, 397–452. New York: Academic Press

Korhola, A., J. Alm, K. Tolonen, J. Turunen and H. Jungner 1996. Three-dimensional reconstruction of carbon accumulation and CH_4 emission during nine millennia in a raised mire. *Journal of Quaternary Science* 11, 161–165

Kuna, M. and D. Adelsbergerová 1995. Prehistoric location preferences: an application of GIS to the Vinorský potok project, Bohemia, the Czech Republic, in G. Lock and Z. Stančič (ed.) *Archaeology and geographical information systems: a European perspective*, 117–131. London: Taylor and Francis

Lamb, H. 1977. *Climate: past, present and future volume 2: Climatic History and the Future*. London: Methuen

Langdon, P. G. and K. E. Barber 2001. New Holocene tephras and a proxy climate record from a blanket mire in northern Skye, Scotland. *Journal of Quaternary Science* 16(8), 753–759

Lawson, I. T., M. J. Church, K. J. Edwards, G. T. Cook and A. J. Dugmore 2007. Peat initiation in the Faroe Isles: climate change, pedogenesis or human activity? *Earth and Environmental Science Proceedings of the Royal Society of Edinburgh* 98, 15–28

Leah, M. D., C. E. Wells, E. Huckerby, E. Stamper and C. Welch 1998. *The wetlands of Shropshire and Staffordshire*. Lancaster: Lancaster Imprints 7

Leary, J. 2009. Perceptions of and responses to the Holocene flooding of the North Sea lowlands. *Oxford Journal of Archaeology* 28(3), 227–237

Lillie, M. C. 1997. Alluvium and warping in the Humberhead levels: the identification of factors obscuring palaeo-land surfaces and the archaeological record, in R. Van de Noort and S. Ellis (ed.) *Wetland heritage of the Humberhead levels, an archaeological survey*, 191–218. Hull: Humber Wetlands Project, University of Hull

Lillie, M. C. 1998. Alluvium and warping in the lower Trent valley, in R. Van de Noort and S. Ellis (ed.) *Wetland heritage of the Ancholme and lower Trent valleys, an archaeological survey*, 102–22. Hull: Humber Wetlands Project, University of Hull

Lillie, M. C. 1999. The palaeoenvironmental survey of the Humber estuary, incorporating an investigation of the nature of warp deposition in the southern part of the Vale of York, in R. Van de Noort and S. Ellis (ed.) *Wetland heritage of the Vale of York, an archaeological survey*, 78–108. Hull: Humber Wetlands Project, University of Hull

Lillie, M. and H. Chapman 2001. Wetland archaeology on the edge: recent research into the nature of landscape development and settlement on the western margins of the European plain, in B. A. Purdy (ed.) *Enduring records. The environmental and cultural heritage of wetlands*, 287–294. Oxford: Oxbow Books

Lillie, M. C. and B. R. Gearey 1999. The palaeoenvironmental survey of the Rivers Aire, ouse, Whare and Derwent, in R. Van de Noort and S. Ellis (ed.) *Wetland heritage of the Vale of York: an archaeological survey*, 35–78. Hull: Humber Wetlands Project, University of Hull

Lillie, M. C. and B. R. Gearey 2000. The Palaeoenvironmental Survey of the Hull Valley and Research at Routh Quarry, in R. Van de Noort and S. Ellis (ed.) *Wetland Heritage of the Hull Valley*, 31–87. Hull: Humber Wetlands Project, University of Hull

Lillie, M. C. and B. R. Gearey 2001. *Peat extraction at Thorne and Hatfield Moors – a project design for archaeological work*. Unpublished report, Wetland Archaeology and Environments Research Centre, University of Hull

Limbert, M. 1985. A history of Hatfield Moors. *Old West Riding Local History Publications* 5, 1–6

Limbert, M. 1986. The exploitation of peat at Thorne. *Old West Riding Local History Publications* 6, 9–16

Limbert, M. 1987. Some notes on the landscape history of Thorne Moors. *Thorne Moors Papers* 1, 42–9

Lindroth, C. H. 1945. *Die Fennoskandischen Carabidae I–II*. Göteborgs K. Vetensk. och VitterhSamh. Handl. (6) B4. Göteborg (reprinted as English translation, 1992)

Lindroth, C. H. 1986. *The Carabidae (Coleoptera) of Fennoscandia and Denmark*. Leiden: SFauna Entomologica Scandinavica 15(2)

Lindsay, R. 2010. *Peatbogs and carbon: a critical synthesis to inform policy development in oceanic peat bog conservation and restoration in the context of climate change*. University of East London

Lindsay, R. A., D. J. Charman, F. Everingham, R. M. O'Reilly, M. A. Palmer, T. A. Rowell and D. A. Stroud 1988. *The Flow Country: The Peatlands of Caithness and Sutherland*. Peterborough: NCC

Lindsay, R. A., F. Everingham, Y. Fanden-Lilja, P. Mayer, R. Nicholls, E. Reid, T. A. Rowell and S. Ross 1992. *Raised bogs in Great Britain, the National Peatland Resource Inventory (NPRI)*. Edinburgh: Scottish Natural Heritage

Lindsay, R. A. and P. Immirzi 1996. *An inventory of lowland raised bogs in Great Britain*. Edinburgh: Scotting Natural Heritage Research, Survey and Monitoring Report 78

Long, A. J., J. B. Innes, J. R. Kirby, J. M. Lloyd, M. M. Rutherford, I. Shennan and M. J. Tooley 1998. Holocene sea-level change and coastal evolution of the Humber estuary, eastern England: an assessment of rapid coastal change. *The Holocene* 8, 229–247

Loveday, R. 2002. Duggleby Howe revisited. *Oxford Journal of Archaeology* 21, 135–146

Lowe, J. J. and M. J. C. Walker 2000. Radiocarbon dating the last glacial–interglacial transition (^{14}C ka BP) in terrestrial and marine records the need for new quality assurance protocols. *Radiocarbon*, 42, 53–68

Lucas, A. T. 1985 Toghers or causeways: some evidence from archaeological, literary, historical and place-name sources. *Proceedings of the Royal Irish Academy* 85(C), 37–60

Lucas, G. 2012. *Understanding the archaeological record.* Cambridge: University Press.

Lucas, G. 2005. *The archaeology of time.* London: Routledge

Lucht, W. H. 1987. *Die Käfer Mitteleuropas. Katalog.* Krefeld: Goeke and Evers

Lucquin, A., R. J. March and S. Cassen 2006. Analysis of adhering organic residues of two 'coupes-à-socles' from the Neolithic funerary site 'La Hougue Bie' in Jersey: evidences of birch bark tar utilization. *Journal of Archaeological Science* 34, 704–710

Mackey, R. 2003. The Iron Age in East Yorkshire: a summary of current knowledge and recommendations for future research, in T. G. Manby, S. Moorhouse and P. Ottaway (ed.) *The archaeology of Yorkshire. An assessment at the beginning of the 21st century*, 117–121. Leeds

Macklin, M., E. Johnstone and J. Lewin 2005. Pervasive and long-term forcing of Holocene river instability and flooding by centennial-scale climate change. *The Holocene* 15, 937–943

Magilton, J. R. 1977. *The Doncaster district: an archaeological survey.* Doncaster: Doncaster Museums and Arts Service

Magny, M. 2004. Holocene climatic variability as reflected by mid-European lake-level fluctuations, and is probable impact on prehistoric human settlements. *Quaternary International* 113, 65–79

Mäkilä, M. 1997. Holocene lateral expansion, peat growth and carbon accumulation on Haukkasuo, a raised bog in southeastern Finlad. *Boreas* 26, 1–14

Mallik, A. U., C. H. Gimmingham and A. A. Rahman 1984. Ecological affects of heather burning 1. Water infiltration, moisture retention and porosity of surface soil. *Journal of Ecology* 72, 676–776

Mansell, L. 2011a. Recent research on Hatfield Moors and the River Torne floodplain. *Thorne and Hatfield Moors Papers* 8, 32–51

Mansell, L. 2011b. *Floodplain-mire interactions and palaeoecology: implications for wetland ontology and Holocene climate change.* Belfast: Unpublished PhD thesis, Queen's University Belfast

Manby, T. 1976. Excavation of the Kilham long barrow, East Riding of Yorkshire. *Proceedings of the Prehistoric Society* 42, 111–159

Manby, T. 1988. The Neolithic in East Yorkshire, in T. Manby (ed.) *Archaeology in eastern Yorkshire*, 35–88. Sheffield: Department of Archaeology and Prehistory, University of Sheffield

Marozas, B. A. and J. A. Zack 1990. GIS and archaeological site location, in K.M.S. Allen, S.W. Green and E. B. W. Zubrow (ed.) *Interpreting space: GIS and archaeology*: 165–172. London: Taylor and Francis

Massey, D. 1999. Space-time. 'Science' and the relationship between physical geography and human geography. *Transactions of the Institute of British Geographers* 24, 261–276

Matthews, G. V. T. 1993. *The Ramsar Convention on Wetlands: its history and development.* Gland, Switzerland: Ramsar Convention Bureau

Matthews, I. 2008 *Roman Lode, Exmoor, Devon, Tephrochronology.* English Heritage Research Department Reports Series 26

Mauquoy, D., B. van Geel, M. Blauww and J. van der Pilcht. 2004. Changes in solar activity and Holocene climate shifts derived from C^{14} wiggle-match dated peat deposits. *The Holocene* 14, 45–52

Mauquoy, D., D. Yeloff, B. Van Geel, D. J. Charman and A. Blundell, A. 2008. Two decadally resolved records from north-west European peat bogs show rapid climate changes associated with solar variability during the mid–late Holocene. *Journal of Quaternary Science* 23, 745–763

McCoy, M. D. and T. N. Ladefoged 2009. New developments in the use of spatial technology in archaeology. *Journal of Archaeological Research* 17, 263–295

McInnes, I. J. 1964. A class II henge in the East Riding of Yorkshire. *Antiquity* 38, 218–219

McTaggart, J. E. M. 1908. The Unreality of Time. *Mind* 17, 457–474

Mehrer, M. W. and K. L. Wescott (ed.) 2006. *GIS and archaeological site location modeling.* Boca Raton: Taylor and Francis

Menotti, F. 2012. *Wetland archaeology and beyond.* Oxford: Oxford University Press

Merrony, C. J. N. 1993. The archaeological assessment in advance of the Dearne Towns link road (stage 4) development at Goldthorpe. *Archaeology in South Yorkshire 1992–1993*, 43–52. Sheffield: South Yorkshire Archaeology Service

Metcalfe, S. E., S. Ellis, B. P. Horton, J. B. Innes, J. McArthur, A. Mitlehner, A. Parkes, J. S. Pethick, J. Rees, J. Ridgway, M. M. Rutherford, I. Shennan and M. J. Tooley 2000. The Holocene evolution of the Humber Estuary: reconstructing change in a dynamic environment. in I. Shennan and J. Andrews (ed.) *Holocene land–ocean interaction and environmental change around the North Sea*, 97–118. Bath: Geological Society Special Publication 166

Metzler, A. 1993. Zwei urgeschichtliche Moorwege im Carnpemoor, Ldkr. Vechta Berichte zur *Denkmalpflege in Niedersachsen* 3(93), 114–116

Middleton, R. and C. Wells 1990. *North-West Wetlands Survey annual report 1990.* Lancaster: Lancaster University Archaeological Unit

Midgley, S. 2010. Monuments and monumentality: the cosmological model of the world of the megaliths. *Documenta Praehistorica* 37, 2–57

Mitsch, W. J. and J. G. Gosselink 1993. *Wetlands* (2nd edition). New York: John Wiley

Mitsch, W. J. and J. G. Gosselink 2000. The value of wetlands: importance of scale and landscape setting. *Ecological Economics* 35, 25–33

Mook, W. G. 1986. Business meeting: recommendations/resolutions adopted by the Twelfth International Radiocarbon Conference. *Radiocarbon* 28, 799

Mook, W. G. and H. T. Waterbolk 1985. *Radiocarbon dating.* Strasbourg: European Science Foundation Handbook for Archaeologists 3

Moore, P. D. 1984. The classification of mires: an Introduction, in P. D. Moore (ed.) *European Mires*, 1–10. London: Academic Press

Moore, P. D. 1986. Hydrological change in mires, in B. E. Berglund (ed.) *Handbook of holocene palaeoecology and palaeohydrology*, 91–107. London: John Wiley

Moore, P. D. 1990. Soils and ecology: temperate wetlands, in M. Williams (ed.) *Wetlands. A threatened landscape*, 95–114. Oxford: Blackwell

Moore, P. D. 1993. The origin of blanket mire revisited, in F. M. Chambers (ed.) *Climate change and human impact on the landscape*, 217–224. London: Chapman and Hall

Moore, P. D. and D. J. Bellamy 1973. *Peatlands.* London: Elek Science

Mortimer, J. R. 1905. *Forty years researches in British and Saxon burial mounds of East Yorkshire.* London

Muir, R. 2000. *The new reading the landscape: fieldwork in landscape history.* Exeter: University of Exeter Press

Neumann, H. 1998. The palaeoenvironmental survey of the Ancholme valley, in R. Van de Noort and S. Ellis (ed.)

Wetland heritage of the Ancholme and lower Trent valleys: an archaeological survey, 75–102. Hull: Humber Wetlands Project, University of Hull

Newnham, R. M., P. J. de Lange and D. J. Lowe 1995. Holocene vegetation, climate and history of a raised bog complex, northern New Zealand based on palynology, plant macrofossils and tephrochronology. *The Holocene* 5, 267–282

Nilsson, A. N. and M. Holmen 1995. *The aquatic adephaga (Coleoptera) of Fennoscandia and Denmark II. Dytiscidae.* Leiden: Faura Entomologica Scandinavica 32

Nilsson, M., M. Klarqvist, E. Bohlin and G. Possnert 2001. Variation in ^{14}C age of macrofossils and different fractions of minute peat samples dated by AMS. *The Holocene* 11, 579–586

Noakes, J. E., S. M. Kim and J. J. Stipp 1965. Chemical and counting advances in Liquid Scintillation Age dating, in E. A. Olsson and R. M. Chatters (ed.) *Proceedings of the Sixth International Conference on Radiocarbon and Tritium Dating*, 68–92. Washington DC: clearinghouse for Federal Scientific and Technical Information, National Bureau of Standards, U.S. Department of Commerce

Northern Ireland Habitat Action Plan 2003. *Lowland raised bog.* Belfast: DOE and Northern Ireland Environment Agency

O'Brien, S. R., P. A. Mayewski, L. D. Mecker, D. A. Meese, M. S. Twickler and S. I. Whitlow 1995. Complexity of Holocene climate as reconstructed from a Greenland icecore. *Science* 270, 1962–1964

Ó Floinn, R. 1995. A gazetteer of bog bodies in Ireland, in R. C. Turner and R. G. Scaife (ed.) *Bog bodies. New discoveries and new perspectives*, 221–234. London: British Museum Press

Ordnance Survey 1973. *Field archaeology in Great Britain.* Southampton: Ordnance Survey

Paasio, I. 1933. Über die Vegetation der Hochmoore Finnland. *Acta Forestalia Fennica* 39, 1–190

Paasio, I. 1934. Soita koskevista morfologis-kasvitopografisista nimityksistä. *Terra* 46, 84–90

Palm, T. 1959. *Die Holz und Rindenkäfer der Sud- und Mittelschwedischen Laubbäume.* (Opuscula Entomologica Supplementum 16). Lund: Entomologiska Sallskapet

Parker Pearson, M. and R. Sydes 1997. The Iron Age enclosures and prehistoric landscape at Sutton Common, South Yorkshire. *Proceedings of the Prehistoric Society* 63, 221–259

Parnell, A., C. Buck and T. Doan 2011. A review of statistical chronological models for high-resolution, proxy-based Holocene palaeoenvironmental reconstruction. *Quaternary Science Review* 30, 2948–60

Parnell, A., J. Haslett, J. Allen, C. Buck and B. Huntley 2008. A flexible approach to assessing synchroneity of past events using Bayesian reconstructions of sedimentation history. *Quaternary Science Reviews* 27, 1872–85

Pfadenhauer, J., H. Schneekloth, R. Schneider and S. Schneider 1993. Mire Distribution, in A. L. Heathwaite (ed.) *Mires. Process, Exploitation and Conservation*, 77–122. Chichester: John Wiley

Piotrowska, N., M. Blaauw, D. Mauquoy and F. M. Chambers 2011. Constructing deposition chronologies for peat deposits using radiocarbon dating. *Mires and Peat* 7, 1–14

Plunkett, G. 2006. Tephra linked peat humification records from Irish raised bogs question nature of solar forcing at 850 cal. BC. *Journal of Quaternary Science* 21, 9–16

Plunkett, G. 2009. Land-use patterns and cultural change in the Middle to Late Bronze Age in Ireland: inferences from pollen records. *Vegetation History and Archaeobotany*, 18, 273–295.

Pollard, J. 2001. The aesthetics of depositional practice. *World Archaeology* 33, 315–333

Pollard, A. M. and C. Heron 1996. *Archaeological chemistry.* Cambridge: Royal Society of Chemistry

Pryor, F. (ed.) 1998. *Etton – excavations at a Neolithic causewayed enclosure near Maxey, Cambs 1982–7.* London: English Heritage

Pryor, F. 2001. *The Flag Fen Basin. Archaeology and environment of a Fenland landscape.* London: English Heritage.

Pryor, F., C. French, D. Crowther, D. Burney, G. Simpson and M. Taylor 1985. *The Fenland Project, No. 1: Archaeology and environment in the lower Welland Valley.* Cambridge: East Anglian Archaeology 27

Rackham, O. 1986. *The history of the countryside.* London: J.M. Dent

Raftery, B. 1990 *Trackways through time. Archaeological investigations on Irish bog roads, 1985–1989.* Dublin: Headline Publishing

Raftery, B. 1996. *Trackway excavations in the Mountdillon Bogs, Co. Longford 1985–1991.* Dublin: Crannog Publications

Raftery, B. 1999. The milling fields, in B. Coles, J. Coles and M. Schou Jørgensen (ed.) *Bog bodies, sacred sites and wetland archaeology*, 191–202. Exeter: WARP Occasional Paper 12

Ramm, H. G. 1971. 70/9 Rudston, E.R. TA07586779, in Yorkshire Archaeological Register. *Yorkshire Archaeological Journal* 43, 193

Ramsar Convention Secretariat 2006. *The Ramsar Convention Manual: a guide to the Convention on Wetlands (Ramsar, Iran, 1971)*, 4th edition. Gland, Switzerland: Ramsar Convention Secretariat

Reimer, P. J., M. G. L. Baillie, E. Bard, A. Bayliss, J. W. Beck, C. Bertrand, P. G. Blackwell, C. E. Buck, G. Burr, K. B. Cutler, P. E. Damon, R. L. Edwards, R. G. Fairbanks, M. Friedrich, T. P. Guilderson, K. A. Hughen, B. Kromer, F. G. McCormac, S. Manning, C. Bronk Ramsey, R. W. Reimer, S. Remmele, J. R. Southon, M. Stuiver, S. Talamo, F. W. Taylor, J. van der Plicht and C. E. Weyhenmeyer 2004. IntCal04 Terrestrial radiocarbon age calibration, 0–26 Cal Kyr BP. *Radiocarbon* 46, 1029–1058

Richards, C. 1996. Henges and water: towards an elemental understanding of monumentality and landscape in late Neolithic Britain. *Journal of Material Culture* 1, 313–336

Ricoeur, P. 1980. Narrative time. *Critical Inquiry* 7, 169–190

Riley, D. N. 1944. The technique of air-archaeology. *Archaeological Journal* 101, 1–16

Riley, D. N. 1980. *Early landscape from the air – studies of crop marks in South Yorkshire and North Nottinghamshire.* Sheffield: Department of Prehistory and Archaeology, University of Sheffield

Riley, D. N. 1988. Air survey of Neolithic sites on the Yorkshire Wolds, in T. G. Manby (ed.) *Archaeology in eastern Yorkshire, essays in honour of T. C. M. Brewster*, 89–93. Sheffield: Department of Archaeology and Prehistory, University of Sheffield

Rippon, S. 2003. *Historic landscape analysis: deciphering the countryside.* York: Council for British Archaeology

Rippon, S. 2012. Historic landscape character and sense of place. *Landscape Research* 2012, 1–24

Roberts, O. T. P. 2007. The Brigg 'raft' reassessed as a bilge Bronze Age boat. *International Journal of Nautical Archaeology* 21, 245–258

Robinson, J. M. and E. Zubrow 1999. Between spaces: interpolation in archaeology, in M. Gillings, D. Mattingly and J. van Dalen (ed.) *Geographic information systems and landscape archaeology*, 65–83. Oxford: Oxbow Books

Rogers, S. and D. J. Bellamy 1972. Peat exploitation and

conservation – a case history. *Proceedings of the Fourth International Peat Symposium* 1, 219–232

Roper, T. 1993. A waste of resources: the origins, growth and decline of Thorne Moors. Sheffield: Unpublished M. Phil Thesis, University of Sheffield.

Roper, T. 1996. Fossil insect evidence for the development of raised mire at Thorne Moors, near Doncaster. *Biodiversity and Conservation* 5, 503–521

Rowley-Conwy, P. 2010. From great bog to sedge fen: a note on Grahame Clark's interpretation of Star Carr in its landscape context, in A. Marchiniak and J. Coles (ed.) *Grahame Clark and his legacy*, 68–84. Newcastle upon Tyne: Cambridge Scholars

Rynne, E. 1965a. A Togher and a bog road in Lullymore Bog. *Journal of the Kildare Archaeological Society* 14, 34–40

Rynne, E. 1965b. Toghers in Littleton Bog, Co. Tipperary. *North Munster Antiquarian Journal* 9, 138–144

Sahlins, M. 1985. *Islands of history*. Chicago: Chicago University Press

Schleifer, N., A. Weller, S. Schneider and A. Junge 2002. Investigation of a Bronze Age plankway by Spectral Induced Polarization. *Archaeological Prospection* 9, 243–53

Schofield, J. E. 2001. *Vegetation succession in the Humber wetlands*. Hull: Unpublished PhD thesis, University of Hull

Schofield, J. E. and M. J. Bunting 2005. Mid-Holocene presence of water chestnut (*Trapa natans* L.) in the meres of Holderness, East Yorkshire, UK. *The Holocene* 15(5), 687–697

Seppa, H. 2002. The mires of Finland: regional and local controls of vegetation, land forms and long term dynamics. *Fennia* 180 1–2, 43–60

Shore, J. S., D. D. Bartley, and D. D. Harkness 1995. Problems encountered with the ^{14}C dating of peat. *Quaternary Science Reviews* 14, 373–383

Slota, Jr. P. J., A. J. T. Jull, T. W. Linick, and L. J. Toolin 1987. Preparation of small samples for ^{14}C accelerator targets by catalytic reduction of CO^2. *Radiocarbon* 29, 303–306

Smith, A, and J. Piltcher 1973. Radiocarbon dates and vegetational history of the British Isles. *New Phytologist*, 72, 903–914

Smith, A. G. 1958. The context of some late Bronze Age and early Iron Age remains from Lincolnshire. *Proceedings of the Prehistoric Society* 224, 78–84

Smith, A. G. and E. W. Cloutman 1988. Reconstruction of Holocene vegetation history in three dimensions at Waun-Fignen-Felen, an upland site in south Wales. *Philosophical Transactions of the Royal Society of London Series B* 322, 159–219

Smith, B. M. 1985. *A palaeoecological study of raised mires in the Humberhead Levels*. Unpublished PhD thesis, University of Wales.

Smith, B. M. 2002. *A palaeoecological study of raised mires in the Humberhead Levels* (ed. P. C. Buckland and M. Limbert). Oxford: Thorne and Hatfield Moors Monograph 1, British Archaeological Report 336

Stainforth, T. 1944. Reed beetles of the genus *Donacia* and its allies in Yorkshire (Col. Chrysomelidae). *Naturalist* 81–91, 127–139

Stead, I. M., J. B. Bourke and D. Brothwell (ed.) 1986. *Lindow Man. The body in the bog*. London: Trustees of the British Museum

Stoertz, C. 1997. *Ancient landscapes of the Yorkshire Wolds: aerial photographic transcription and analysis*. Swindon: RCHME

Stonehouse, W.B. 1839. *History and topography of the Isle of Axholme: being that part of Lincolnshire that is west of the Trent*. London

Stovin Ms. (1747), in C. Jackson (ed.) 1882. The Stovin manuscript. *Yorkshire Archaeological and Topographical Journal* 7, 194–238

Sturt, F. 2006. Local knowledge is required: a rhythmananalytical approach to the late Mesolithic and early Neolithic of the East Anglian Fenland, UK. *Journal of Maritime Archaeology* 1, 119–139

Stradling, W. 1849. The turbaries between Glaston and the sea. *Somerset Archaeology and Natural History* 1, 48–62

Streefkerk, J. G. and W. A. Casparie 1989. *The hydrology of bog ecosystems. Guidelines for management*. Staatsbosbeheer

Stenhouse, M. J. and M. S. Baxter 1983. ^{14}C dating reproducibility: evidence from routine dating of archaeological samples. *PACT* 8, 147–161

Stuiver, M. and R. S. Kra 1986. Editorial comment. *Radiocarbon* 28(2B), ii

Stuiver, M. and H. A. Polach 1977. Reporting of ^{14}C data. *Radiocarbon* 19, 355–63

Stuiver, M. and P. J. Reimer 1986. A computer program for radiocarbon age calculation. *Radiocarbon* 28, 1022–1030

Stuiver, M. and P. J. Reimer 1993. Extended ^{14}C data base and revised CALIB 3.0 ^{14}C age calibration program. *Radiocarbon* 35, 215–230

Swindles, G. T., P. J. Morris, A. J. Baird, M. Blaauw and G. Plunket 2012. Ecohydrological feedbacks confound peat based climate reconstructions. *Geophysical Research Letters* 39(11), L11401

Swindles, G. T. and G. Plunkett 2009. Testing the palaeoclimatic significance of the Northern Irish bog-oak record. *The Holocene* 20, 155–159

Swindles, G. T., G. Plunkett and H. M. Roe 2007. A delayed response to solar forcing at 2800 cal. BP: multi-proxy evidence from three Irish peatlands. *The Holocene* 17, 177–182

Switsur, V. R. and E. V. Wright 1989. Radiocarbon ages and calibrated dates for the boats from North Ferriby, Humberside – a re-appraisal. *Archaeological Journal* 146, 58–67

Taylor, C. 1984. *Village and farmstead. A history of rural settlement in England*. London: George Philip

Taylor, J. A. 1983 The peatlands of Great Britain and Ireland, in A. J. P. Gore (ed.) *Mires: swamp, bog, fen and moor. Ecosystems of the world* 4b, 1–46. Amsterdam: Elsevier

Taylor, M. 1998. Wood and bark from the enclosure ditch, in F. Pryor (ed.) *Etton – excavations at a Neolithic causewayed enclosure near Maxey, Cambs 1982–7*. London: English Heritage

Telford, R. J., E. Heegaard and H. J. B. Birks 2004. All age-depth models are wrong, but how badly? *Quaternary Science Reviews* 23, 1–5

Thomas, J. 1993. The politics of vision and the archaeologies of landscape, in B. Bender (ed.) *Landscape: politics and perspectives*, 19–48. Oxford: Berg

Thomas, J. 2004. *Archaeology and modernity*. London: Routledge

Tilley, C. 1994. *A phenomenology of landscape. Places, paths and monuments*. Oxford: Berg

Tipping, R. 2008. Blanket peat in the Scottish Highlands: timing, cause, spread and the myth of environmental determinism. *Biodiversity and Conservation* 17, 2097–2113

Tipping, R., A. Haggart, P. Milburn and J. Thomas 2004. Landscape perception in the early Bronze Age: henge construction at Picts Knowe, southern Scotland: a palaeoecological perspective, in E. Carver and O. Delong (ed.) *Modern views – ancient lands*, 41–51. Oxford: Archaeopress

Tohall, P. and W. Van Zeist 1955. A Trackway in Corlona Bog, Co. Leitrim. *Journal of the Royal Society of Antiquaries of Ireland* 85, 77–83

Tomlinson, J. 1882. *The level of Hatfield Chase and parts adjacent*. Doncaster

Tuan, Y-F. 1977. *Space and place. The perspective of experience*. London: Edward Arnold

Tuittila, E.-S., M. Väliranta, J. Laine and A. Korhola 2007.

Controls of mire vegetation succession in a southern boreal bog. *Journal of Vegetation Science* 18, 891–902

Tuittila, E-S., V-M. Komulainen, J. Vasander and J. Laine 1999. Restored and cut-away peatland as a sink for atmospheric CO 2. *Oecologia* 120, 563–574

Turner, J. 1962. The *Tilia* decline an anthropogenic interpretation. *New Phytologist* 61, 328–341

Turner, R. C. and M. Rhodes 1992. A bog body and its shoes from Amcotts, Lincolnshire. *Antiquaries Journal* 72, 76–90

Turner, R. C. and R. G. Scaife (ed.) 1995. *Bog bodies: new discoveries and new perspectives*. London: British Museum Press

Turner, V. W. 1966. Colour classification in Ndembu ritual, in M. Banton (ed.) *Anthropological approaches to the study of religion*, 47–84. London: Tavistock Publications

Turney, C. S. M. and H. Brown 2007. Catastrophic early Holocene sea level rise, human migration and the Neolithic transition in Europe. *Quaternary Science Reviews* 26, 2036–2041

Tweddle, J. 2001. Regional vegetational history, in M. D. Bateman, P. C. Buckland, C. D. Friedrick and N. J. Whitehouse (ed.) *The Quaternary of east Yorkshire and north Lincolnshire field guide*, 35–47. London: Quaternary Research Association

Ucko, P. J., M. Hunter, A. J. Clark and A. David 1991. *Avebury reconsidered: from the late 1660s to the 1990s*. London: Unwin Hyman

Utsi, E. 2004. Ground-penetrating radar time-slices from North Ballachulish Moss. *Archaeological Prospection* 11(2), 65–75

Väliranta, M., A. Korhola, H. Seppä, E. S. Tuittila, K. Sarmaja-Korjonen, J. Laine and J. Alm 2007. High resolution reconstruction of wetness dynamics in a southern Boreal raised bog, Finland, during the late Holocene: a quantitative approach. *The Holocene* 17, 1093–1107

Van de Noort, R. 2002. Flat, flatter, flattest – the English Heritage wetlands surveys in retrospect, in T. Lane and J. Coles (ed.) *Through wet and dry: essays in honour of David Hall*, 87–95. Sleaford and Exeter: Lincolnshire Archaeology and Heritage Report Series 5/WARP Occassional Paper 17

Van de Noort, R. 2004. *The Humber Wetlands. The archaeology of a dynamic landscape*. Cambridge: Cambridge University Press

Van de Noort, R. 2011. Conceptualising climate change archaeology. *Antiquity* 85, 1039–1048.

Van de Noort, R., H. P. Chapman and J. L. Cheetham 2001. In situ preservation as a dynamic process: the example of Sutton Common, UK. *Antiquity* 75, 94–100

Van de Noort, R., H. P. Chapman and J. R. Collis 2007. *Sutton Common. The excavation of an Iron Age 'marsh-fort'*. York: Council for British Archaeology Research Report 154

Van de Noort, R. and P. Davies 1993. *Wetland heritage – an archaeological assessment of the Humber wetlands*. Hull: Humber Wetlands Project, University of Hull

Van de Noort, R. and S. Ellis (ed.) 1997. *Wetland heritage of the Humberhead Levels, an archaeological survey*. Hull: Humber Wetlands Project

Van de Noort, R, H. Chapman, R. Head and M. Dinnin 1997. The archaeological survey of West, Hatfield and Thorne Moors, in R. Van de Noort and S. Ellis (ed.) *Wetland Heritage of the Humberhead Levels – an archaeological survey*, 369–81. Hull: Humber Wetlands Project, University of Hull

Van de Noort, R., W. Fletcher, G. Thomas, I. Carstairs and D. Patrick 2002. *Monuments at risk in England's wetlands*. Swindon: English Heritage and the University of Exeter

Van de Noort, R., R. Middleton, A. Foxton and A. Bayliss 1999. The 'Kilnsea-boat', and some implications from the discovery of England's oldest plank boat remains. *Antiquity* 73, 131–143

Van de Noort, R. and A. O'Sullivan 2006. *Rethinking wetland archaeology*. London: Duckworth

Van der Plicht, J., W. A. B. van der Sanden, A. T. Aerts and H. J. Streurman 2004. Dating bog bodies by means of ^{14}C-AMS. *Journal of Archaeological Science* 31, 471–491

Vandeputte, K., L. Moens and R. Dams 1996. Improved sealed-tube combustion of organic samples to CO2 for stable isotopic analysis, radiocarbon dating and percent carbon determinations. *Analytical Letters* 29, 2761–2774

Van der Sanden, W. 1996. *Through nature to eternity: the bog bodies of northwest Europe*. Amsterdam: Batavian Lion International

Van Geel, B., J. Buurman, and H. T. Waterbolk 1996. Archaeological and palaeoecological indications of an abrupt climate change in The Netherlands and evidence for climatological teleconnections around 2650 BP. *Journal of Quaternary Science* 11, 451–460

Van Leusen, M. and H. Kamermans 2005. *Predictive modelling for archaeological heritage management: a research agenda*. NAR Report 29

Wachowicz, M. and R. G. Healey 1994. Towards temporality in GIS, in M. F. Worboys (ed.) *Innovations in GIS 1*, 105–118. London: Taylor and Francis

Waddington, J. M., K. Harrison, E. Kellner and A. J. Baird 2009. Effect of atmospheric pressure and temperature on entrapped gas content in peat. *Hydrological Processes* 23, 2970–2980

Walker, D. 1970. Direction and rate in some British post-glacial hydroseres, in D. Walker and R. G. West (ed.) *Studies in the Vegetational History of the British Isles*, 117–139. Cambridge: Cambridge University Press

Walker M. J. C., C. Bryant, G. R. Coope, D. D. Harkness, J. J. Lowe and E. M. Scott 2001. Towards a radiocarbon chronology of the Late-Glacial sample selection strategies. *Radiocarbon* 43, 1007–1021

Waller, M. P., A. J. Long, D. J. Long and J. B. Innes 1999. Patterns and processes in the development of coastal mire vegetation: Multi-site investigations from Walland Marsh, Southeast England. *Quaternary Science Reviews* 18, 1419–1444

Ward, G. K. and S. R. Wilson 1978. Procedures for comparing and combining radiocarbon age determinations a critique. *Archaeometry* 20(1), 19–31

Warren, R. E. 1990. Predictive modelling in archaeology: a primer, in K. M. S. Allen, S. W. Green and E. B. W. Zubrow (ed.) *Interpreting space: GIS and archaeology*, 90–111. London: Taylor and Francis

Watts, W. 1973. rates of change and stability in vegetation in the perspective of long periods of time, in H. Birks and R. West (ed.) *Quaternary Palaeoecology*, 195–206, Blackwell Scientific Publishing

Westcott, K. and R. J. Brandon 2000. *Practical applications of GIS for archaeologists*. London: Taylor and Frances

Wheatley, D. and M. Gillings 2000. Vision, perception and GIS: developing enriched approaches to the study of archaeological visibility, in G. Lock (ed.) *Beyond the map. Archaeology and spatial technologies*, 1–27. Amsterdam: IOS Press

Wheatley, D. and M. Gillings 2002. *Spatial technology and archaeology. The archaeological applications of GIS*. London: Taylor and Francis

Whitehouse, N. J. 1993. *A mid-Holocene forested site from Thorne Moors: the fossil insect evidence*. Sheffield: Unpublished MSc Thesis, University of Sheffield

Whitehouse, N. J. 1997a. Silent witnesses: a Urwald fossil insect assemblage from Thorne Moors. *Thorne and Hatfield Moors Papers* 4, 19–54

Whitehouse, N. J. 1997b. Insect fauna associated with *Pinus sylvestris* from the mid-Holocene of the Humberhead levels,

Yorkshire, UK, in A. C Ashworth, P. C. Buckland and J. Sadler (ed.) *Studies in Quaternary Entomology: An inordinate fondness for Insects. Quaternary Proceedings* 5, 293–303

Whitehouse, N. J. 1998. The evolution of the Holocene wetland landscape of the Humberhead levels from a fossil insect perspective. Sheffield: Unpublished PhD Thesis, University of Sheffield

Whitehouse, N. J. 2000. Forest fires and insects: palaeo-entomological research from a sub-fossil burnt forst. *Palaeogeography, Palaeoclimatology, Palaeoecology* 164, 247–262

Whitehouse, N. J. 2004. Mire ontogeny, environment and climatic change inferred from fossil beetle successions from Hatfield Moors, eastern England. *The Holocene* 14, 79–94

Whitehouse, N. J. 2006. The Holocene British and Irish ancient forest fossil beetle fauna: implications for forest history, biodiversity and faunal colonisation. *Quaternary Science Reviews* 25, 1755–1789

Whitehouse, N. J., G. Boswijk and P. C. Buckland 1997. Peatlands, past, present and future: some comments from the fossil record, in L. Parkyn, R. E. Stoneman and H. A. P. Ingram (ed.) *Conserving peatlands*, 54–63. Wallingford: CAP International 7

Whitehouse, N. J, G. Boswijk and P. C. Buckland 2001. The Humberhead Peatlands, in M. D. Bateman, P. C. Buckland, C. D. Friedrick, N. J. Whitehouse (ed.) *The Quaternary of east Yorkshire and north Lincolnshire field guide*, 167–189. London: Quaternary Research Association

Whitehouse, N. J., P. C. Buckland, G. Boswijk and B. M. Smith 2001. Hatfield Moors, in, M. D. Bateman, P. C. Buckland, C. D. Frederick and N. J. Whitehouse (ed.) *The Quaternary of east Yorkshire and north Lincolnshire field guide*, 179–184. London: Quaternary Research Association

Whitehouse, N. J., P. C. Buckland, P. Wagner and B. M. Smith 2001. Lindholme Island (SE708063), in M. D. Bateman, P. C. Buckland, C. D. Frederick and N. J. Whitehouse (ed.) *The Quaternary of east Yorkshire and north Lincolnshire field guide*, 185–193. London Quaternary Research Association

Whitehouse, N. J., P. G. Langdon, R. Bustin and S. Galsworthy 2008. Fossil insects and ecosystem dynamics in wetlands: implications for biodiversity and conservation. *Biodiversity and Conservation* 17, 2055–2078

Whittle, A., F. Healey and A. Bayliss 2011. *Gathering time: dating the early Neolithic enclosures of southern Britain and Ireland*. Oxford: Oxbow Books

Willems, W. 1998. *The future of European archaeology*. Oxford: Oxbow Lecture Series 3

Wilson, D. R. 1982. *Air photo interpretation for archaeologists*. London: B T Batsford

Wolhfarth, B., G. Skog, G. Possert and B. Holqvist 1998. Pitfalls in the radiocarbon-dating of terrestrial macrofossils. *Journal of Quaternary Science* 13, 137–145

Worrall, F., G. D. Clay, R. Marrs and M. S. Reed 2010. *IUCN technical reviews 5 – impacts of burning on the management of peatlands*. York: IUCN Peatland Commission

Wright, E. V. 1990. *The Ferriby boats: seacraft of the Bronze Age*. London and New York: Routledge

Wright, E. V., R. E. M. Hedges, A. Bayliss and R. Van de Noort 2001. New AMS radiocarbon dates for the North Ferriby boats: a contribution to dating prehistoric seafaring in northwestern Europe. *Antiquity* 75, 726–724

Yeloff, D. E., B. van Geel, P. Broekens, J. Bakker and D. Mauquoy 2007. Mid–late Holocene vegetation and land-use history in the Hadrian's Wall region of northern England: the record from the Butterburn Flow. *The Holocene* 17, 527–538

Yu, G. and S. P. Harrison 1995. Holocene changes in atmospheric circulation patterns as shown by lake status changes in northern Europe. *Boreas* 24, 260–268

Zimmerman, D. W. 2005. The A-theory of time, the B-theory of time, and 'taking tense seriously'. *Dialectica* 59(4), 401–457

Zoltai, S. C. 1988. Wetland Environments and Classification, in National Wetlands Working Group (ed.) *Wetlands of Canada. Ecological Land Classification Series, No. 24*, 1–26. Montreal: Polyscience Publications

Index

Hatfield and Thorne Moors are not individually indexed. Numbers in italics denote pages with illustrations

aerial photography 6, 17, 41, 153
age-altitude model 100–101
archaeological record/remains 21, 26–29, *28*, 37, 41, 42, 119, 134, 139, 141, 152, 161; *see also* by period
agriculture/cultivation 17, 20, 32, 35, 93, 95–96, 138, 140, 147, 152, 159, 162
Alnus (alder) 32, 33–34, 35, 40, 80, 87, 88–89, 90, 96, 97, 133, 134, 141, 142, 148
avenue 137, 150

bark, birch 121, *124*, 129, 135, 136, 137
Betula (birch) 28–29, 33–34, 35, 40, 67, 68, 81, 86, 87, 88–89, 90, 97, 132, 133, 134, 135, 139, 141, 142, 168
    bark 121, *124*, 129, 135, 136, 137
boat 152, 159
bog body 6, *7*, 13, 27, 28, 141, 163
    Tollund Man *7*
bog burst 15, *17*, 159
bog oak 14, 27
Bog Surface Wetness (BSW) 14, 32, 33, 35, 47, 77, 96–97, 148, 149, 159, 164
Bronze Age
    artefacts 15, 27, 28, 29, 140
    barrow 152
    boat 152
    clearance 31, 35, 95, 97–98, 140, 145–147, 148, 149
    cremation cemetery/burial 15, 80
    cultivation 95–96
    metalwork 13, 138
    peat surface and accumulation 109, 110, 114, 118, 138–140, *139*, 141–144, *145*, 155, 156, 161, 162
    settlement 15
    Thorne trackway 15, 28–29, *28*, *29*, 31, 129, 144–145, 162, 173, 184, *184*
    woodland 33, 141, 149

Calluna vulgaris (heather) 33, 34, 78, 86, 87, 88, 89, 90, 97, 124, 127–128, 131–132, 133, 134, 135, 139, 140, 142, 147, 148, 169
charcoal 30, 33, 88, 124, 132, 148, 167
chronology 1–2, 4–5, 6, 8, 35, 37, 40, 41, 99, 164, *and see under* model *and* peat
clearance 30, 31, 80, 93, 94–95, 140, 145–147, 148, 149, 156
    Bronze Age 31, 35, 95, 97–98, 140, 145–147, 148, 149
    Iron Age 30, 31, 32, 33, 35, 94–95, 96, 97, 140–141, 147, 152
    Neolithic 32, 35, 145, 148
    Romano-British 30, 33, 35, 96
climate change 5, 9–10, 14, 15–16, 30, 37, 40, 90–97, *91*, *93–94*, 96–97, 148, 149, 153, 159–160 *and see* environmental change
coin, Roman 13, 27
Coleoptera/beetles 10, 29, 31, 32–33, 39, 86, 87, 89, 90, 132, 135, 153,
    associated with Neolithic trackway/platform 124–129, *127*, 185–187
Corlea Bog Co. Longford IA trackways 13
Corylus avellana (hazel) 33, 88, 97, 132, 133
Cottage Dyke 68, 70, 75, *75*, 116
crops 32, 96, 147
Crowle Moor 19, 20, 22, 43, 69, 90, 96, 112, 133, 141, 143, 148, 149, 156, 157, 172, *175*, 176, *177*

dendrochronology 5, 11–13, 14, *17*, 29, 33, 39, 132, 135, 142
Derryville Bog Co. Tipperary, development of *16*, *17*, 65, 130, 140, 158
    trackways 15
diatoms 47, 77

Digital Elevation Model (DEM) 26, 41–42, *42*, 46, 47, 50, 52, 53, *54–61*, 65–67, *66*, 68–76, *68*, *71–74*, 100, 102, *104*, 108–110, 112–114, 117
Digital Surface Model (DSM) 22, 26, 50
Digital Terrain Model (DTM) 22, 26, 50, 62, 66
ditch/drain survey 6, 119, 120
drainage, of mire/peatland 6, 13, 17–19, 35 17–19, 22, 26–29, 37, 41, 49, *49*, 65, 67, 99, 117, 134, 140, 153
driver (of change) 8, 15, 39, 96, 114, 117, 118, 147
dune, sand sand dune *49*, 50, 52, 65, 67, 86, 87, 89, 101, 124, 128, 131, 133, 155, 161

elevation 24, 26, 40, 42, 46, 47, 49, 50, 52, 53, 55, 57, 62, 63, 65, 67–70, 75, *82*, 86, 87, 100–2, 105, 107, 108–112, 114, 117, 118, 154, 155; *see also under* model
Elm Decline 30, 93, *93*, 145
Environment Agency LIDAR data 21, 22, 26, *26*, 41, 42, 50, 60, 108
Environmental change 1, 5, 6, 8, 9, 15–17, 99, 118, 138, 150, 153, 154, 156, 160, 161, 164
Eriophorum spp. (sedges) 11, 33, 65, 75, 76, 86, 87, 88, 89, 90, 97, 127, 133, 134, 135, 139, 140, 142, 147, 148
excavation 8, 29, 33, 39
 of Neolithic trackway and platform 119–130, *120–121*, *124–125*, *129*

fen 9, 15, 87, 97, 139, 147, 149
fen carr/woodland 11, 33, 34, 40, 90, 96, 97, 129, 130, 134, 139, 141, 142, 160
fieldwalking 27, 28, 50, 119, 120, 139, 150, 162
fieldwork 8, 20, 41, 42, 49, 77, 100, 153, 154
flooding/flood-warping of mires 19–20
floodplain/floodplain deposits 17, *18*, 33, 35–38, 64, 100–1, *102*, *112*, 138, 140–141, 147–9, *151*, 152, 158
flow-accumulation model 101–107, *103–107*, 112–114, *113*, 117, 154, 155, 156
Fraxinus (ash) 32, 127, 128

geology 22–26, *25*, 42, 147, 163
geophysics 6, 17, 41, 42, 50, 153, 163, 165; *see also* Ground Penetrating Radar
GIS (Geographical Information System) 1, 4, 5, 40, 41–42, 46, 48, 49, 50, 67, 77, 100, 104, 107, *108*, *115*, 129, 131, 139, 147, 149, 154, 155, 156, 164, 165
Glastonbury and Meare lake villages 13, 27
Goole Moor 19, 20, 27, 43, 90, 92, 95, 96, 97, 111, 113, 143, 144, 146, 147, 148, 149, 173, 176, *178–180*
grassland/pasture 31, 67, 96
Ground Penetrating Radar (GPR) 6, 17, 21, 41, 42, 45–46, *45*, 49, 50–52, 55–57, 65–66, 67, 76, 153, 154, 160, 161, 163, 165; *see also* geophysics *and* remote sensing

heath 32, 33, 34, 86, 87, 88–89, 90, 97, 127–128, 131–134, 135, 139, 142, 148, 156
henge monument 138, 150

heritage/landscape management 1, 4, 6, 14, 37, 39, 41, 100, 118, 131, 152, 153, 154, 161–163
historic landscape characterisation 4
Holocene
 climate 14, 15, 40, 42, 47, 159, 164
 landscape 1, 25, 165
 sea level 5, 147–148, 160
 sequence 11, 35–37, 39
 woodland 33, 89
human activity 1, 4, 8, 13–15, 27, 29, 32, 37, 38, 39, 40, 42, 47, 89, 90–97, 131, 132, 140, 141, 143–146, 148, 149–152, *151*, 153, 154, 159, 160, 162; *see also* by period
 bone 27
 impact 33, 39, 40, 47, 77, 80, 93, 95, 97, 98, 148, 154, 159
Humber Wetlands Project 8, 27, 33, 35, 109
Humberhead Levels 18, 32, 89, 90–93, *91*, 96, 135, 147, 149
 'recurrence surfaces' 89, 90–93, *91*, 96, 98, 140, 149, 159
humin/humic acid fractions 78–80, 82–84, *85*, 87, 103, *103*, *106*, 107, 158–159, *158*
hydrology/palaeohydrology 9–11, 14, 15, 33–35, 41, 47, 96–97, 149, 153, 156, 159, 160, 164

Iron Age
 boat 152, 159
 clearance 30, 31, 32, 33, 35, 94–95, 96, 97, 140–141, 147, 152
 settlement and material culture 13, 15
 trackway 13, *13*, 15

lacustrine deposits 19–20, 23–24
Lake Humber 19–20, 23, 24–26
landscape archaeology 1, 8, *36*, 38, 41
legacy datasets 8, 21, 22, 26, *26*, 40, 41–2, 45–7, 49–50, 53–57, 60, 65, 67–68, 70, 76, 108, 116, 117, 155, 167–184, *169–184*, *see also* Environment Agency *and* Natural England
Light Detection and Ranging (LIDAR) 6, 21, 22, 26, *26*, 41, 42, 46, 47, 49, 50, 53–55, 60, 66–67, 76, 108, 109, 154, 161, 163, 165
Lindholme Island and surrounding peat 17, 19–20, 22, 27, 28, 42–43, 49–50, 65, 67, 90, 96, 100, 104–105, *105*, 109, 118, 120, 131, 132, 134, 135, 138, 139, 140, 162
 borehole survey 49–52, 85, 119
 hermitage 27
 Neolithic trackway and platform 8, 11–14, 28, 51, 86, 87, 89, 97, 109, 110, 119–130, *121–122*, *124–125*, *129*, 131, 135–138, *136–137*, 150, 152, 160, 162
 radiocarbon sequences/model 82–83, 85–87, 88, 97
Lisheen Archaeological Project 15, 158
lithics 132, 141, 150
 associated with trackway and platform 120, 122–123, *123–124*, 152

map/mapping
  geological 22–26, *25*, 42
  historical 21, 22, *23–24*
  Ordnance Survey 21, 22, *23–24*, 41, 49–50, 51, 52, 117
  regression 6
megalithic monument 138
Mesolithic 28, 110, 118, 131–133, *132*, 141, 148, 150, *151*, 155, 156, 161
mesotrophic mire/peatland 34, 39–40, 90, 92, 127–128, 139, 142
Middle Moor
  borehole survey 33, 35, 67, 70, *71–73*, 87, 90, 112
  palaeochannel 76, 97, 118, 141, 149, 161
  peat development 75, 112, 113, 117–118, 141, 143–144, 148, 149, 155, 156
  pollen sequence 35, 141, 148
  radiocarbon sequence 33, 81, 87, 90, 98, 117, 141, 156, 158, 161
mire, blanket 8
mire, burning of surface 33, 132, 134, 148
mire raised (general) 6–8
  archaeological and palaeoenvironmental investigation 11–17, 29–37, *30–31*
  finds from 13–17
  inception of 8, 30, 33, 37, 39–40, 76, 89–90, 97, 100–106, *101–103*
  lagg zone/vegetation 11, 87
  location formation and development of 9–11, *12*, 15, 33, *36*
  mesotrophic 34, 39–40, 90, 92, 127–128, 139, 142
  ombotrophic 8, 9–11, 14, 17, 33, 37, 39–40, 85, 86, 87, 88, 89–90, 96, 97, 119, 127, 128, 132, 135, 139, 140–141, 143, 147, 148, 149, 150, 155, 156–158, *157*, 160, 162
  reclamation of 17–18
model
  age-altitude 100–101
  chronological 8, 39, 40, 48, 77–97, *78–79*, *82*, *84–85*, *91*, *93–94*, 100–110, *101–110*, 111–114, *111–115*, 117, 139, 141, 149, 154, 164, 167–184, *169–184*
  cut/surviving peat surface 108–9, *109*, 117, 118, 139, 161
  deposit 41, 42, 45, 49, 66, 67, 77, 81, 100, 146, 160–161
  Digital Elevation (DEM) 26, 41–42, *42*, 46, 47, 50, 52, 53, *54–61*, 65–67, *66*, 68–76, *68*, *71–74*, 100, 102, *104*, 108–110, 112–114, 117
  Digital Surface (DSM) 22, 26, 50
  Digital Terrain (DTM) 22, 26, 50, 62, 66
  elevation elevation 40, 42, 46, 47, 49, 50, 52, 53, 55, 57, 62, 63, 65, 67–70, 75, 100–2, 105, 107, 108–112, 114, 117, 118
  flow-accumulation 101–107, *103–107*, 112–114, *113*, 117, 154, 155, 156
  of Neolithic trackway site 129–130, *129*, 135
  palaeoenvironmental 39, 40, 47, 80, 154

peat depth 41–2, 45, 43–5, 47–8, 49, 50–1, 62–5, *63–4*, 75–80, *75*, 108–10, 114–7, *116*, 161, 163, 167
peat/mire inception and spread *35*, 37, 40, 46–47, 49, 77–90, 99–118, *100–108*, *113–115*, *115–116*, 117–118, 131–135, *132*, *134–135*, 138–147, *139*, *143–146*, 154, 155–158, *157*, 161
pre-peat surface 41, 42–46, *43–46*, 48, 49–76, *49*, *54–61*, *63–64*, *66*, *68*, *71–75*, 85, 97, 99–100, 104, 117, 141, 154, 155
'recurrence surfaces' 89, 90–93, *91*, 96, 98, 140, 149, 159
relative sea level 147–148
spatial 1, 8, 99–100, 106–107
wetland inception related to watercourses and surface run-off 101–106, *102–106*, 112–114, *112*, 154
Mountdillon Co. Longford mires 14–15

Natural England 1, 19, 20, 53, 66, 120, 162
Natural England, datasets 21, 42, *43–4*, 45–6, 47, 51–57, 62–65, 67, 68, 70, 75, 76, 114–5, 117, 160
Neolithic
  artefacts 27, 28, 29, 134, 150
  clearance 32, 35, 145, 148
  fen 15
  Hatfield platform 8, 28, 51, 86, 109, 110, 119–130, *121–122*, *124–125*, *129*, 135–138, *136–137*, 150, 152, 160, 162
  Hatfield trackway 8, 11–14, 28, 51, 86, 87, 89, 97, 109, 110, 119–130, *121–122*, *124–125*, *129*, 131, 135–138, *136–137*, 150, 152, 160, 162
  monument 138, 150
  peat surface and accumulation 49–67, *54–61*, *63–64*, *66*, *68*, 81, 85, 97, 109, 114, 117, 118, 133–135, *133–134*, 141, 143, *143–144*, 155, 156, 161
  woodland 33, 141, 150

Packards South/Southwest
  borehole survey 50–51, 52, 55, *59*, 65, 87
  peat inception and growth 87, 100–102, 104, 109
  radiocarbon dates/model 87, 101
palaeochannel add 141, 155
palaeoenvironmental
  analysis/research 8, 14–17, 29–37, 39, 42, 65, 107, 158, 163, 164, 165
  data/record 6, 8, 17, 38–39, 80, 97, 108, 133, 138–139, 140, 149, 153, 154, 159, 160
  model 39, 40, 47, 80, 154
  previous research 29–37, *30–31*, 39–40, 42–44, 65, 67, 77, 80, 90, 97, 135, 140–142, 155, 156, 159, 165
  reconstruction 6, 124
  see also individual coleoptera, diatoms, enviromental change, palynology/pollen, plant remains *and* vegetation succession
paludification 11, 30, 37, 86, 97, 100, 105, 111, 117, 129, 134, 135, 142, 147, 148, 154, 156

palynology/pollen 5, 10, 11, 14, 15–16, 29–37, 39, 47, 77, 80, 88, 89, 93–95, 96, 97, 135, 140, 145, 147, 148, 149, 153, 159

peat/peatland
    cutting/extraction/milling 6, 9, 13, 17–19, *19*, 27, 28–29, 37–39, 91–93, 96, 99, 108–109, *109*, 117, 118, 119, 134, 140, 153, 162, 164–165
    depth 41–2, 45, *43–5*, 47–8, 49, 50–1, 62–5, *63–4*, 75–80, *75*, 108–10, 114–7, *116*, 161, 163, 167
    drainage of 6, 13, 17–19, 35 17–19, 22, 26–29, 37, 41, 49, *49*, 65, 67, 99, 117, 134, 140, 153
    extent (general) 1, 9, *10*
    growth/spread (general) 1, 8, 14, 15, 30, 147–149, 161
    humification studies 29, 32
    humin/humic acid fraction 78–80, 82–84, *85*, 87, 103, *103*, *106*, 107, 158–159, *158*
    inception and spread *35–6*, 37–40, 42, 46–47, 49, 77–90, 99–118, *100–108*, 110–118, *111–113*, *115–116*, 131–135, *132–136*, 138–150, *139*, 141–147, *143–146*, 148, 154, 155–158, *157*, 161
    mesotrophic 34, 39–40, 90, 92, 127–128, 139, 142
    ombotrophic 8, 9–11, 14, 17, 33, 37, 39–40, 85, 86, 87, 88, 89–90, 96, 97, 119, 127, 128, 132, 135, 139, 140–141, 143, 147, 148, 149, 150, 155, 156–158, *157*, 160, 162
    restoration/rewetting 1, *3*, 9, 19, 20, 39, 45, 47, 49, 67, 82, 153, 162
    surviving resource 108–110, *108–110*, 114–117, *115–117*, 118, 139, 154, 161–163
    surviving surface 41, 42, 47, 54, 63, 108–9, *109*, 117, 118, 139, 161

Phragmites australis (common reed) 75, 86, 87, 141

Pinus sylvestris (Scots' pine) 28–29, 31, 32, 33, 34, 40, 81, 86, 87, 90, 97, *100*, 121, 127, 128, 129, 131–132, 134, 135, *138*, 142, 156, 168

plant remains/macrofossils 10, 14, 29–30, 32, 33–35, 39, 65, 78–79, 81, 84, 86, 87, 88, 89, 96–97, *103*, *106*, 107, 132, 140
    associated with trackway/platform 124–129

Plantago lanceolata (ribwort plantain) 32, 93–96, *94*, 140, 145, 146, 149

platform, Neolithic wooden 8, 15, 51, 86, 109, 110, 119–130, *121–122*, *124–125*, *129*, 135–137, *136–137*, 150, 152, 160, 162
    approach and threshold 137–138
    function 136–138
    see also trackway

Porters Drain 100, 118
    borehole survey 50–52, 63–65, *64*
    palaeoenvironmental sequence 33–35, 97, 134, 158
    radiocarbon sequence 80–81, 85, 88, 89, 90, 97, 169, 174, *174*

pre-peat landscape/surface add from models 1, 8, 9, 29, 33, 37, 39, 41, 42–46, *43–46*, 48, 49–76, *49*, *54–61*, *63–64*, *66*, *68*, *71–75*, 81, 85, 87, 97, 99–100, 104, 110, 117, 120–121, *121*, 127, 133, 141–142, 153, 154, 155, 156, 163

preservation
    in situ 8, 162
    of archaeological remains 9, 10, 41, 100, 108, 161
    of organic remains 8, 10, 41, 108, 164
    of peatland resource 108–110, *108–110*, 114–117, *115–117*, 118, 139, 154, 161–163

previous palaeoenvironmental research 29–37, *30–31*, 39–40, 42–44, 65, 67, 77, 80, 90, 97, 135, 140–142, 155, 156, 159, 165

proxy data 9–10, 14, 33, 37, 47, 96, 149, 154, 157, 159, 163, 164

Quercus (oak) 11, 31, 32, 33, 35, 40, 87, 88, 97, 127, 128, 131–132, 133, 135, 141, 142; *see also* woodland/forest
    bog oak 14, 27

radiocarbon date/dating 1, 5, 8, 14, *17*, 29, 33, 35, 37, 39, 40, 46, 47, 65, 77–98, *78–79*, *82*, *84–85*, *91*, *93–94*, 99, 100, 101–118, *101–103*, *105–107*, *110*, *112–113*, 120, 121–122, 125, *126*, 132–133, 135, 139, 140, 141, 142, 145–152, 155–160, *157–158*, 163, 167–184, *169–184*, see also *individual sites* and *model, chronological*
    Bayesian modelling 1, 5, 41, 47, 80–98, *84*, *91*, *93–94*, 107, 122, 140, 154, 156–160, *157*, 163, 164
    humin/humic acid/fulvic acid fractions 78–80, 82–84, *85*, 87, 103, *103*, *106*, 107, 158–159, *158*
    reservoir ('old wood') effect 77–78

rainfall/precipitation 9–11, 14, 17; *and see* under model

Rawcliffe Moors 20, 40, 43, 67–69, *74*, 90, 95, 96, 97, 112, 114, 141, 143, 144, 145, 147, 149, 155, 156, 157, 161, 173, *181*, 183

reedswamp 11, 79

remote sensing
    geophysics 6, 17, 41, 42, 50, 153, 163, 165
    Ground Penetrating Radar (GPR) 6, 17, 21, 41, 42, 45–46, *45*, 49, 50–52, 55–57, 65–66, 67, 76, 153, 154, 160, 161, 163, 165
    Light Detection and Ranging (LIDAR) 6, 26, *26*, 41, 42, 46, 47, 49, 50, 53–55, 66–67, 76, 108, 109, 154, 161, 163, 165

ritual activity and performance 136–138

River Don/Old River Don 17, 25, 112, 141, 149, 150

River Idle 17, 25, 35–36, 100, 101, 147, 150

River Torne 17, 22, 25, 33, 35, 37, 100, 101, 138, 140, 141, 147, 150

River Trent 17, 28, 150

Roman
    artefacts 27, 29
    clearance 30, 33, 35, 96
    coins 27

sand dune *49*, 50, 52, 65, 67, 86, 87, 89, 101, 124, 128, 131, 133, 155, 161

sea level change/curve 5, 37, 47, 77, 90, 147–148, 160
    Holocene 5, 147–148, 160
    relationship with peat formation 147–148

site location analysis 4
soil 29, 89, 131–132, 141
    micromorphology 29, 131
    pre-peat 33, 83, 97, 132
Somerset Levels 6, 11–13, 27, 128, 129, 138, 162
Somerset Levels Project 14, 15
space
    concept of 4, 13
    neutrality of 4
Sphagnum 11, 30, 33, 34, 65, 75, 76, 78, 86, 87, 89, 90, 91, 124, 140, 142–143, 148
stratigraphy 1, 15, 30, 32, 33, 65, 75–76, 89, 96–97, 100, 101, 141, 164
    palimpsest 1–2, 153
survey 6, 20–21, 27–28
    borehole 6, 39, 42–47, *46*, 49, 50–65, *54–61*, *63–64*, 67–76, *69–75*, 81, 108, 110, 114, 119, 121, *121*, 154, 160–161
    ditch/drain 6, 119, 120
    fieldwalking 27–28, 50, 119, 120, 139, 150, 162
Sweet Track, Somerset 11–13, 128

testate amoebae 14, 29–30, 32, 33, 35, 39, 47, 77, 88, 96–97, 140
Thorne Bronze Age trackway 28–29, *28*, *29*, 31, 144–145, 162, 173, 184, *184*
Tilia (lime) 31, 32, 88, 89, 93–5, *93*, 97, 132, 133, 142, 145, 146, *see also* elm decline
Tollund Man 7
topography 4, 29, 39, 42, 49, 51, 52, 65, 69, 76, 81, 86, 100, 105, 110, 160; *see also* under model

trackway 6, 8, 11–14, *13*, 15, 28, *28*, *29*, 31, 51, 86, 87, 89, 97, 109, 110, 119–130, *121–122*, *124–125*, *129*, 131, 134, 135–137, *136–137*, 144–145, 150, 152, 160, 162
    Corlea Bog Co. Longford 13
    evidence of woodworking 121, *125*
    function 136–138
    Hatfield, Neolithic 8, 11–14, 28, 51, 86, 87, 89, 97, 109, 110, 119–130, *121–122*, *124–125*, *129*, 131, 135–138, *136–137*, 150, 152, 160, 162
    Iron Age 13, *13*, 15
    radiocarbon dating 121–122
    structure and composition 120–121, *121*, 129, 135–136
    Sweet Track, Somerset 11–13, 128
Thorne, Bronze Age 28–29, *28*, *29*, 31, 144–145, 162, 173, 184, *184*

urtica spp. (nettles) 32, 95, 145, 146

vegetation succession/change 32–35, 40, 47, 77, 80, 89, 90, 93, 97, 121, 133–143, 160; *and see* environmental change
    hydroseres 11, *12*, 37, 47, 100, 101, 105
visualisation 4, 48, 76, 155

'warping' 17–18, 19, 20, 27, 140
woodland/forest 11, 27, 31, 32, 33, 35, 40, 49, 65, 67, 68, 86, 87, 88–89, 93, 97, 129, 130, 131–132, 133–135, 139, 141, 150
    ancient 128
    burning of 31–32, 132
    Romano-British and Medieval regeneration 35, 97–98